SI UNITS USED IN MECHANICS

Quantity	Unit	SI Symbol
(Base Units)		
Length	meter°	m
Mass	kilogram	kg
Time	second	s
(Derived Units)		
Acceleration, linear	meter/second^2	m/s^2
Acceleration, angular	radian/second^2	rad/s^2
Area	meter^2	m^2
Density	kilogram/meter^3	kg/m^3
Force	newton	N $(= $ kg$\cdot$m/s$^2)$
Frequency	hertz	Hz $(= 1/$s$)$
Impulse, linear	newton-second	N$\cdot$s
Impulse, angular	newton-meter-second	N$\cdot$m$\cdot$s
Moment of force	newton-meter	N$\cdot$m
Moment of inertia, area	meter^4	m^4
Moment of inertia, mass	kilogram-meter^2	kg$\cdot$m^2
Momentum, linear	kilogram-meter/second	kg$\cdot$m/s $(= $ N$\cdot$s$)$
Momentum, angular	kilogram-meter^2/second	kg$\cdot$m^2/s $(= $ N$\cdot$m$\cdot$s$)$
Power	watt	W $(= $ J/s $= $ N$\cdot$m/s$)$
Pressure, stress	pascal	Pa $(= $ N/m$^2)$
Product of inertia, area	meter^4	m^4
Product of inertia, mass	kilogram-meter^2	kg$\cdot$m^2
Spring constant	newton/meter	N/m
Velocity, linear	meter/second	m/s
Velocity, angular	radian/second	rad/s
Volume	meter^3	m^3
Work, energy	joule	J $(= $ N$\cdot$m$)$
(Supplementary and Other Acceptable Units)		
Distance (navigation)	nautical mile	$(= 1.852$ km$)$
Mass	ton (metric)	t $(= 1000$ kg$)$
Plane angle	degrees (decimal)	°
Plane angle	radian	—
Speed	knot	$(1.852$ km/h$)$
Time	day	d
Time	hour	h
Time	minute	min

° Also spelled *metre*.

SI UNIT PREFIXES

Multiplication Factor	Prefix	Symbol
1 000 000 000 000 $= 10^{12}$	terra	T
1 000 000 000 $= 10^{9}$	giga	G
1 000 000 $= 10^{6}$	mega	M
1 000 $= 10^{3}$	kilo	k
100 $= 10^{2}$	hecto	h
10 $= 10$	deka	da
0.1 $= 10^{-1}$	deci	d
0.01 $= 10^{-2}$	centi	c
0.001 $= 10^{-3}$	milli	m
0.000 001 $= 10^{-6}$	micro	μ
0.000 000 001 $= 10^{-9}$	nano	n
0.000 000 000 001 $= 10^{-12}$	pico	p

SELECTED RULES FOR WRITING METRIC QUANTITIES

1. (a) Use prefixes to keep numerical values generally between 0.1 and 1000.
   (b) Use of the prefixes hecto, decka, deci, and centi should be generally avoided except for certain areas or volumes where the numbers would be otherwise awkward.
   (c) Use prefixes only in the numerator of unit combinations. The one exception is the base unit kilogram. (*Example:* write kN/m not N/mm; J/kg not mJ/g)
   (d) Avoid double prefixes. (*Example:* write GN not kMN)
2. Unit designations
   (a) Use a dot for multiplication of units. (*Example:* write N$\cdot$m not Nm)
   (b) Avoid ambiguous double solidus. (*Example:* write N/m^2 not N/m/m)
   (c) Exponents refer to entire unit. (*Example:* mm^2 means (mm)^2)
3. Number grouping
   Use a space rather than a comma to separate numbers in groups of three, counting from the decimal point in both directions. (*Example:* 4 607 321.048 72) Space may be omitted for numbers of four digits. (*Example:* 4296 or 0.0476)

ENGINEERING MECHANICS
VOLUME 1
STATICS

ENGINEERING MECHANICS
VOLUME 1
STATICS

J. L. MERIAM

Professor of Mechanical Engineering
California Polytechnic State University

JOHN WILEY & SONS
New York   Santa Barbara   London   Sydney   Toronto

This book was printed and bound by Quinn and Boden.
It was set in Laurel by York Graphic Services.
The drawings were designed and executed by John Balbalis
with the assistance of the Wiley Illustration Department.
Text design by Jerome Wilke and Edward A. Butler.

Cover design by Edward A. Butler.

Library of Congress Cataloging in Publication Data:

Author's name entry.
Brief title entry: Subtitle, if any xxxxxxxxxxxxxxxxxx
xxxxxxxxxxxxxx

(Series Statement)
Notes xxxxxxxxxxxxxxx
L.C. Subject Heading   Joint Author Entry
Title and Series added entries

L.C. Call No.   Dewey Classification No.   L.C. Card No.
ISBN 0-471-xxxxx-x
ISBN 0-471-xxxxx-x (pbk)

Printed in the United States of America

10 9 8 7 6 5 4 3 2 1

FOREWORD

The innovations and contributions of Dr. James L. Meriam to the field of engineering mechanics cannot be overstated. He has undoubtedly had more influence on instruction in mechanics during the last quarter century than any other individual.

No one who commenced the study of engineering mechanics after 1950 can fully appreciate the apprehension and lack of understanding that the average engineer once experienced when facing a problem in mechanics. Professor Meriam did much to bring clarity and comprehension to this subject. His first book on mechanics in 1951 literally reconstructed undergraduate mechanics and became the definitive textbooks for the next decade. Hist texts were logically organized, easy to read, directed to the average engineering undergraduate and were packed with exciting examples of real-life engineering problems superbly illustrated. These books became the model for other engineering mechanics texts in the 1950s and 1960s.

Dr. Meriam began his work in mechanical engineering at Yale University where he earned his B.E., M. Eng., and Ph.D. degrees. He had early industrial experience with Pratt and Whitney Aircraft and the General Electric Company, which stimulated his first contributions to mechanics in mathematical and experimental stress analysis. In the Second World War he served in the U.S. Coast Guard. These early experiences influenced Professor Meriam in two ways. First, he discovered his profound interest in the practical applications of mechanics and, second, he developed a lifelong interest in ships and the sea.

Dr. Meriam was a member of the faculty of the University of California, Berkeley, for twenty-one years where he served as Professor of Engineering Mechanics, Assistant Dean of Graduate Studies, and Chairman of the Division of Mechanics and Design. In 1963 he became Dean of Engineering at Duke University where he devoted his full energies to the development of its School of Engineering. In 1972 Professor Meriam followed his desire to return to full-time teaching and accepted appointment as Professor of Mechanical Engineering at the California Polytechnic State University. Dr. Meriam has always placed great emphasis on teaching, and this trait has been recognized by his students wherever he has taught. For example, at Berkeley in 1963 he was the first recipient of the Outstanding Faculty Award of Tau Beta Pi, given primarily for excellence in teaching.

The free-body diagram is the foundation for mechanics. This was not a new concept with Dr. Meriam, but the emphasis and rigor with which he developed the free body in mechanics were new and highly successful. They permeate his writing and teaching. He was the first author to show clearly how the method of virtual work in statics can be employed to solve a class of problems largely neglected by previous authors. In dynamics, plane motion became understandable, and in his later editions, three-dimensional kinematics and kinetic received the same treatment. He is credited with original developments in the theory of variable-mass dynamics, which are contained in his *Dynamics, 2nd Edition*. More recently, Professor Meriam has been a leader in promoting the use of SI units, and his *SI Versions of Statics* and *Dynamics* published in 1975 were the first mechanics textbooks in SI units in this country.

Professor Meriam's new book promises to meet and even to exceed the high standards that he has set in the past. Without question it contains one of the most outstanding collections of problems yet assembled. This new text is especially designed to assist students in the preliminary stages of each new topic, and then to lead them to more challenging engineering applications as well. The new book will appeal to a wide audience of students, teachers, and engineers and will further extend the author's contributions to mechanics.

Robert F. Steidel
Professor of Mechanical Engineering
University of California, Berkeley

PREFACE
To the Student

As you undertake the study of engineering mechanics, first statics and then dynamics, you will be building a foundation of analytical capability for the solution of a great variety of engineering problems. Modern engineering practice demands a high level of analytical capability, and you will find that your study of mechanics will help you immensely in developing this capacity.

In engineering mechanics we learn to construct and solve mathematical models which describe the effects of force and motion on a variety of structures and machines that are of concern to engineers. In applying our principles of mechanics we formulate these models by incorporating appropriate physical assumptions and mathematical approximations. Both in the formulation and solution of mechanics problems you will have frequent occasion to use your background in plane and solid geometry, scalar and vector algebra, trigonometry, analytic geometry, and calculus. Indeed, you are likely to discover new significance to these mathematical tools as you make them work for you in mechanics.

Your success in mechanics (and throughout engineering) will be highly contingent upon developing a well-disciplined method of attack from hypothesis to conclusion in which the applicable principles are applied rigorously. From many years of experience as a teacher and an engineer I know the importance of developing the ability to represent one's work in a clear, logical, and concise manner. Mechanics is an excellent place in which to develop these habits of logical thinking and effective communication.

ENGINEERING MECHANICS contains a large number of sample problems in which the solutions are presented in detail. Also included in these examples are helpful observations with mention made of common errors and pitfalls to be avoided. In addition, the book contains a large selection of simple, introductory problems and problems of intermediate difficulty to help you gain initial confidence and understanding of each new topic. Also included are many problems which illustrate significant and contemporary engineering situations. These will stimulate your interest and help you to develop an appreciation for the many applications of mechanics in engineering.

I am pleased to extend my encouragement to you as a student of mechanics. I hope this book will provide both help and stimulation as you develop your background in engineering.

J. L. Meriam

Santa Barbara, California
January 1978

PREFACE
To the Instructor

The primary purpose of the study of engineering mechanics is to develop capacity to predict the effects of force and motion in the course of carrying out the creative design function of engineering. Successful prediction requires more than a mere knowledge of the physical and mathematical principles of mechanics. Also required is the ability to visualize physical configurations in terms of real materials, actual constraints, and practical limitations which govern the behavior of machines and structures. One of our primary objectives in teaching mechanics should be to help the student develop this ability to visualize, which is so vital to problem formulation. Indeed, the construction of a meaningful mathematical model is often a more important experience than its solution. Maximum progress is made when the principles and their limitations are learned together within the context of engineering application.

Courses in mechanics are often regarded by students as a difficult requirement and frequently as an uninteresting academic hurdle as well. The difficulty stems from the extent to which reasoning from fundamentals, as distinguished from rote learning, is required. The disinterest that is frequently felt is due primarily to the extent to which mechanics is presented as an academic discipline largely lacking in engineering purpose and challenge. This attitude is traceable to the frequent tendency in the presentation of mechanics to use problems mainly as a vehicle to illustrate theory rather than to develop theory for the purpose of solving problems. When the first view is allowed to predominate, problems tend to become overly idealized and unrelated to engineering with the result that the exercise becomes dull, academic, and uninteresting. This approach deprives the student of much of the valuable experience in formulating problems and thus of discovering the need for and meaning of theory. The second view provides by far the stronger motive for learning theory and leads to a better balance between theory and application. The crucial role of interest and purpose in providing the strongest possible motive for learning cannot be overemphasized. Furthermore, we should stress the view that, at best, theory can only approximate the real world of mechanics rather than the view that the real world approximates the theory. This difference in philosophy is indeed basic and distinguishes the *engineering* of mechanics from the *science* of mechanics.

During the past twenty years there has been a strong trend in engineering education to increase the extent and level of theory in the engineering-science courses. Nowhere has this trend been felt more than in mechanics courses. To the extent that students are prepared to handle the accelerated treatment, the trend is beneficial. There is evidence and justifiable concern, however, that a significant disparity has more recently appeared between coverage and comprehension. Among the contributing factors we should note three trends. First, emphasis on the geometric and physical meanings of prerequisite mathematics appears to have diminished. Second, there has been a significant reduction and even elimination of instruction in graphics which in the past served to enhance the visualization and representation of mechanics problems. Third, in advancing the mathematical level of our treatment of mechanics there has been a tendency to allow the notational manipulation of vector operations to mask or replace geometric visualization. Mechanics is inherently a subject which depends on geometric and physical perception, and we should increase our efforts to develop this ability.

One of our responsibilities as teachers of mechanics is to use the mathematics which is most appropriate for the problem at hand. The use of vector notation for one-dimensional problems is usually trivial; for two-dimensional problems it is often optional; but for three-dimensional problems it is quite essential. As we introduce vector operations in two-dimensional problems, it is especially important that their geometric meaning be emphasized. A vector equation is brought to life by a sketch of the corresponding vector polygon, which often discloses through its geometry the shortest solution. There are, of course, many mechanics problems where the complexity of variable interdependence is beyond the normal powers of visualization and physical perception, and reliance on analysis is essential. Nevertheless, our students become better engineers when their abilities to perceive, visualize, and represent are developed to the fullest.

As teachers of engineering mechanics we have the strongest obligation to the engineering profession to set reasonable standards of performance and to uphold them. In addition, we have a serious responsibility to encourage our students to think for themselves. Too much help with details that students should be reasonably able to handle from prerequisite subjects can be as bad as too little help and can easily condition them to becoming overly dependent on others rather than to exercising their own initiative and ability. Also, when mechanics is subdivided into an excessive number of small compartments, each with detailed and repetitious instructions, students can have difficulty seeing the "forest for the trees" and, consequently, fail to perceive the unity of mechanics and the far-reaching applicability of its few basic principles and methods.

ENGINEERING MECHANICS is written with the foregoing philosophy in mind. It is intended primarily for the first engineering course in mechanics, generally taught in the second year of study. The book omits a number of the more advanced topics contained in the author's more extensive treatments, *Statics* and *Dynamics, 2nd Edition* and *SI Version,* and is designed especially to facilitate self-study. To this end a major feature of the book is a greatly expanded treatment of sample problems, which are presented in a single-page format for more convenient study. In addition to presenting the solution in detail, each sample problem also contains comments and cautions keyed to salient points in the solution and printed in colored type. These comments alert students to common pitfalls and should provide a valuable aid to their self-study efforts.

Volume 1, Statics, contains 66 sample problems and 740 unsolved problems from which a wide choice of assignments can be made. Of these problems over 50 percent are totally new with the balance selected from the author's *Statics, 2nd Edition* and *SI Version.* In recognition of the need for the predominant emphasis on SI units there are two problems in SI units for every one in U.S. customary units. This apportionment between the two sets of units permits classroom use anywhere from a 50–50 emphasis to a 100 percent SI treatment. Each problem set begins with relatively simple, uncomplicated problems to help students gain confidence with the new topic. Many practical problems and examples of interesting engineering situations drawn from a wide range of applications are represented in the problem collection. Simple numerical values have been used throughout, however, so as not to complicate the solutions and divert attention from the principles. The problems are arranged generally in order of increasing difficulty, and the answers to most of them are given. The more difficult problems are identified by a ▶ and may often be used to provide a comprehensive classroom experience when solved by the instructor. All numerical solutions have been carried out and checked with an electronic computer without rounding intermediate values. Consequently, the final answers should be correct to within the number of significant figures cited. The author is confident that the book is exceptionally free from error.

ENGINEERING MECHANICS is written in a style which is both concise and friendly. The major emphasis is focused on basic principles and methods rather than on a multitude of special cases. Strong effort has been made to show both the cohesiveness of the relatively few fundamental ideas and the great variety of problems which these few ideas will solve.

In Chapter 2 the properties of forces, moments, couples, and resultants are developed so that the student may proceed directly to the equilibrium of noncurrent force systems in Chapter 3 without belaboring unnecessarily the relatively trivial problem of the equilibrium of concurrent forces acting on a particle. In both Chapters 2 and 3 analysis of two-dimensional problems is presented before three-dimensional problems are treated. The vast majority of students acquire a greater physical insight and understanding of mechanics by first gaining confidence in two-dimensional analysis before coping with the third dimension.

Application of equilibrium principles to simple trusses and to frames and machines is presented in Chapter 4 with primary attention given to two-dimensional systems. A sufficient number of three-dimensional examples are included, however, to enable students to exercise more general vector tools of analysis.

The concepts and categories of distributed forces are introduced at the beginning of Chapter 5 with the balance of the chapter divided into two main sections. Section A treats centroids and mass centers where detailed examples are presented to help students master early applications of calculus to physical and geometrical problems. Section B includes the special topics of beams, flexible cables, and fluid forces, which may be omitted without loss of continuity of basic concepts.

Chapter 6 on friction is divided into Section A on the phenomenon of dry friction, and Section B on selected machine applications. Although Section B may be omitted if time is limited, this material does provide a valuable experience for the student in dealing with distributed forces.

Chapter 7 presents a consolidated introduction to virtual work with application limited to single-degree-of-freedom systems. Special emphasis is placed on the advantage of the virtual-work and energy method for interconnected systems and stability determination. Virtual work provides an excellent opportunity to convince the student of the power of mathematical analysis in mechanics.

Moments and products of inertia of areas are presented in Appendix A. This topic helps to bridge the subjects of statics and solid mechanics. Appendix B contains a summary review of selected topics of elementary mathematics that the student should be prepared to use in mechanics.

It is a pleasure for me to recognize again the continuing contribution of Dr. A. L. Hale of the Bell Telephone Laboratories for his invaluable suggestions and careful checking of the manuscript. The critical reviews of Professor Andrew Pytel of The Pennsylvania State University and Professor John Biddle of the California State Polytechnic University have also been of great assistance and are acknowledged with gratitude. In addition, appreciation is expressed to Professor J. M. Henderson of the University of California, Davis, for helpful comments and suggestions of selected problems. Contribution by the staff of John Wiley & Sons during the planning and production of the book reflects a high degree of professional competence and is duely recognized. The support of the California Polytechnic State University in granting me a leave of absence in which to prepare this book is likewise acknowledged. Finally, I acknowledge the patience and forebearance of my wife, Julia, during the many hours required to prepare this manuscript.

J. L. Meriam

Santa Barbara, California
January 1978

CONTENTS

INTRODUCTION TO STATICS

1

1/1 MECHANICS. Mechanics is that physical science which deals with the state of rest or motion of bodies under the action of forces. No one subject plays a greater role in engineering analysis than does mechanics. The early history of this subject is synonymous with the very beginnings of engineering. Modern research and development in the fields of vibrations, stability and strength of structures and machines, rocket and spacecraft design, automatic control, engine performance, fluid flow, electrical machines and apparatus, and molecular, atomic, and subatomic behavior are highly dependent upon the basic principles of mechanics. A thorough understanding of this subject is an absolute prerequisite for work in these and many other fields.

Mechanics is the oldest of the physical sciences. The earliest recorded writings in this field are those of Archimedes (287–212 B.C.) which concern the principle of the lever and the principle of buoyancy. Substantial progress awaited the formulation of the laws of vector combination of forces by Stevinus (1548–1620), who also formulated most of the principles of statics. The first investigation of a dynamic problem is credited to Galileo (1564–1642) in connection with his experiments with falling stones. The accurate formulation of the laws of motion, as well as the law of gravitation, was made by Newton (1642–1727), who also conceived the idea of the infinitesimal in mathematical analysis. Substantial contributions to the development of mechanics were also made by da Vinci, Varignon, D'Alembert, Lagrange, Laplace, and others.

The principles of mechanics as a science embody the rigor of mathematics upon which they are highly dependent. On the other hand, the purpose of engineering mechanics is the application of these principles to the solution of practical problems. In this book we shall be concerned both with the rigorous development of principles and their application. The basic principles of mechanics are relatively few in number, but they have exceedingly wide application, and the methods employed in mechanics will carry over into many fields of engineering endeavor.

The subject of mechanics is logically divided into two parts, *statics*, which concerns the equilibrium of bodies under the action of forces, and *dynamics,* which concerns the motion of bodies. ENGINEERING MECHANICS is divided into these two parts, Volume 1 *Statics* and Volume 2 *Dynamics.*

1/2 BASIC CONCEPTS.   Certain concepts and definitions are basic to the study of mechanics, and they should be understood at the outset.

Space is the geometric region occupied by bodies whose positions are described by linear and angular measurements relative to a coordinate system. For three-dimensional problems our space will require three independent coordinates. For two-dimensional problems only two coordinates will be required.

Time is the measure of the succession of events and is a basic quantity in dynamics. Time is not directly involved in the analysis of statics problems.

Mass is a measure of the inertia of a body, which is its resistance to a change of motion. Of more importance to us in statics, mass is also the property of every body by which it experiences mutual attraction to other bodies.

Force is the action of one body on another. A force tends to move a body in the direction of its action. The action of a force is characterized by its *magnitude,* by the *direction* of its action, and by its *point of application.* Force is a vector quantity, and its properties are discussed in detail in Chapter 2.

Particle. A body of negligible dimensions is called a particle. In the mathematical sense a particle is a body whose dimensions approach zero so that it may be analyzed as a point mass. Frequently a particle is chosen as a differential element of a body. Also, when the dimensions of a body are irrelevant to the description of its position or the action of forces applied to it, the body may be treated as a particle.

Rigid body. A body is considered rigid when the relative movements between its parts are negligible for the purpose at hand. For instance, the calculation of the tension in the cable which supports the boom of a mobile crane under load is essentially unaffected by the small internal strains (deformations) in the structural members of the boom. For the purpose, then, of determining the external forces which act on the boom, we may treat it as a rigid body. Statics deals primarily with the calculation of external forces which act on rigid bodies which are in equilibrium. To determine the internal stresses and strains, the deformation characteristics of the material of the boom would have to be analyzed. This type of analysis belongs in the study of the mechanics of deformable bodies which comes after the study of statics.

1/3 SCALARS AND VECTORS.   In statics we deal with two kinds of quantities, scalars and vectors. Scalar quantities are those with which a magnitude alone is associated. Examples of scalar quantities in mechanics are time, volume, density, speed, energy, and mass. Vector quantities, on the other hand, possess direction as well as magnitude and must obey the parallelogram law of addition as described in this article. Examples of vectors are displacement, velocity, acceleration, force, moment, and momentum.

Physical quantities that are vectors fall into one of three classifications, free, sliding, or fixed.

A *free vector* is one whose action is not confined to or associated with a unique line in space. For example, if a body moves without rotation, then the movement or displacement of any point in the body may be taken as a vector, and this vector will describe equally well the direction and magnitude of the displacement of every point in the body. Hence we may represent the displacement of such a body by a free vector.

A *sliding vector* is one for which a unique line in space must be maintained along which the quantity acts. When we deal with the external action of a force on a rigid body, the force may be applied at any point along its line of action without changing its effect on the body as a whole° and hence may be considered a sliding vector.

A *fixed vector* is one for which a unique point of application is specified, and therefore the vector occupies a particular position in space. The action of a force on a deformable or nonrigid body must be specified by a fixed vector at the point of application of the force. In this problem the forces and deformations internal to the body will be dependent on the point of application of the force as well as its magnitude and line of action.

A vector quantity $\mathbf{V}$ is represented by a line segment, Fig. 1/1, having the direction of the vector and having an arrowhead to indicate the sense. The length of the directed line segment represents to some convenient scale the magnitude $|\mathbf{V}|$ of the vector and is written with lightface italic type V. In scalar equations and frequently on diagrams where only the magnitude of a vector is labeled the symbol will appear in lightface italic type. Boldface type is used for vector quantities whenever the directional aspect of the vector is a part of its mathematical representation. When writing vector equations we must preserve the mathematical distinction between vectors and scalars. It is recommended that in all handwritten work a distinguishing mark be used for each vector quantity, such as an underline, $\underline{V}$, or an arrow over the symbol, $\vec{V}$, to take the place of boldface type in print. The direction of the vector $\mathbf{V}$ may be measured by an angle θ from some known reference direction as indicated. The negative of $\mathbf{V}$ is a vector $-\mathbf{V}$ directed in the sense opposite to $\mathbf{V}$ as shown.

Figure 1/1

° This is the so-called *principle of transmissibility*, which is discussed in Art. 2/2.

(a)

(b)   *(c)*

Figure 1/2

Figure 1/3

(a)

(b)

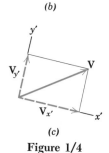

(c)

Figure 1/4

In addition to possessing the properties of magnitude and direction, vectors must also obey the parallelogram law of combination. This law requires that two vectors $\mathbf{V}_1$ and $\mathbf{V}_2$, treated as free vectors, Fig. 1/2a, may be replaced by their equivalent $\mathbf{V}$ which is the diagonal of the parallelogram formed by $\mathbf{V}_1$ and $\mathbf{V}_2$ as its two sides, as shown in Fig. 1/2b. This combination or vector sum is represented by the vector equation

$$\mathbf{V} = \mathbf{V}_1 + \mathbf{V}_2$$

where the plus sign used in conjunction with the vector quantities (boldface type) means *vector* and not *scalar* addition. The scalar sum of the magnitudes of the two vectors is written in the usual way as $V_1 + V_2$, and it is clear from the geometry of the parallelogram that $V \neq V_1 + V_2$.

The two vectors $\mathbf{V}_1$ and $\mathbf{V}_2$, again treated as free vectors, may also be added head-to-tail by the triangle law, as shown in Fig. 1/2c, to obtain the identical vector sum $\mathbf{V}$. We see from the diagram that the order of addition of the vectors does not affect their sum, so that $\mathbf{V}_1 + \mathbf{V}_2 = \mathbf{V}_2 + \mathbf{V}_1$.

The difference $\mathbf{V}_1 - \mathbf{V}_2$ between the two vectors is easily obtained by adding $-\mathbf{V}_2$ to $\mathbf{V}_1$ as shown in Fig. 1/3 where either the triangle or parallelogram procedure may be used. The difference $\mathbf{V}'$ between the two vectors is expressed by the vector equation

$$\mathbf{V}' = \mathbf{V}_1 - \mathbf{V}_2$$

where the minus sign is used to denote *vector subtraction*.

Any two or more vectors whose sum equals a certain vector $\mathbf{V}$ are said to be the *components* of that vector. Hence the vectors $\mathbf{V}_1$ and $\mathbf{V}_2$ in Fig. 1/4a are the components of $\mathbf{V}$ in the directions 1 and 2, respectively. It is usually most convenient to deal with vector components that are mutually perpendicular, and these are called *rectangular components*. The vectors $\mathbf{V}_x$ and $\mathbf{V}_y$ in Fig. 1/4b are the x- and y-components, respectively, of $\mathbf{V}$. Likewise, in Fig. 1/4c, $\mathbf{V}_{x'}$ and $\mathbf{V}_{y'}$ are the x'- and y'-components of $\mathbf{V}$. When expressed in rectangular components, the direction of the vector with respect to, say, the x-axis is clearly specified by

$$\theta = \tan^{-1}\frac{V_y}{V_x}$$

For some problems, particularly three-dimensional ones, it is convenient to express the rectangular components of $\mathbf{V}$, Fig. 1/5, in terms of unit vectors $\mathbf{i}$, $\mathbf{j}$, $\mathbf{k}$, which are vectors in the x-, y-, and z-directions, respectively, with magnitudes of unity. The vector sum of the components is written

$$\boxed{\mathbf{V} = \mathbf{i}V_x + \mathbf{j}V_y + \mathbf{k}V_z}$$

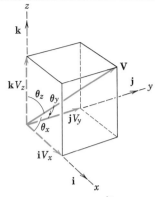

Figure 1/5

When we substitute the *direction cosines* l, m, and n of $\mathbf{V}$ given by

$$l = \cos\theta_x \qquad m = \cos\theta_y \qquad n = \cos\theta_z$$

we may write the magnitudes of the components of $\mathbf{V}$ as

$$\boxed{\begin{array}{ccc} V_x = lV & V_y = mV & V_z = nV \\ \multicolumn{3}{c}{V^2 = V_x^2 + V_y^2 + V_z^2} \end{array}}$$

with

Note also that $l^2 + m^2 + n^2 = 1$.

1/4 NEWTON'S LAWS. Sir Isaac Newton was the first to state correctly the basic laws governing the motion of a particle and to demonstrate their validity.° Slightly reworded to use modern terminology, these laws are:

Law I. A particle remains at rest or continues to move in a straight line with a uniform velocity if there is no unbalanced force acting on it.

Law II. The acceleration of a particle is proportional to the resultant force acting on it and is in the direction of this force.

Law III. The forces of action and reaction between interacting bodies are equal in magnitude, opposite in direction, and collinear.

The correctness of these laws has been verified by innumerable accurate physical measurements. Newton's second law forms the basis for most of the analysis in dynamics. As applied to a particle of mass m it may be stated as

$$\boxed{\mathbf{F} = m\mathbf{a}} \tag{1/1}$$

where $\mathbf{F}$ is the resultant force acting on the particle and $\mathbf{a}$ is the resulting acceleration. This equation is a *vector* equation since the direction of $\mathbf{F}$ must agree with the direction of $\mathbf{a}$ in addition to the equality in magnitudes of $\mathbf{F}$ and $m\mathbf{a}$. Newton's first law contains the principle of the equilibrium of forces, which is the main topic of concern in statics. Actually this law is a consequence of the second law, since there is no acceleration when the force is zero, and the particle either is at rest or is moving with a constant velocity. The first law adds nothing new to the description of motion but is included here since it was a part of Newton's classical statements.

° Newton's original formulations may be found in the translation of his *Principia* (1687) revised by F. Cajori, University of California Press, 1934.

The third law is basic to our understanding of force. It states that forces always occur in pairs of equal and opposite forces. Thus the downward force exerted on the desk by the pencil is accompanied by an upward force of equal magnitude exerted on the pencil by the desk. This principle holds for all forces, variable or constant, regardless of their source and holds at every instant of time during which the forces are applied. Lack of careful attention to this basic law is the cause of frequent error by the beginner. In analyzing bodies under the action of forces it is absolutely necessary to be clear about which of the pair of forces is being considered. It is necessary first of all to *isolate* the body under consideration and then to consider only the one force of the pair which acts *on* the body in question.

1/5 UNITS. The International System of Units, abbreviated SI (from the French, Système International d'Unités), has been accepted throughout the world and in the United States is rapidly replacing the U.S. customary system of units. During the years of transition U.S. engineers will need to be familiar with both systems, although their primary use will be for the SI metric system. In this book both SI and U.S. units are used, although the principal emphasis is on SI units.

The SI system is an absolute system of units based on the quantities of length, time, and mass with the units of force derived from Eq. 1/1. The main characteristics of SI units are set forth inside the front cover of the book along with the numerical conversions between U.S. customary and SI units. In addition, charts which give the approximate conversions between selected quantities in the two systems appear inside the back cover of the book for convenient reference. In statics we are primarily concerned with the units of length and force, with mass being involved only when we compute gravitational force, as explained in the next article. In SI units the unit of force is the newton (symbol N) which, by definition, is that force required to give a one-kilogram mass an acceleration of one meter per second squared. The kilogram is a unit of mass and *not* a unit of force. In the MKS gravitational system, which has been in use for years in many countries, the kilogram, like the pound in the U.S. customary system, has been used both as a unit of mass and a unit of force. It is necessary to guard against this practice when using the SI system.

Primary standards for the measurements of mass, length, and time have been established by international agreement and are as follows:

Mass. The kilogram is defined as the mass of a certain platinum-iridium cylinder which is kept at the International Bureau of Weights and Measures near Paris, France. An accurate copy of this cylinder is kept at the National Bureau of Standards in the United States and serves as the standard of mass for this country.

Length. The meter, originally defined as one ten millionth of the distance from the pole to the equator along the meridian through Paris, was later defined as the length of a certain platinum-iridium bar kept at the International Bureau of Weights and Measures. The difficulty of accessibility and accuracy of reproduction of measurement prompted the adoption of a more accurate and reproducible standard of length for the meter, which is now defined as 1 650 763.73 wavelengths of a certain radiation of the krypton-86 atom.

Time. The second was originally defined as the fraction 1/(86 400) of the mean solar day. Irregularities in the Earth's rotation have led to difficulties with this definition, and a more accurate and reproducible standard has been adopted. The second is now defined as the duration of 9 192 631 770 periods of the radiation of a certain state of the cesium-133 atom.

Clearly, for most engineering work, and for our purpose in studying mechanics, the accuracy of these standards is considerably beyond our needs.

In the U.S. customary system the unit of force is the pound (symbol lb). The standard pound is the force required to give a one-pound mass an acceleration of 32.1740 ft/sec^2, which is the standard value of gravitational acceleration g at sea level and at a latitude of 45°. Other units of force in use with the U.S. customary system are the *kilopound* (kip), which equals 1000 lb and the *ton*, which equals 2000 lb. In SI units the corresponding standard value of g is $9.806 \ 65 \text{ m/s}^2$. By applying Eq. 1/1 to free-fall conditions for the standard one-pound mass, we find that one pound of force is equivalent to $(0.453 \ 592 \ 37)(9.806 \ 65) = 4.4482$ newtons of force. To three significant figures, then, we have the equivalences

$$1 \text{ lb} = 4.45 \text{ N} \quad \text{or} \quad 1 \text{ N} = 0.225 \text{ lb}$$

The unit of mass in the U.S. customary system is derived from Eq. 1/1 and is the mass which has an acceleration of 1 ft/sec^2 when acted upon by a 1-lb force. This mass is called a *slug* and equals 32.1740 pounds mass. Thus from Eq. 1/1

$$\text{force (lb)} = \text{mass (slugs)} \times \text{acceleration (ft/sec}^2)$$

From the gravitational experiment where W is the weight we derive the mass m from

$$m \text{ (slugs)} = \frac{W \text{ (lb)}}{g \text{ (ft/sec}^2)}$$

In Fig. 1/6 are depicted examples of force, mass, and length in the two systems of units to aid in visualizing their relative magnitudes.

FORCE

LENGTH

Figure 1/6

1/6 LAW OF GRAVITATION.

In statics as well as dynamics we have frequent need to compute the weight of (gravitational force acting on) a body. This computation depends on the law of gravitation, which was also formulated by Newton. The law of gravitation is expressed by the equation

$$F = K \frac{m_1 m_2}{r^2} \qquad (1/2)$$

where F = the mutual force of attraction between two particles
K = a universal constant known as the constant of gravitation
m_1, m_2 = the masses of the two particles
r = the distance between the centers of the particles

The mutual forces F obey the law of action and reaction, since they are equal and opposite and are directed along the line joining the centers of the particles. By experiment the gravitational constant is found to be $K = 6.673(10^{-11})\text{m}^3/(\text{kg·s}^2)$. Gravitational forces exist between every pair of bodies. On the surface of the earth the only gravitational force of appreciable magnitude is the force due to the earth's attraction. Thus each of two iron spheres 100 mm in diameter is attracted to the earth with a gravitational force of 37.9 N, which is called its weight. On the other hand the force of mutual attraction between the spheres if they are just touching is 0.000 000 099 4 N. This force is clearly negligible compared with the earth's attraction of 37.9 N, and consequently the gravitational attraction of the earth is the only gravitational force of any appreciable magnitude which need be considered for most engineering experiments conducted on the earth's surface.

The gravitational attraction of the earth on a body is known as the *weight* of the body. This force exists whether the body is at rest or in motion. Since this attraction is a force, the weight of a body should be expressed in newtons (N) in SI units. Unfortunately in common practice the mass unit kilogram (kg) has been used extensively as a measure of weight. When expressed in kilograms the word "weight" technically means mass. To avoid confusion the word "weight" in this book shall be restricted to mean the force of gravitational attraction, and it will always be expressed in newtons.

For a body of mass m on the surface of the earth the earth's gravitational attraction on the body as specified by Eq. 1/2 may be calculated from the results of the simple gravitational experiment. If the gravitational force or weight has a magnitude W, then, since the body falls with an acceleration g, Eq. 1/1 gives

$$W = mg \tag{1/3}$$

The weight W will be in newtons (N) when m is in kilograms (kg) and g is in meters per second squared (m/s²). The standard value $g = 9.81$ m/s² will be sufficiently accurate for our calculations in statics. The corresponding value of g in U.S. customary units is 32.2 ft/sec².

The true weight (gravitational attraction) and the apparent weight (as measured by a spring scale) are slightly different. The difference, which is due to the rotation of the earth, is quite small and will be neglected. This effect will be discussed in *Vol. 2 Dynamics*.

1/7 ACCURACY, LIMITS, AND APPROXIMATIONS. The number of significant figures shown in an answer should be no greater than the number of figures which can be justified by the accuracy of the given data. Hence the cross-sectional area of a square bar whose side, 24 mm, say, was measured to the nearest half millimeter should be written as 580 mm² and not as 576 mm², as would be indicated if the numbers were multiplied out.

When calculations involve small differences in large quantities, greater accuracy in the data is required to achieve a given accuracy in the results. Hence it is necessary to know the numbers 4.2503 and 4.2391 to an accuracy of five significant figures in order that their difference 0.0112 be expressed to three-figure accuracy. It is often difficult in somewhat lengthy computations to know at the outset the number of significant figures needed in the original data to ensure a certain accuracy in the answer. Accuracy to three significant figures is considered satisfactory for the majority of engineering calculations.

The *order* of differential quantities is the subject of frequent misunderstanding. Higher-order differentials may always be neglected compared with lower-order differentials when the mathematical limit is approached. As an example the element of volume ΔV of a right circular cone of altitude h and base radius r may be taken to be a circular slice a distance x from the vertex and of thickness Δx. It can be verified that the complete expression for the volume of the element may be written as

$$\Delta V = \frac{\pi r^2}{h^2}[x^2 \, \Delta x + x(\Delta x)^2 + \tfrac{1}{3}(\Delta x)^3]$$

It should be recognized that, when passing to the limit in going from ΔV to dV and from Δx to dx, the terms in $(\Delta x)^2$ and $(\Delta x)^3$ drop out, leaving merely

$$dV = \frac{\pi r^2}{h^2}x^2 \, dx$$

which gives an exact expression when integrated.

When dealing with small angles we can usually make use of simplifying assumptions. Consider the right triangle of Fig. 1/7 where the angle θ, expressed in radians, is relatively small. With the hypotenuse taken as unity, we see from the geometry of the figure that the arc length $1 \times \theta$ and $\sin \theta$ are very nearly the same. Also $\cos \theta$ is close to unity. Furthermore, $\sin \theta$ and $\tan \theta$ have almost the same values. Thus for small angles we may write

$$\sin \theta \approx \tan \theta \approx \theta \qquad \cos \theta \approx 1$$

These approximations amount to retaining only the first terms in the series expansions for these three functions. As an example of these approximations, for an angle of $1°$

$$1° = 0.017\ 453 \text{ rad}$$
$$\sin 1° = 0.017\ 452$$
$$\tan 1° = 0.017\ 455$$
$$\cos 1° = 0.999\ 848$$

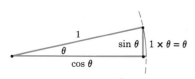

Figure 1/7

If a closer approximation is desired, the first two terms may be retained, and they are

$$\sin \theta = \theta - \theta^3/6 \qquad \tan \theta = \theta + \theta^3/3 \qquad \cos \theta = 1 - \theta^2/2$$

The error in replacing the sine by the angle for $1°$ is only 0.005 percent. For $5°$ the error is 0.13 percent, and for $10°$ the error is still only 0.51 percent. As the angle θ approaches zero, it should now be clear that the following relations are true in the mathematical limit:

$$\sin d\theta = \tan d\theta = d\theta$$

$$\cos d\theta = 1$$

The angle $d\theta$ is, of course, expressed in radian measure.

1/8 DESCRIPTION OF STATICS PROBLEMS.

The study of statics is directed toward the quantitative description of forces that act on engineering structures in equilibrium. Mathematics establishes the relations between the various quantities involved and makes it possible for us to predict effects from these relations. A dual thought process is required in formulating this description. It is necessary to think in terms of the physical situation and in terms of the corresponding mathematical description. Analysis of every problem will require the repeated transition of thought between the physical and the mathematical. Without question, one of the most common problems encountered by the student is the difficulty in making this transition of thought freely. We should recognize that the mathematical formulation of a physical problem represents an ideal limiting description, or model, which approximates but never quite matches the actual physical situation.

When constructing the idealized mathematical model for a given engineering problem certain approximations will always be involved. Some of these approximations may be mathematical, whereas others will be physical. For instance, it is often necessary for us to neglect small distances, angles, or forces compared with large distances, angles, or forces. A force which is actually distributed over a small area of the body upon which it acts may be considered a concentrated force if the dimensions of the area involved are small compared with other pertinent dimensions. The weight of a steel cable per unit length may be neglected if the tension in the cable is many times greater than its total weight, whereas the cable weight may not be neglected if the problem calls for a determination of the deflection or sag of a suspended cable due to its weight. Thus the degree of assumption involved depends on what information is desired and on the accuracy required. We must be constantly alert to the various assumptions called for in the formulation of real problems. The ability to understand and make use of the appropriate assumptions in the formulation and solution of engineering problems is certainly one of the most important characteristics of a successful engineer. One of the major aims of this book is to provide a maximum of opportunity to develop this ability through the formulation and analysis of many practical problems involving the principles of statics.

Graphics is an important analytical tool which serves us in three capacities. First, it makes possible the representation of a physical system on paper by means of a sketch or diagram. Geometrical representation is vital to physical interpretation and aids greatly in visualizing the three-dimensional aspects of many problems. Second, graphics often affords a means of solving physical relations where a direct mathematical solution would be awkward or difficult. Graphical solutions not only provide us with a practical means for obtaining results, but they also aid greatly in making the transition of thought between the physical situation and the mathematical expression because both are represented simultaneously. A third use of graphics is in the display of results in charts or graphs which become a valuable aid to representation.

An effective method of attack on statics problems, as in all engineering problems, is essential. The development of good habits in formulating problems and in representing their solutions will prove to be an invaluable asset. Each solution should proceed with a logical sequence of steps from hypothesis to conclusion, and its representation should include a clear statement of the following parts, each clearly identified:

1. Given data
2. Results desired
3. Necessary diagrams
4. Calculations
5. Answers and conclusions

In addition it is well to incorporate a series of checks on the calculations at intermediate points in the solution. We should observe reasonableness of numerical magnitudes, and the accuracy and dimensional homogeneity of terms should be frequently checked. It is also important that all work be neat and orderly. Careless solutions that cannot be easily read by others are of little or no value. The discipline involved in adherence to good form will in itself be an invaluable aid to the development of the abilities for formulation and analysis. Many problems that at first may seem difficult and complicated become clear and straightforward once they are begun with a logical and disciplined method of attack.

The subject of statics is based on surprisingly few fundamental concepts and involves mainly the application of these basic relations to a variety of situations. In this application the *method* of analysis is all-important. In solving a problem it is essential that the laws which apply be carefully fixed in mind and that we apply these principles literally and exactly. In applying the principles which define the requirements for forces acting on a body it is essential that we *isolate* the body in question from all other bodies so that complete and accurate account of all forces which act on this body may be taken. This *isolation* should exist mentally as well as be represented on paper. The diagram of such an isolated body with the representation of *all* external forces acting on it is called a *free-body diagram*. It has long been established that the *free-body-diagram method* is the key to the understanding of mechanics. This is so because the *isolation* of a body is the tool by which *cause* and *effect* are clearly separated and by which our attention to the literal application of a principle is accurately focused. The technique of drawing free-body diagrams is covered in Chapter 3 where they are first used.

In applying the laws of statics, we may use numerical values of the quantities directly in proceeding toward the solution, or we may use algebraic symbols to represent the quantities involved and leave the answer as a formula. With numerical substitution the magnitude of each quantity expressed in its particular units is evident at each stage of the calculation. This approach offers advantage when the practical significance of the magnitude of each term is important. The symbolic solution, however, has several advantages over the numerical solution. First, the abbreviation achieved by the use of symbols aids in focusing our attention on the connection between the physical situation and its related mathematical description. Second, a symbolic solution allows us to make a dimensional check at every step, whereas dimensional homogeneity may be lost when numerical values only are used. Third, we may use a symbolic solution repeatedly for obtaining answers to the same problem when different sets and sizes of units are used. Facility with both forms of solution is essential, and ample practice with each should be sought in the problem work.

The student will find that solutions to the problems of statics may be obtained in one of three ways. First, we may utilize a direct mathematical solution by hand calculation where answers appear either as algebraic symbols or as numerical results. The large majority of problems come under this category. Second, we may approximate the results of certain problems by graphical solutions. Third, the modern digital computer is of particular advantage where a large number of equations or repeated data are involved in numerical form. Students who have ready access to digital computation facilities may wish to solve a selected few problems by this means. In order to reduce computation time in the problem work the data for most of the problems are given in simple numbers. The choice of the most expedient method of solution is an important aspect of the experience to be gained from the problem work.

FORCE SYSTEMS

2

(a)

Cable tension
P

(b)

Figure 2/1

2/1 INTRODUCTION. In this chapter and in the chapters that follow the properties and effects of various kinds of forces as they act on engineering structures and mechanisms will be examined. The experience gained through this examination will prove to be of fundamental use throughout the study of mechanics and in the study of other subjects such as stress analysis, design of structures and machines, and fluid flow. The foundation for a basic understanding of not only statics but also of the entire subject of mechanics is laid in this chapter, and the student should master this material thoroughly.

2/2 FORCE. Before dealing with a group or *system* of forces it is necessary for us to examine the properties of a single force in some detail. A force has been defined as the action of one body on another. We find that force is a vector quantity, since its effect depends on the direction as well as on the magnitude of the action and since forces may be combined according to the parallelogram law of vector combination. The action of the cable tension on the bracket in Fig. 2/1*a* is represented in Fig. 2/1*b* by the force vector **P** of magnitude *P*. The effect of this action on the bracket will depend on *P*, the angle θ, and the location of the point of application *A*. Changing any one of these three specifications will alter the effect on the bracket, as could be detected, for instance, by the force in one of the bolts which secure the bracket to the base or the internal stress and strain in the material of the bracket at any point. Thus the complete specification of the action of a force must include its *magnitude, direction,* and *point of application,* in which case it is treated as a fixed vector.

Force is applied either by direct mechanical contact or by remote action. Gravitational and magnetic forces are applied by remote action. All other forces are applied through direct physical contact.

The action of a force on a body can be separated into two effects, *external* and *internal*. For the bracket of Fig. 2/1 the effects of **P** external to the bracket are the reactions or forces (not shown) exerted on the bracket by the foundation and bolts because of the action of **P**. Forces external to a body are then of two kinds, *applied* forces and *reactive* forces. The effects of **P** internal to the bracket are the resulting internal stresses and strains distributed throughout the material of the bracket. The relation between internal forces and internal strains involves the material properties of the body and is studied in strength of materials, elasticity, and plasticity.

In dealing with the mechanics of rigid bodies, where concern is given only to the net *external* effects of forces, experience shows us that it is not necessary to restrict the action of an applied force to a given point. Hence the force **P** acting on the rigid plate in Fig. 2/2 may be applied at *A* or at *B* or at any other point on its action line, and the net external effects of **P** on the bracket will not change. The external effects are the force exerted on the plate by the bearing support at *O* and the force exerted on the plate by the roller support at *C*. This conclusion is described by the *principle of transmissibility*, which states that a force may be applied at any point on its given line of action without altering the resultant effects of the force *external* to the *rigid* body on which it acts. When only the resultant external effects of a force are to be investigated, the force may be treated as a *sliding* vector, and it is necessary and sufficient to specify the *magnitude, direction,* and *line of action* of the force. Since this book deals essentially with the mechanics of rigid bodies, we will treat almost all forces as sliding vectors for the rigid body on which they act.

Forces may be either *concentrated* or *distributed*. Actually every contact force is applied over a finite area and is therefore a distributed force. When the dimensions of the area are very small compared with the other dimensions of the body, we may consider the force to be concentrated at a point with negligible loss of accuracy. Force may be distributed over an area, as in the case of mechanical contact, or it may be distributed over a volume when gravity or magnetic force is acting. The *weight* of a body is the force of gravitational attraction distributed over its volume and may be taken as a concentrated force acting through the center of gravity. The position of the center of gravity is frequently obvious from considerations of symmetry. If the position is not obvious, then a separate calculation, explained in Chapter 5, will be necessary to locate the center of gravity.

A force may be measured either by comparison with other known forces, using a mechanical balance, or by the calibrated movement of

Figure 2/2

(a)

(b)

(c)

(d)

Figure 2/3

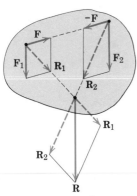

Figure 2/4

an elastic element. All such comparisons or calibrations have as their basis a primary standard. The standard unit of force in SI units is the newton (N) and in the U.S. customary system is the pound (lb), as defined in Art. 1/5.

The characteristic of a force expressed by Newton's third law must be carefully observed. The *action* of a force is always accompanied by an *equal* and *opposite reaction*. It is essential for us to fix clearly in mind which force of the pair is being considered. The answer is always clear when the body in question is *isolated* and the force exerted *on* that body (not *by* the body) is represented. It is very easy to make a careless mistake and consider the wrong force of the pair unless we distinguish carefully between action and reaction.

Two forces $\mathbf{F}_1$ and $\mathbf{F}_2$ that are concurrent may be added by the parallelogram law in their common plane to obtain their sum or *resultant* $\mathbf{R}$ as shown in Fig. 2/3a. If the two concurrent forces lie in the same plane but are applied at two different points as in Fig. 2/3b, by the principle of transmissibility we may move them along their lines of action and complete their vector sum $\mathbf{R}$ at the point of concurrency. The resultant $\mathbf{R}$ may replace $\mathbf{F}_1$ and $\mathbf{F}_2$ without altering the external effects on the body upon which they act. The triangle law may also be used to obtain $\mathbf{R}$, but it will require moving the line of action of one of the forces as shown in Fig. 2/3c. In Fig. 2/3d the same two forces are added, and although the correct magnitude and direction of $\mathbf{R}$ are preserved, we lose the correct line of action, since $\mathbf{R}$ obtained in this way does not pass through A. This type of combination should be avoided. Mathematically the sum of the two forces may be written by the vector equation

$$\mathbf{R} = \mathbf{F}_1 + \mathbf{F}_2$$

In addition to the need for combining forces to obtain their resultant, we often have occasion to replace a force by its *components* which act in two specified directions. Thus the force $\mathbf{R}$ in Fig. 2/3a may be replaced by or *resolved* into two components $\mathbf{F}_1$ and $\mathbf{F}_2$ with these specified directions merely by completing the parallelogram as shown to obtain the magnitudes of $\mathbf{F}_1$ and $\mathbf{F}_2$.

A special case of addition is presented when the two forces $\mathbf{F}_1$ and $\mathbf{F}_2$ are parallel, Fig. 2/4. They may be combined by first adding two equal, opposite, and collinear forces $\mathbf{F}$ and $-\mathbf{F}$ of convenient magnitude which taken together produce no external effect on the body. Adding $\mathbf{F}_1$ and $\mathbf{F}$ to produce $\mathbf{R}_1$ and combining with the sum $\mathbf{R}_2$ of $\mathbf{F}_2$ and $-\mathbf{F}$ yield the resultant $\mathbf{R}$ correct in magnitude, direction, and line of action. This procedure is also useful in obtaining a graphical combination of two forces that are almost parallel and hence have a point of concurrency which is remote and inconvenient.

It is usually helpful to master the analysis of force systems in two dimensions before undertaking three-dimensional analysis. To this end the remainder of the chapter is subdivided into these two categories. However, for students who have a good command of vector analysis, these sections may be studied simultaneously if preferred.

SECTION A. TWO-DIMENSIONAL FORCE SYSTEMS

2/3 RECTANGULAR COMPONENTS. The most common two-dimensional resolution of a force **F** is resolution into *rectangular components* $\mathbf{F}_x$ and $\mathbf{F}_y$ as shown in Fig. 2/5. It should be immediately evident from the figure that

$$\boxed{\begin{aligned} F_x &= F \cos\theta & F &= \sqrt{F_x^2 + F_y^2} \\ F_y &= F \sin\theta & \theta &= \tan^{-1}\frac{F_y}{F_x} \end{aligned}} \qquad (2/1)$$

Figure 2/5

where F is the magnitude of **F** and where F_x and F_y are the magnitudes of $\mathbf{F}_x$ and $\mathbf{F}_y$. If we introduce unit vectors **i** and **j** in the *x*- and *y*-directions as shown in Fig. 2/5, we may write the vector equation

$$\boxed{\mathbf{F} = \mathbf{F}_x + \mathbf{F}_y = \mathbf{i}F_x + \mathbf{j}F_y} \qquad (2/2)$$

To eliminate any ambiguity it is desirable to show the components of a force in dotted lines, as in Fig. 2/5, and the force in a full line, or vice versa. With either of these conventions it will always be clear that a force and its components are being represented and not three separate forces as would be implied by three solid-line vectors.

Actual problems do not come with reference axes, so their assignment is a matter of arbitrary convenience, and the choice is frequently up to the student. The logical choice is usually indicated by the manner in which the geometry of the problem is specified. When the principal dimensions of a body are given in the horizontal and vertical directions, for example, then assignment of reference axes in these directions is generally convenient. However, dimensions are not always given in horizontal and vertical directions, angles need not be measured counterclockwise from the *x*-axis, and the origin of coordinates need not be on the line of action of a force. Therefore, it is essential that we be able to determine the correct components of a force no matter how the axes are oriented or how the angles are measured. Figure 2/6 suggests a few typical examples of resolution situations in two dimensions, the results of which should be readily apparent. Thus it is seen that memorization of Eqs. 2/1 is not a substitute for an understanding of the parallelogram law and for the correct projection of a vector onto a reference axis. A neatly drawn sketch always helps to clarify the geometry and avoid error.

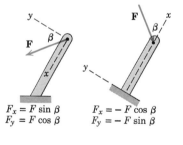

$$\begin{array}{ll} F_x = F \sin\beta & F_x = -F \cos\beta \\ F_y = F \cos\beta & F_y = -F \sin\beta \end{array}$$

$$\begin{array}{ll} F_x = F \sin(\pi - \beta) & F_x = F \cos(\beta - \alpha) \\ F_y = -F \cos(\pi - \beta) & F_y = F \sin(\beta - \alpha) \end{array}$$

Figure 2/6

Sample Problem 2/1

The 100-N force **F** is applied to the fixed bracket as shown. Determine the magnitudes of the rectangular components of **F** in (1) the x- and y-directions and (2) the x'- and y'-directions. Also (3) find the components of **F** in the x'- and y-directions.

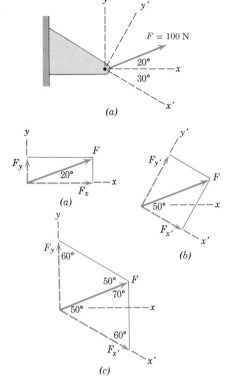

(a)

Solution. Part (1). The x- and y-components of **F** are shown in Fig. a and are

$$F_x = F\cos\theta_x = 100\cos 20° = 94.0 \text{ N}$$

$$F_y = F\cos\theta_y = 100\cos 70° = 34.2 \text{ N} \qquad Ans.$$

Part (2). The x'- and y'-components of **F** are the projections on these axes as shown in Fig. b and are

$$F_{x'} = F\cos\theta_{x'} = 100\cos 50° = 64.3 \text{ N}$$

$$F_{y'} = F\cos\theta_{y'} = 100\cos 40° = 76.6 \text{ N} \qquad Ans.$$

Part (3). The components of **F** in the x'- and y-directions are non-rectangular and are obtained by completing the parallelogram as shown in Fig. c. The components may be calculated by the law of sines which gives

$$\frac{F_{x'}}{\sin 70°} = \frac{F}{\sin 60°} \qquad F_{x'} = \frac{0.940}{0.866}100 = 108.5 \text{ N}$$

$$\frac{F_y}{\sin 50°} = \frac{F}{\sin 60°} \qquad F_y = \frac{0.766}{0.866}100 = 88.5 \text{ N} \qquad Ans.$$

① Obtain $F_{x'}$ and F_y graphically and compare your results with the calculated values.

Sample Problem 2/2

Combine the two forces **P** and **T**, which act on the fixed structure at B, into a single force **R**.

Solution. The parallelogram addition of **T** and **P** is shown in the figure. We must first find the angle α before R can be found by the law of cosines. From the given figure

$$\tan\alpha = \frac{\overline{BD}}{\overline{AD}} = \frac{6\sin 60°}{3 + 6\cos 60°} = 0.866, \quad \alpha = 40.9°$$

The law of cosines applied to the parallelogram of vectors gives

$$R^2 = 600^2 + 800^2 - 2(600)(800)\cos 40.9° = 274\,300$$

$$R = 524 \text{ lb} \qquad Ans.$$

The angle θ gives the direction of **R**, and from the law of sines

$$\frac{600}{\sin\theta} = \frac{524}{\sin 40.9°}, \quad \sin\theta = 0.750, \quad \theta = 48.6° \qquad Ans.$$

① Note carefully the repositioning of P to permit parallelogram addition at B.

PROBLEMS

2/1 Calculate the x- and y-components of the force **P** acting on the structural member if the magnitude of the force is 20 kN.

Ans. $P_x = -19.70$ kN, $P_y = 3.47$ kN

Problem 2/1

2/2 Find the rectangular component of the 600-lb tension **T** of Sample Problem 2/2 along the line BC.

2/3 When the load L is 7 m from the pivot at C, the tension **T** in the cable has a magnitude of 15 kN. Express **T** as a vector using the unit vectors **i** and **j**.

Ans. **T** $= 12.86$**i** $+ 7.72$**j** kN

Problem 2/3

2/4 A cable exerts a force **F** on the bracket of the structural member to which it is attached. If the magnitude of the x-component of **F** is 900 N, calculate the y-component and the magnitude of **F**.

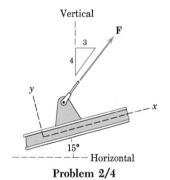

Problem 2/4

2/5 The turnbuckle C is tightened until the light cable which is stretched across the frame is under a tension of 900 N. Compute the x- and y-components of the force exerted *on* the frame at B *by* the cable.

Ans. $F_x = -879$ N, $F_y = -195.2$ N

Problem 2/5

Problem 2/6

2/6 The tension **T** in the vertical cable equals the weight of the crate. Calculate the components T_t and T_n along and normal to the boom, respectively, of the force **T** applied to the boom at A by the crate.

Problem 2/7

2/7 The circular cam has an offset $e = 2$ in. and a radius $r = 4$ in. For the position where $\theta = 30°$ the smooth undersurface of the plunger exerts a downward force of 400 lb on the cam normal to the contacting surfaces. Calculate the rectangular component F' of this force along the line joining the contact point and the center of the shaft which turns the cam.

Ans. $F' = 378$ lb

Problem 2/8

2/8 The hydraulic cylinder exerts a force of 40 kN in the direction of its shaft against the load that it is hoisting. Determine the components F_n and F_t normal and tangent to AB for the position $\theta = 30°$.

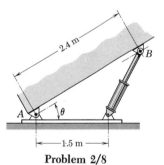

Problem 2/9

2/9 Calculate the magnitude of the single force **R** which is equivalent to the two forces shown. Also calculate the angle θ measured counterclockwise from the positive x-axis to **R**.

Ans. $R = 9.17$ kN, $\theta = 109.1°$

2/10 Solve Problem 2/9 graphically.

2/11 At what angle θ must the 400-lb force be applied in order that the resultant **R** of the two forces will have a magnitude of 1000 lb? For this condition what will be the angle β between **R** and the horizontal?

Ans. $\theta = 51.3°$, $\beta = 18.2°$

Problem 2/11

2/12 The 10-kN vertical force is to be replaced by two forces, $\mathbf{F}_1$ directed along the 45° line *a-a* and $\mathbf{F}_2$ which has a magnitude of 8 kN. Calculate the magnitude of $\mathbf{F}_1$ and the counterclockwise angle θ made by $\mathbf{F}_2$ with the *x*-axis.

Ans. $F_1 = 10.81$ kN with $\theta = 17.1°$,
or $F_1 = 3.33$ kN with $\theta = 72.9°$

Problem 2/12

2/13 Solve Prob. 2/12 graphically.

2/14 The rigid member *ABC* is supported by the pin *A* and the hinged link *D* and is subjected to a force **F** at *C*. Can it be concluded from the principle of transmissibility that the force exerted by the pin at *A* on member *ABC* would be the same if **F** were applied either at *D* or at *E* rather than at *C*?

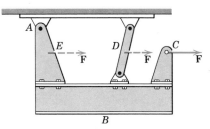

Problem 2/14

2/15 Resolve the 4-kN force into two components, one along *AB* and the other along *BC*.

Ans. $F_{AB} = 2.93$ kN, $F_{BC} = 3.59$ kN

Problem 2/15

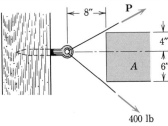

Problem 2/16

2/16 It is desired to remove the spike from the timber by applying force along its horizontal axis. An obstruction *A* prevents direct access, so that two forces, one 400 lb and the other **P**, are applied by cables as shown. Compute the magnitude of **P** necessary to ensure tension *T* along the axis of the spike. Also find *T*.

Problem 2/17

2/17 Replace the 6-kN and 4-kN forces by a single equivalent force **R** expressed as a vector using unit vectors in the *x*- and *y*-directions. Compute the angle θ made by **R** with the *x*-axis.
Ans. **R** $= 2.63\mathbf{i} + 6.06\mathbf{j}$ kN, $\theta = 66.6°$

Problem 2/18

2/18 The resultant of the 100-kN load and the accompanying tension *T* in member *B* passes through point *A* and results in a certain force on the pin which supports the truss at this point. If the 100-kN load is replaced by a force **P** applied along the dotted line shown, determine the magnitude of **P** which will result in the same effect on the pin at *A* as when the 100-kN force was applied. Specify the corresponding increment ΔT in the tension of member *B*. All interior angles of the truss are either 45° or 90°. (*Hint:* The resultant of *P* and the new *T* would be the same vector as the resultant of the 100-kN force and the original *T*.)

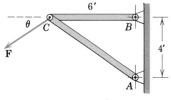

Problem 2/19

2/19 At what maximum angle θ should the force **F** be directed so that the magnitude of its component along *CA* does not exceed 80 percent of the magnitude of its component along *BC*?   *Ans.* $\theta = 53.0°$

2/20 The force $\mathbf{F} = -40\mathbf{i} + 60\mathbf{j}$ lb is to be replaced by two forces, one along the *y*-axis and one in the horizontal *h*-direction, which produce the same combined effect as **F**. Determine the magnitudes of these components.

Problem 2/20

2/21 When resolved into rectangular components along the *x*- and *y*-axes, the force **P** has an *x*-component of 500 N. For resolution along the *x'*- and *y'*-axes, $P_{x'} = 800$ N. Express **P** as a vector using unit vectors with the *x-y* axes and calculate the *y'*-component of **P**. (With the aid of the geometry of the vector figure, carry out the solution without involving simultaneous solutions.)

     Ans. $\mathbf{P} = 500\mathbf{i} + 965\mathbf{j}$ N, $P_{y'} = 736$ N

Problem 2/21

2/22 Solve Prob. 2/21 graphically.

2/23 It is known that the resultant of the two forces passes through point *A*. Determine the magnitude of **P**.

     Ans. $P = 675$ N

Problem 2/23

2/24 The two forces which act on the rigid frame are to be replaced by a single equivalent force **R** applied at point *A*. Determine the magnitude of **R** and the distance *x* to point *A*. Solve graphically or algebraically.

     Ans. $R = 898$ lb, $x = 4.05$ ft

Problem 2/24

2/4 MOMENT.   In addition to the tendency to move a body in the direction of its application, a force also tends to rotate the body about any axis which does not intersect the line of action of the force and which is not parallel to it. This tendency is known as the *moment* **M** of the force about the given axis. The moment of a force is also frequently referred to as *torque*.

Figure 2/7*a* shows a two-dimensional body acted upon by a force **F** in its plane. The magnitude of the moment or tendency of the force to rotate the body about the axis *O-O* normal to the plane of the body is, clearly, proportional both to the magnitude of the force and to the *moment arm d*, which is the perpendicular distance from the axis to the line of action of the force. Therefore the magnitude of the moment is defined as

$$M = Fd \qquad\qquad (2/3)$$

The moment is a vector **M** perpendicular to the plane of the body. The sense of **M** depends on the direction in which **F** tends to rotate the body. The right-hand rule, Fig. 2/7*b*, is used to identify this sense, and the moment of **F** about *O-O* may be represented as a vector pointing in the direction of the thumb with the fingers curled in the direction of the tendency to rotate. The moment **M** obeys all the rules of vector combination and may be considered a sliding vector with a line of action coinciding with the moment axis. The basic units of moment in SI units are newton-meters (N·m) and in the U.S. customary system are pound-feet (lb-ft).

When dealing with forces all of which act in a given plane we customarily speak of the moment about a point. Actually the moment with respect to an axis normal to the plane and passing through the point is implied. Thus the moment of force **F** about point *O* in Fig. 2/7*c* has the magnitude $M_O = Fd$ and is counterclockwise. Vector representation of moments for coplanar forces is unnecessary, since the vectors are either out from the paper (counterclockwise) or

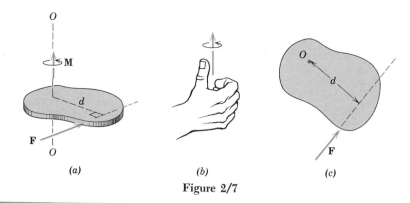

(*a*)   (*b*)   (*c*)

Figure 2/7

into the paper (clockwise). Since the addition of parallel free vectors may be accomplished with *scalar* algebra, the moment directions may be accounted for by using a plus sign (+) for counterclockwise moments and a minus sign (−) for clockwise moments, or vice versa. It is necessary only for us to be consistent within a given problem in using either sign convention.

    Principle of moments. One of the most important principles of mechanics is *Varignon's theorem*, or the *principle of moments,* which for coplanar forces states that the moment of a force about any point is equal to the sum of the moments of the components of the force about the same point. To prove this statement we consider a force R and two equivalent components P and Q acting at point A, Fig. 2/8.

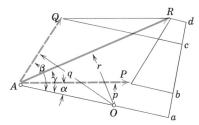

Figure 2/8

Point O is selected arbitrarily as the moment center. Construct the line AO and project the three vectors onto the normal to this line. Also construct the moment arms p, q, r of the three forces to point O and designate the angles of the vectors to the line AO by α, β, γ as shown in the figure. Since the parallelogram whose sides are $\overline{P}$ and $\overline{Q}$ requires that $\overline{ac} = \overline{bd}$, it is evident that

$$\overline{ad} = \overline{ab} + \overline{bd} = \overline{ab} + \overline{ac}$$

or

$$R \sin \gamma = P \sin \alpha + Q \sin \beta$$

Multiplying by the distance $\overline{AO}$ and substituting the values of p, q, r give

$$Rr = Pp + Qq$$

which proves that the moment of a force about any point equals the sum of the moments of its two components about the same point. Varignon's theorem need not be restricted to the case of only two components but applies equally well to three or more, since it is always possible by direct combination to reduce the number of components to two for which the theorem was proved.

Sample Problem 2/3

Calculate the moment about the base point O of the 600-N force in four different ways.

Solution. (*I*) The moment arm to the 600-N force is

$$d = 4 \cos 40° + 2 \sin 40° = 4.35 \text{ m}$$

By $M = Fd$ the moment is clockwise and has the magnitude

$$M_O = 600(4.35) = 2610 \text{ N} \cdot \text{m} \qquad Ans.$$

(*II*) Replace the force by its rectangular components at A

$$F_1 = 600 \cos 40° = 460 \text{ N}, \quad F_2 = 600 \sin 40° = 386 \text{ N}$$

By Varignon's theorem, then, the moment becomes

$$M_O = 460(4) + 386(2) = 2610 \text{ N} \cdot \text{m} \qquad Ans.$$

① The required geometry here and in similar problems should not cause difficulty if the sketch is carefully drawn.

(*III*) By the principle of transmissibility move the 600-N force along its line of action to point B, which eliminates the moment of the component F_2. The moment arm of F_1 becomes

$$d_1 = 4 + 2 \tan 40° = 5.68 \text{ m}$$

and the moment is

$$M_O = 460(5.68) = 2610 \text{ N} \cdot \text{m} \qquad Ans.$$

② This procedure is frequently the shortest approach.

(*IV*) Moving the force to point C eliminates the moment of the component F_1. The moment arm of F_2 becomes

$$d_2 = 2 + 4 \text{ ctn } 40° = 6.77 \text{ m}$$

and the moment is

$$M_O = 386(6.77) = 2610 \text{ N} \cdot \text{m} \qquad Ans.$$

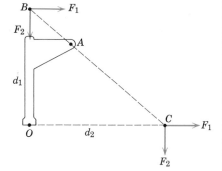

③ The fact that points B and C are not on the body proper should not cause concern, as the mathematical calculation of the moment of a force does not require that the force be on the body.

PROBLEMS

2/25 In Sample Problem 2/3 calculate the magnitude of the smallest force **P** which can be applied at point A and produce the same moment about O as that produced by the 600-N force. Find the corresponding angle θ between **P** and the horizontal.
                    Ans. $P = 584$ N, $\theta = 26.6°$

2/26 The moment of the force **P** about point A is 30 N·m. Calculate the magnitude of **P**. The plate upon which the force acts is divided into 0.1-m squares.

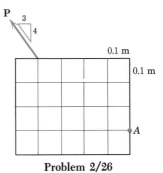

Problem 2/26

2/27 The rectangular plate is made up of 1-m squares as shown. A 10-kN force is applied at point A in the direction shown. Calculate the moment M_B of the force about point B.   *Ans.* $M_B = 27.7$ kN·m
                                        clockwise

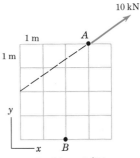

Problem 2/27

2/28 Compute the moment of the 400-N force about point O.

Problem 2/28

1000 lb

A | 3' | 5' | B

Problem 2/29

2/29 It is desired to replace the 1000-lb force acting on the beam by two downward forces, R_A at A and R_B at B, which will not alter the overall external effect on the beam. Calculate these forces.

1 m

→800 N

2 m

300 N ←

1 m

Problem 2/30

2/30 Determine the distance y down from the top of the pole at which a single horizontal force **P** must be applied in order to duplicate the external effects of the two given forces.   *Ans.* $y = 0.4$ m

250 N

15°

← 200 mm →

30 mm

Problem 2/31

2/31 Calculate the moment of the 250-N force on the handle of the monkey wrench about the center of the bolt.

200 N

450 mm

650 mm

20°

O

125 mm

Problem 2/32

2/32 A force of 200 N is applied to the end of the wrench to tighten a flange bolt which holds the wheel to the axle. Determine the moment M produced by this force about the center O of the wheel for the position of the wrench shown.   *Ans.* $M = 78.3$ N·m

2/33 Calculate the moment about point A exerted by the 40,000-lb force supported by the hoisting cable of the tractor crane for the position shown.

Problem 2/33

2/34 In raising the flagpole from the position shown, the tension T in the cable must supply a moment about O of 73 kN·m. Determine T.   *Ans. T = 8.65 kN*

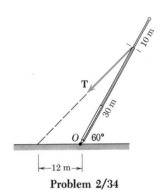

Problem 2/34

2/35 Determine the angle θ which will maximize the moment M_O of the 50-lb force about the shaft axis at O. Also compute M_O.

Problem 2/35

2/36 If the combined moment about point A of the 50-kN force and the force **P** is zero, determine both graphically and algebraically the magnitude of **P**. The plate upon which the forces act is divided into squares.   *Ans. P = 51.5 kN*

Problem 2/36

60 lb

Problem 2/37

2/37 If the combined moment about C of the two forces is zero, determine the magnitude R of their resultant.

Problem 2/38

2/38 In the slider-crank mechanism shown, the connecting rod AB of length l supports a variable compressive force $\mathbf{C}$. Derive an expression for the moment of $\mathbf{C}$ about the crank axis O in terms of C, r, l, and the variable angle θ.

Problem 2/39

2/39 The masthead fitting supports the two forces shown. Determine the magnitude of $\mathbf{T}$ which will cause no bending (zero moment) of the mast at point C.

2/40 For Prob. 2/39 calculate the magnitude of $\mathbf{T}$ so that there will be no bending in the mast at point O.
Ans. $T = 3.23$ kN

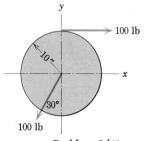

Problem 2/41

2/41 Determine graphically or algebraically the coordinates of the point A on the rim of the wheel about which the combined moment M of the two forces is a maximum. Find M for this point.
Ans. $A(-8.66$ in., -5 in.$)$, $M = 2000$ lb-in.

2/5 COUPLE. The moment produced by two equal and opposite and noncollinear forces is known as a *couple*. Couples have certain unique properties and have important applications in mechanics.

Consider the action of two equal and opposite forces **F** and −**F** a distance d apart, Fig. 2/9a. These two forces cannot be combined into a single force, since their sum in every direction is zero. Their effect is entirely to produce a tendency of rotation. The combined moment of the two forces about an axis normal to their plane and passing through any point such as O in their plane is the *couple* **M.** It has a magnitude

$$M = F(a + d) - Fa$$

or

$$M = Fd$$

and is in the counterclockwise direction when viewed from above for the case illustrated. Note especially that the magnitude of the couple contains no reference to the dimension a which locates the forces with respect to the moment center O. It follows that the moment of a couple has the same value for *all* moment centers. We may therefore represent a couple by a *free* vector **M,** as shown in Fig. 2/9b, where the direction of **M** is normal to the plane of the couple and the sense of the vector is established by the right-hand convention.

Since the couple vector **M** will always be perpendicular to the plane of the forces which constitute the couple, in two-dimensional analysis we can represent the sense of a couple vector as clockwise or counterclockwise by one of the conventions shown in Fig. 2/9c. Later when we deal with couple vectors in three-dimensional problems, we will retain the vector notation for their representation.

A couple is unchanged as long as the magnitude and direction of its vector remain constant. Consequently a given couple will not be altered by changing the values of F and d as long as their product remains the same. Likewise a couple is not affected by allowing the

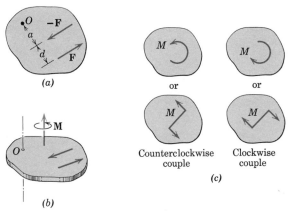

(a)

(b)

M or M

Counterclockwise couple   Clockwise couple

(c)

Figure 2/9

forces to act in any one of parallel planes. Figure 2/10 shows four different configurations of the same couple **M.** In each of the four cases the couple is described by the identical free vector that represents the identical tendencies to rotate the bodies in the direction shown.

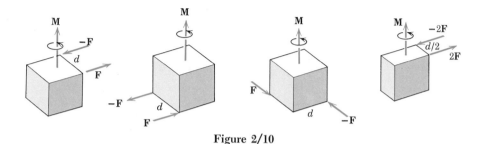

Figure 2/10

The effect of a force acting on a body has been described in terms of the tendency to push or pull the body in the direction of the force and to rotate the body about any axis which does not intersect the line of the force. The representation of this dual effect is often facilitated by replacing the given force by an equal parallel force and a couple to compensate for the change in the moment of the force. This resolution of a force into a force and a couple is illustrated in Fig. 2/11, where

Figure 2/11

the given force **F** acting at point A is replaced by an equal force **F** at some point B and the counterclockwise couple $M = Fd$. The transfer is seen from the middle figure, where the equal and opposite forces **F** and $-\mathbf{F}$ are added at point B without introducing any net external effects on the body. We now see that the original force at A and the equal and opposite one at B constitute the couple $M = Fd$, which is counterclockwise for the sample chosen, as shown in the right-hand part of the figure. Thus we have replaced the original force at A by the same force acting at a different point B and a couple without altering the external effects of the original force on the body. It follows also that a given couple and a force which lies in the plane of the couple (normal to the couple vector) may be combined to produce a single force by reversing the procedure. The resolution of a force into an equivalent force and couple is a step that finds repeated application in mechanics and should be thoroughly mastered.

Sample Problem 2/4

The rigid bracket is subjected to a couple consisting of the 200-N forces. Replace this couple by an equivalent couple consisting of the two forces **P** and −**P** which have a magnitude of 500 N. Find the proper angle θ.

Dimensions in Millimeters

Solution. The given couple is counterclockwise when viewed in the plane of the forces from above, and its magnitude is

$[M = Fd]$ $M = 200(0.200 + 0.100) = 60 \text{ N·m}$

The forces **P** and −**P** produce a counterclockwise couple

$[M = Fd]$ $M = 500(0.100 + 0.060) \cos\theta$

Equating the two expressions gives

$$60 = 500(0.160)\cos\theta$$

$$\theta = \cos^{-1}\frac{60}{80} = 41.4° \qquad Ans.$$

① Note that the only dimensions that are relevant are those that give the perpendicular distances between the forces of the couples.

Sample Problem 2/5

Replace the horizontal 80-lb force acting on the lever by an equivalent system consisting of a force at O and a couple.

Solution. We apply two equal and opposite 80-lb forces at O and identify the counterclockwise couple

$[M = Fd]$ $M = 80(9 \sin 60°) = 624 \text{ lb-in.}$ *Ans.*

① Thus the original force is equivalent to the 80-lb force at O and the 624-lb-in. couple as shown in the third of the three equivalent figures.

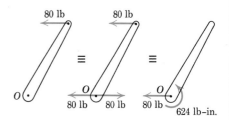

① The reverse of this problem is often encountered, namely, the replacement of a force and a couple by a single force. Proceeding in reverse is the same as replacing the couple by two forces one of which is equal and opposite to the 80-lb force at O. The moment arm to the second force would be $M/F =$

$624/80 = 7.79$ in. which is $9 \sin 60°$, thus determining the line of action of the single resultant force of 80 lb.

Problem 2/42

F

Problem 2/43

Problem 2/44

Problem 2/45

PROBLEMS

2/42 The action of the 10-kN force on the steel column can be analyzed by considering it to produce a compression along the centerline and a couple. If the couple has a magnitude of 800 N·m, determine the offset d.   *Ans.* $d = 80$ mm

2/43 In designing the lifting hook we note that the action of the applied force **F** at the critical section of the hook is a pull at B and a couple. If the magnitude of the couple is 4000 lb-ft, determine the magnitude of **F**.

2/44 Each propeller of the twin-screw ship develops a full-speed thrust F of 300 kN. In maneuvering the ship one propeller is turning full speed ahead and the other full speed in reverse. What thrust P must each tug exert on the ship to counteract the turning effect of the ship's propellers?   *Ans.* $P = 51.4$ kN

2/45 Express the combined moment of the two forces about the y-axis and about the y'-axis in vector notation using the unit vectors shown.

2/46 The simple truss supports a load of 40 kN. The vertical wall exerts a horizontal force against the supporting roller at *A*, and the hinged connection at *B* exerts the additional force on the truss required to maintain equilibrium. The 40-kN load and the vertical component of the reaction at *B* constitute a couple that is equal and opposite to the couple due to the two horizontal forces. Calculate the magnitude *B* of the force acting on the pin connection at *B*.

Ans. $B = 75.5$ kN

Problem 2/46

2/47 The angle plate is subjected to the two 250-N forces shown. It is desired to replace these forces by an equivalent set consisting of the 200-N force applied at *A* and a second force applied at *B*. Determine the *y*-coordinate of *B*.

Problem 2/47

2/48 The rear wheel of an accelerating car is acted upon by a friction force *F* of 2.4 kN and a torque on the axle which amounts to a couple *M*. If the force and the couple are replaced by an equivalent force which acts through a point 12 mm directly above the center of the wheel, find *M*.   Ans. $M = 929$ N·m

Problem 2/48

2/49 The 300-lb-in. couple is applied to the vertical shaft which is welded to the horizontal rectangular plate. If the couple and 60-lb force are to be replaced by a single equivalent force at *B*, determine the distance *x*.

Problem 2/49

Problem 2/50

2/50 The wrench is subjected to the 200-N force and the force **P** as shown. If the equivalent of the two forces is a force **R** at O and a couple $\mathbf{M} = 20\mathbf{k}$ N·m, determine **P** and **R**.

$$Ans. \; \mathbf{P} = 40\mathbf{j} \; N, \; \mathbf{R} = -160\mathbf{j} \; N$$

Problem 2/51

2/51 The control lever is subjected to a clockwise couple of 80 N·m exerted by its shaft at A and is to be designed to operate with a 200-N pull as shown. If the resultant of the couple and the force passes through A, determine the proper dimension x of the lever.

Problem 2/52

2/52 Replace the couple and force shown by a single force **F** applied at a point D. Locate D by determining the distance b.   Ans. $b = 213$ mm

1.5 kN
20°
20°
2.4 kN
200 mm
120 mm
O
x
y

Problem 2/53

2/53 The figure represents two integral gears subjected to the tooth-contact forces shown. Replace the two forces by an equivalent single force **R** at the rotation axis O and a corresponding couple M. If the gears start from rest under the action of the tooth loads shown, in what direction would rotation occur?

$$Ans. \; R = 3.56 \; kN, \; \theta_x = 51.1°$$
$$M = 11.28 \; N·m \; counterclockwise$$

2/54 The bracket is fastened to the girder by means of the two rivets A and B and supports the 520-lb force. Replace this force by a force acting along the horizontal centerline between the rivets and a couple. Then redistribute this force and couple by replacing them by two forces, one at A and one at B, and thus determine the forces supported by the rivets.

Problem 2/54

2/55 If the force system $\mathbf{F}_1$ and $\mathbf{F}_2$ is to be equivalent to the system of the two 200-N forces, determine θ and the magnitudes of $\mathbf{F}_1$ and $\mathbf{F}_2$.

Ans. $\theta = 26.6°$, $|\mathbf{F}_1| = |\mathbf{F}_2| = F = 335$ N

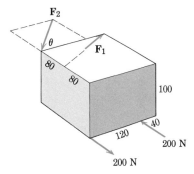

Dimensions in Millimeters

Problem 2/55

2/6 RESULTANTS. In the previous four articles we have developed the properties of force, moment, and couple. With the aid of these descriptions we are now ready to describe the resultant action of a group or *system* of forces. Most problems in mechanics deal with a system of forces, and it is generally necessary to reduce the system to its simplest form in describing its action. The resultant of a system of forces is the simplest force combination that can replace the original forces without altering the external effect of the system on the rigid body to which the forces are applied. The equilibrium of a body is the condition where the resultant of all forces that act on it is zero. When the resultant of all forces on a body is not zero, the acceleration of the body is described by equating the force resultant to the product of the mass and acceleration of the body. Thus the determination of resultants is basic to both statics and dynamics.

The most common type of force system occurs when the forces all act in a single plane, say the *x-y* plane, as illustrated by the system of three forces $\mathbf{F}_1$, $\mathbf{F}_2$, and $\mathbf{F}_3$ in Fig. 2/12*a*. The resultant force $\mathbf{R}$ is

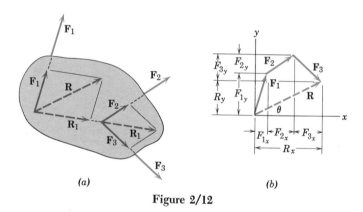

(a)   (b)

Figure 2/12

obtained in magnitude and direction by forming the *force polygon* in the *b*-part of the figure where the forces are added head-to-tail in any sequence. Thus for any system of coplanar forces we may write

$$\mathbf{R} = \mathbf{F}_1 + \mathbf{F}_2 + \mathbf{F}_3 + \cdots = \Sigma\mathbf{F}$$
$$R_x = \Sigma F_x \qquad R_y = \Sigma F_y \qquad R = \sqrt{(\Sigma F_x)^2 + (\Sigma F_y)^2}$$
$$\theta = \tan^{-1}\frac{R_y}{R_x} = \tan^{-1}\frac{\Sigma F_y}{\Sigma F_x}$$

(2/4)

Graphically the correct line of action of $\mathbf{R}$ may be obtained by preserving the correct lines of action of the forces and adding them by the parallelogram law as indicated in the *a*-part of the figure for the case of three forces where the sum $\mathbf{R}_1$ of $\mathbf{F}_2$ and $\mathbf{F}_3$ is added to $\mathbf{F}_1$ to obtain $\mathbf{R}$. In this process the principle of transmissibility has been used.

Algebraically, we may locate the resultant force by using Varignon's moment principle with the selection of some convenient point O as a moment center, Fig. 2/13. Thus the unknown arm d is computed from

$$Rd = \Sigma Fd = F_1 d_1 - F_2 d_2 + F_3 d_3$$

or simply

$$\boxed{Rd = \Sigma M_O} \qquad (2/5)$$

which is a restatement of Varignon's theorem or the *principle of moments,* which says that the moment of the resultant force about any point O equals the sum of the moments of the forces of the system about that same point. This principle is one of the most widely used of all principles in mechanics.

We see that it requires three equations, $R_x = \Sigma F_x$, $R_y = \Sigma F_y$, and $Rd = \Sigma M_O$, to determine completely the resultant $\mathbf{R}$ of a general coplanar force system. For a concurrent force system, the moment equation about the point of concurrency is automatically satisfied, and only the two force equations are needed to determine $\mathbf{R}$. For a parallel force system, one force equation in the direction of the forces and one moment equation will be sufficient to determine $\mathbf{R}$.

If the resultant force $\mathbf{R}$ for a given force system is zero, the resultant of the system need not be zero as it may be a couple M. In this event $\Sigma M_O = M$, where O is any convenient moment center. The three forces in Fig. 2/14, for instance, have a zero resultant force but have a resultant clockwise couple $M = F_3 d$.

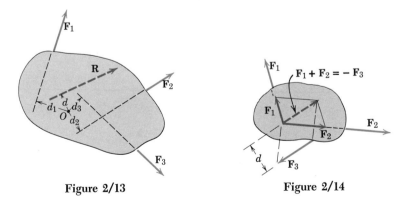

Figure 2/13   Figure 2/14

Sample Problem 2/6

Determine the resultant of the four forces and one couple that act on the plate shown.

___

Solution. Point O is selected arbitrarily as a convenient origin of coordinates and moment center. The components R_x and R_y, the magnitude of the resultant $\mathbf{R}$, and the angle θ made by $\mathbf{R}$ with the x-axis become

$$[R_x = \Sigma F_x] \qquad R_x = 40 + 80\cos 30° - 60\cos 45° = 66.9 \text{ N}$$

$$[R_y = \Sigma F_y] \qquad R_y = 50 + 80\sin 30° + 60\sin 45° = 132.4 \text{ N}$$

$$[R = \sqrt{R_x{}^2 + R_y{}^2}] \qquad R = \sqrt{(66.9)^2 + (132.4)^2} = 148.3 \text{ N} \qquad Ans.$$

$$\left[\theta = \tan^{-1}\frac{R_y}{R_x}\right] \qquad \theta = \tan^{-1}\frac{132.4}{66.9} = 63.2° \qquad Ans.$$

Although the couple has no influence on the magnitude and direction of $\mathbf{R}$, it does influence the moment of the resultant, which will now be determined. The position of the line of action of $\mathbf{R}$ is found from the principle of moments (Varignon's theorem). With O as the moment center, with d as the moment arm of $\mathbf{R}$, and with the counterclockwise sense chosen arbitrarily as positive, this principle requires

① $[Rd = \Sigma M_O] \qquad 148.3d = 140 - 50(5) + 60\cos 45°(4) - 60\sin 45°(7)$

$$d = -1.60 \text{ m}$$

The negative sign indicates that the moment of the resultant is acting in a clockwise rather than counterclockwise sense about O. Hence the resultant may be applied at any point on a line, making an angle of 63.2° with the x-axis and tangent to a circle of 1.6-m radius about O as shown in the figure. The clockwise moment of $\mathbf{R}$ requires that the line of action of $\mathbf{R}$ be tangent at point A and not at point B, as would have been the case if the moment had been acting in a counterclockwise sense.

Alternatively, we may locate $\mathbf{R}$ by finding the point C on, say, the x-axis through which $\mathbf{R}$ must pass. If we assumed initially that C was to the left of O, then a clockwise moment sum would give us

② $[Rd = \Sigma M_O] \qquad 132.4b = -140 + 50(5) - 60\cos 45°(4) + 60\sin 45°(7)$

$$b = 1.79 \text{ m}$$

① It is noted that the choice of point O as a moment center eliminated any moments due to the two forces that pass through O. The careful selection of a convenient moment center that eliminates as many terms as possible from the moment equations is an important simplification in mechanics calculations.

② If we had assumed that C was on the positive x-axis, b would have been negative. As a check note that $d = b\sin\theta$, $1.60 = 1.79\sin 63.2°$. Also observe that we could locate $\mathbf{R}$ by finding the point on the y-axis through which $\mathbf{R}$ passes. In this calculation R_y would not appear.

PROBLEMS

2/56 Determine the resultant **R** of the four forces acting on the gusset plate. Also find the magnitude of **R** and the angle θ_x which the resultant makes with the x-axis.

$$Ans. \ \mathbf{R} = 34.9\mathbf{i} + 41.8\mathbf{j} \text{ kN}$$
$$R = 54.5 \text{ kN}$$
$$\theta_x = 50.2°$$

Problem 2/56

2/57 Calculate the magnitude of the tension **T** and the angle θ for which the eye bolt will be under a resultant downward force of 15 kN.

$$Ans. \ T = 12.85 \text{ kN}, \ \theta = 38.9°$$

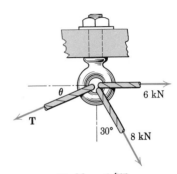

Problem 2/57

2/58 Determine the height h above the base B at which the resultant of the three forces acts.

Problem 2/58

2/59 Where does the resultant of the two forces act?

$$Ans. \ 10.70 \text{ m to the left of } A$$

Problem 2/59

Problem 2/60

Problem 2/61

Problem 2/62

Problem 2/63

2/60 Explain why the resultant of the three parallel forces always passes through point A regardless of the magnitude of **P**.

2/61 Determine the force R that could replace the four forces that act on the cantilever beam and not change the reaction on the end of the beam at the supporting weld at A. Locate R by finding its distance b to the left of A.

   Ans. $R = 1.5$ kN down, $b = 11$ m

2/62 In the equilibrium position shown the resultant of the three forces acting on the bell crank passes through the bearing O. Determine the vertical force **P**. Does the result depend on θ?

2/63 Determine the magnitude of the force **F** applied to the handle which will make the resultant of the three forces pass through O.   *Ans.* $F = 260$ N

2/64 The gear reducer shown is acted upon by the two couples, its weight of 200 N, and a vertical force at each of the mountings A and B. If the resultant of this system of two couples and three forces is zero, determine the vertical forces at A and B.

Problem 2/64

2/65 Determine the resultant $\mathbf{R}$ of the four forces and locate the point A on the upper chord through which $\mathbf{R}$ must pass. *Ans.* $\mathbf{R} = 4000\mathbf{i} - 11,460\mathbf{j}$ lb, 6.11 ft to the right of B

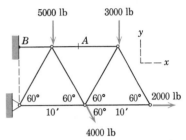

Problem 2/65

2/66 Replace the three forces and couple by an equivalent force $\mathbf{R}$ at A and a couple M. Specify M and the magnitude of $\mathbf{R}$.

Problem 2/66

2/67 Determine the resultant $\mathbf{R}$ of the three forces and two couples shown. Find the coordinate x of the point on the x-axis through which $\mathbf{R}$ passes. *Ans.* $\mathbf{R} = -1.5\mathbf{i} - 2\mathbf{j}$ kN, $x = 290$ mm

Problem 2/67

60 N·m
100 N
200 mm
30°
y
300 mm
600 N
x
400 mm
100 N

Problem 2/68

2/68 Calculate the *y*-coordinate of the point on the *y*-axis through which the resultant of the three forces and couple must pass.   *Ans. y = 109 mm*

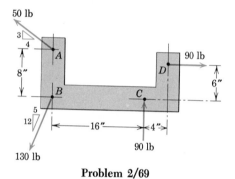

50 lb
3
4
A
90 lb
D
8"
6"
B
C
5
12
16"
4"
90 lb
130 lb

Problem 2/69

2/69 Determine graphically the resultant of the four forces. Check your result by a calculation.

200 N
200 mm
30°
100 mm
100 mm
40 N·m
A
150 N
x
80 N
45°
15 N·m
y

Problem 2/70

2/70 Replace the three forces and two couples acting on the rigid unit by an equivalent single force **R** at point *A* and a couple *M*.

600 lb
3'
7'
5'
4000 lb-ft
A
160°
4'
1000 lb
800 lb

Problem 2/71

2/71 Represent the resultant of the three forces and couple by a force **R** at *A* and a couple *M*. Find *M* and the magnitude of **R**.
Ans. M = 8420 lb-ft counterclockwise, R = 950 lb

2/72 The gear and attached V-belt pulley are turning counterclockwise and are subjected to the tooth load of 1600 N and the 800-N and 450-N tensions in the V-belt. Represent the action of these three forces by a resultant force **R** at O and a couple of magnitude M. Is the unit slowing down or speeding up?   *Ans.* **R** $= 930$**i** $+ 1666$**j** N, speeding up

Problem 2/72

▶2/73 The jet plane has a mass m of 30 Mg and is climbing at the 15° angle at constant velocity under the steady thrust T of 86 kN. Since the velocity is constant, the resultant of all forces acting on the aircraft is zero. Calculate the drag D, the lift L, and the force P acting on the stabilizer for the conditions represented.

Ans. $D = 9.83$ kN, $L = 279.6$ kN, $P = 4.69$ kN

Problem 2/73

SECTION B. THREE-DIMENSIONAL FORCE SYSTEMS

2/7 RECTANGULAR COMPONENTS.   Many problems in mechanics require analysis in three dimensions, and it is often necessary to resolve a force into its three mutually perpendicular components. Thus, the force **F** acting at point O in Fig. 2/15 has the *rectangular components* F_x, F_y, F_z, where

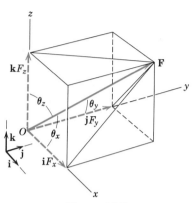

Figure 2/15

$$
\begin{aligned}
F_x &= F \cos \theta_x & F &= \sqrt{F_x^2 + F_y^2 + F_z^2} \\
F_y &= F \cos \theta_y & \mathbf{F} &= \mathbf{i}F_x + \mathbf{j}F_y + \mathbf{k}F_z \\
F_z &= F \cos \theta_z & \mathbf{F} &= F(\mathbf{i} \cos \theta_x + \mathbf{j} \cos \theta_y + \mathbf{k} \cos \theta_z)
\end{aligned} \tag{2/6}
$$

The unit vectors **i, j, k** are in the x-, y-, and z-directions, respectively. If we introduce the *direction cosines* of **F** which are $l = \cos \theta_x$, $m = \cos \theta_y$, $n = \cos \theta_z$, where $l^2 + m^2 + n^2 = 1$, we may write the force as

$$
\mathbf{F} = F(\mathbf{i}l + \mathbf{j}m + \mathbf{k}n) \tag{2/7}
$$

The choice of orientation of the coordinate system is quite arbitrary, with convenience being the primary consideration. But we must use a right-handed set of axes in our three-dimensional work so as to preserve the relative orientations of the axes. When we rotate from the x- to the y-axis, the positive direction for the z-axis in a right-handed system is that of the advancement of a right-handed screw when rotated in the same sense.

Rectangular components of a force **F** (or other vector) may be written alternatively with the aid of the vector operation known as the *dot* or *scalar product*.° By definition, the dot product of two vectors **P** and **Q**, Fig. 2/16*a*, is the product of their magnitudes times the cosine of the angle α between them and is written

$$
\mathbf{P} \cdot \mathbf{Q} = PQ \cos \alpha
$$

This product may be viewed either as the projection (component)

(a)	(b)

Figure 2/16

°See item 6 of section 7 in Appendix B.

$P \cos \alpha$ of **P** in the direction of **Q** multiplied by Q or as the projection (component) $Q \cos \alpha$ of **Q** in the direction of **P** multiplied by P. In either case the dot product of the two vectors is a scalar quantity. Thus the component $F_x = F \cos \theta_x$ of the force **F** in Fig. 2/15, for instance, may be written as $F_x = \mathbf{F} \cdot \mathbf{i}$ where **i** is the unit vector in the x-direction. In more general terms, if **n** is a unit vector in a specified direction, the component of **F** in the **n**-direction, Fig. 2/16*b*, has the magnitude $F_n = \mathbf{F} \cdot \mathbf{n}$. If it is desired to write the component vector in the **n**-direction as a vector quantity, then its scalar magnitude, expressed by $\mathbf{F} \cdot \mathbf{n}$, must be multiplied by the unit vector **n** to give $\mathbf{F}_n = (\mathbf{F} \cdot \mathbf{n})\mathbf{n}$, which may be written merely as $\mathbf{F}_n = \mathbf{F} \cdot \mathbf{nn}$.

If **n** has direction cosines α, β, γ, then we may write **n** in vector component form like any other vector as

$$\mathbf{n} = \mathbf{i}\alpha + \mathbf{j}\beta + \mathbf{k}\gamma$$

where in this case its magnitude is unity. If **F** has the direction cosines l, m, n with respect to reference axes x-y-z, then the component of **F** in the **n**-direction becomes

$$F_n = \mathbf{F} \cdot \mathbf{n} = F(\mathbf{i}l + \mathbf{j}m + \mathbf{k}n) \cdot (\mathbf{i}\alpha + \mathbf{j}\beta + \mathbf{k}\gamma)$$

$$= F(l\alpha + m\beta + n\gamma)$$

since $\qquad \mathbf{i} \cdot \mathbf{i} = \mathbf{j} \cdot \mathbf{j} = \mathbf{k} \cdot \mathbf{k} = 1$

and $\qquad \mathbf{i} \cdot \mathbf{j} = \mathbf{j} \cdot \mathbf{i} = \mathbf{i} \cdot \mathbf{k} = \mathbf{k} \cdot \mathbf{i} = \mathbf{j} \cdot \mathbf{k} = \mathbf{k} \cdot \mathbf{j} = 0$

If the angle between the force **F** and the direction specified by the unit vector **n** is θ, then by virtue of the dot-product relationship we have $\mathbf{F} \cdot \mathbf{n} = Fn \cos \theta = F \cos \theta$ where $|\mathbf{n}| = n = 1$. Thus the angle between **F** and **n** is given by

$$\theta = \cos^{-1} \frac{\mathbf{F} \cdot \mathbf{n}}{F} \qquad (2/8)$$

or, in general, the angle between any two vectors **P** and **Q** is

$$\theta = \cos^{-1} \frac{\mathbf{P} \cdot \mathbf{Q}}{PQ} \qquad (2/8a)$$

If a force **F** is perpendicular to a line whose direction is specified by the unit vector **n**, then $\cos \theta = 0$, and $\mathbf{F} \cdot \mathbf{n} = 0$. Note especially that this relationship does not mean that either **F** or **n** is zero, as would be the case with scalar multiplication where $(A)(B) = 0$ requires that either A or B be zero.

It should be observed that the dot-product relationship applies to nonintersecting vectors as well as to intersecting vectors. Thus the dot product of the nonintersecting vectors **P** and **Q** in Fig. 2/17 is the projection of **P′** on **Q**, which is $P'Q \cos \alpha = PQ \cos \alpha$ since **P′** and **P** are the same free vectors.

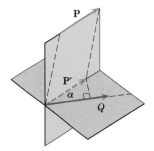

Figure 2/17

Sample Problem 2/7

A force **F** with a magnitude of 100 N is applied at the origin O of the axes x-y-z as shown. The line of action of **F** passes through a point A whose coordinates are 3 m, 4 m, and 5 m. Determine (a) the x-, y-, and z-components of **F**, (b) the projection of **F** on the x-y plane, and (c) the component F_n of **F** in the direction of the line O-n which passes through point B as shown.

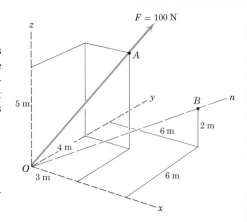

Solution. Part (a). The direction cosines of **F** are

$$l = \frac{3}{7.071} = 0.424 \qquad m = \frac{4}{7.071} = 0.566 \qquad n = \frac{5}{7.071} = 0.707$$

where the diagonal to point A is $\sqrt{3^2 + 4^2 + 5^2} = \sqrt{50} = 7.071$ m. The components become

$$F_x = Fl = 100(0.424) = 42.4 \text{ N}$$
$$F_y = Fm = 100(0.566) = 56.6 \text{ N}$$
$$F_z = Fn = 100(0.707) = 70.7 \text{ N} \qquad\qquad Ans.$$

Part (b). The cosine of the angle θ_{xy} between **F** and the x-y plane is

$$\cos \theta_{xy} = \frac{\sqrt{3^2 + 4^2}}{7.071} = 0.707$$

so that $F_{xy} = F \cos \theta_{xy} = 100(0.707) = 70.7$ N. $\qquad\qquad Ans.$

Part (c). The direction cosines of a unit vector **n** along O-n are

$$\alpha = \beta = \frac{6}{\sqrt{6^2 + 6^2 + 2^2}} = 0.688 \qquad \gamma = \frac{2}{\sqrt{6^2 + 6^2 + 2^2}} = 0.229$$

Thus the component of **F** along O-n becomes

$$F_n = \mathbf{F} \cdot \mathbf{n} = 100(0.424\mathbf{i} + 0.566\mathbf{j} + 0.707\mathbf{k}) \cdot (0.688\mathbf{i} + 0.688\mathbf{j} + 0.229\mathbf{k})$$
$$= 100[(0.424)(0.688) + (0.566)(0.688) + (0.707)(0.229)]$$
$$= 84.4 \text{ N} \qquad\qquad Ans.$$

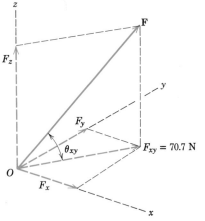

① In this example all components are positive. Be prepared for the case where a direction cosine, and hence the component, is negative.

② The dot product automatically finds the length of the projection or component of **F** on line On as shown. To express this component as a vector, we would write $\mathbf{F} \cdot \mathbf{nn}$ or 84.4**n** N.

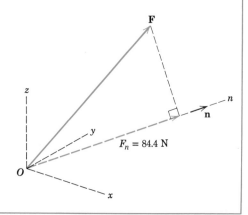

PROBLEMS

2/74 If the *x*-component of **P** equals 60 lb, determine the *y*-component of **P**.   *Ans.* $P_y = 85.9$ lb

Problem 2/74

2/75 The turnbuckle is tightened until the tension in the cable *AB* equals 1.2 kN. Write the vector expression for the tension **T** as a force acting on the lever.

Problem 2/75

2/76 The cable exerts a tension of 2 kN on the fixed bracket at *A*. Write the vector expression for the tension **T**.
          Ans. **T** $= 2(-0.920\mathbf{i} + 0.383\mathbf{j} + 0.077\mathbf{k})$ kN

Problem 2/76

2/77 The 140-lb force **F** is directed along the diagonal of the rectangular parallelepiped whose sides are in the ratios 2:3:6. Express **F** as a vector.

Problem 2/77

Problem 2/78

Problem 2/79

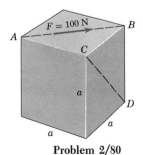

Problem 2/80

2/78 The tension in the supporting cable AB is 10 kN. Express this tension as a force vector $\mathbf{T}$ acting on BC.

$$Ans. \; \mathbf{T} = \frac{10}{\sqrt{389}}(8\mathbf{i} - 15\mathbf{j} + 10\mathbf{k}) \; \text{kN}$$

2/79 The crane exerts a force $\mathbf{T} = -6\mathbf{i} - 8\mathbf{j} + 10\mathbf{k}$ kN on the concrete pillar to which its cable is attached. Calculate the angles which $\mathbf{T}$ makes with the positive x-axis and with the x-y plane.

2/80 A force $\mathbf{F}$ has a magnitude of 1200 N. If its direction cosine with respect to the x-axis is 0.25 and if the ratio of its x-component to its z-component is 0.6, compute F_y. $\qquad$ *Ans.* $F_y = 1049$ N

2/81 The line of action of a force $\mathbf{P}$ makes an angle of 120° with the positive x-axis, and $\mathbf{P}$ has a direction cosine with respect to the z-axis of 0.6. If the y-component of $\mathbf{P}$ is 200 N, write $\mathbf{P}$ as a vector.

2/82 The direction cosines of a certain force vector $\mathbf{F}$ with respect to the x- and y-directions are -0.4 and 0.6, respectively. If $\mathbf{F} \cdot \mathbf{k} = 60$ lb, compute the angle θ_{xy} made by $\mathbf{F}$ with the x-y plane and find the x-component of $\mathbf{F}$.

$$Ans. \; \theta_{xy} = 43.9°, \; F_x = -34.6 \; \text{lb}$$

2/83 Calculate the magnitude of the projection F_{CD} of the 100-N force on the face diagonal CD of the cube.

2/84 Determine the x-coordinate of the small pulley at C which will ensure that there is no component of the tension in cable OC along AB.   *Ans.* $x = 5.33$ ft

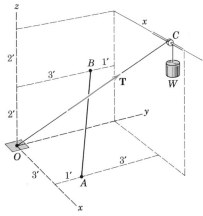

Problem 2/84

2/85 Derive the expression for the component F_{DC} of the force **F** along the line directed from D to C.

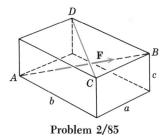

Problem 2/85

2/86 The force **F** has a magnitude of 2 kN and is directed from A to B. Calculate the component F_{CD} of **F** along CD.   *Ans.* $F_{CD} = \sqrt{6/5}$ kN

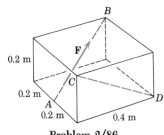

Problem 2/86

▶**2/87** The access door is held in the 30° open position by the chain AB. If the tension in the chain is 100 N, determine the component of the tension along the diagonal axis CD of the door.   *Ans.* $T_{CD} = 46.0$ N

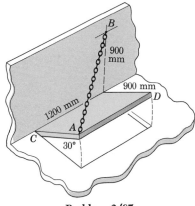

Problem 2/87

2/8 MOMENT AND COUPLE. A somewhat more general formulation of the concepts of moment and couple than was used in two-dimensional analysis will now be developed for our use in three-dimensional problems. It is only when we work in three dimensions that the full advantage of vector analysis becomes apparent.

Moment. Consider a force **F** with a given line of action, Fig. 2/18*a*, and any point *O* not on this line. Point *O* and the line of **F** establish a plane *A*. The moment **M**$_O$ of **F** about an axis through *O* normal to the plane has the magnitude $M_O = Fd$, where *d* is the perpendicular distance from *O* to the line of **F**. This moment is also referred to as the moment of **F** about the *point O*. The vector **M**$_O$ is normal to the plane and along the axis through *O*. Both the magnitude and the direction of **M**$_O$ may be described by the vector operation known as the *cross* or *vector product* (see item 7 in Art. B7 of Appendix B). A vector **r** is introduced that extends from *O* to *any* point on the line of action of **F**. By definition, the cross product of **r** and **F** is written **r** × **F** and has the magnitude $(r \sin \alpha)F$, which is the same as Fd, the magnitude of **M**$_O$. The correct direction and sense of the moment are established by the right-hand rule, described previously in Arts. 2/4 and 2/5. Thus, with **r** and **F** treated as free vectors, Fig. 2/18*b*, the thumb points in the direction of **M**$_O$ if the fingers of the right hand curl in the direction of rotation from **r** to **F**. Therefore we may write the moment of **F** about the axis through *O* as

(a)

(b)

Figure 2/18

$$\boxed{\mathbf{M}_O = \mathbf{r} \times \mathbf{F}} \tag{2/9}$$

The order **r** × **F** of the vectors *must* be maintained, since **F** × **r** would produce a vector with a sense opposite to that of **M**$_O$ or **F** × **r** = −**M**$_O$.

The cross-product expression for **M**$_O$ may be written in determinant form which is

$$\boxed{\mathbf{M}_O = \begin{vmatrix} \mathbf{i} & \mathbf{j} & \mathbf{k} \\ r_x & r_y & r_z \\ F_x & F_y & F_z \end{vmatrix}} \tag{2/10}$$

(Again, refer to item 7 in Art. B7 of Appendix B if you are not already familiar with the expansion of the cross product.) The symmetry and order of the terms should be carefully noted, and care must be exercised to ensure the use of a *right-handed* coordinate system upon which the correct evaluation of vector operations depends. Expansion of the determinant gives

$$\mathbf{M}_O = \mathbf{i}(r_y F_z - r_z F_y) + \mathbf{j}(r_z F_x - r_x F_z) + \mathbf{k}(r_x F_y - r_y F_x)$$

To gain more confidence in the cross-product relationship let us observe the three components of the moment of a force about a point as seen from Fig. 2/19, which shows the three components of a force **F** acting at a point A located from O by the vector **r**. The scalar magnitudes of the moments of these forces about the positive x-, y-, and z-axes through O are seen to be

$$M_x = r_y F_z - r_z F_y, \quad M_y = r_z F_x - r_x F_z, \quad M_z = r_x F_y - r_y F_x$$

which agree with the respective terms in the determinant expansion for the cross product **r × F**.

The moment $\mathbf{M}_\lambda$ of **F** about *any* axis λ through O, Fig. 2/20, may now be written. If **n** is a unit vector in the λ-direction, then by using the dot product expression for the component of a vector as described in Art. 2/7, the component of $\mathbf{M}_O$ in the direction of λ is merely $\mathbf{M}_O \cdot \mathbf{n}$, which is the scalar magnitude of the moment $\mathbf{M}_\lambda$ of **F** about λ. To obtain the vector expression for the moment of **F** about λ, the magnitude must be multiplied by the directional unit vector **n** to give

$$\boxed{\mathbf{M}_\lambda = (\mathbf{r} \times \mathbf{F} \cdot \mathbf{n})\mathbf{n}} \tag{2/11}$$

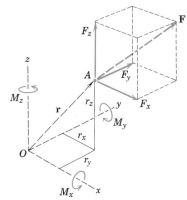

Figure 2/19

where **r × F** replaces $\mathbf{M}_O$. The expression **r × F · n** is known as a *triple scalar product* (see item 8 in Art. B7, Appendix B). It need not be written $(\mathbf{r} \times \mathbf{F}) \cdot \mathbf{n}$, since the association $\mathbf{r} \times (\mathbf{F} \cdot \mathbf{n})$ would have no meaning because a cross product cannot be formed by a vector and a scalar. The triple scalar product may be represented by the determinant

$$\boxed{|\mathbf{M}_\lambda| = M_\lambda = \begin{vmatrix} r_x & r_y & r_z \\ F_x & F_y & F_z \\ \alpha & \beta & \gamma \end{vmatrix}} \tag{2/12}$$

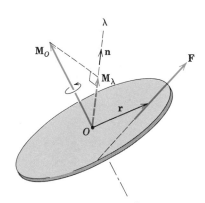

Figure 2/20

where α, β, γ are the direction cosines of the unit vector **n**. This result is easily verified when the indicated operations are carried out.

Principle of moments. In Art. 2/4 on the moment of a force we introduced the *principle of moments* or *Varignon's theorem* for forces in two dimensions. The principle is equally applicable in three dimensions and is easily proved by applying the distributive law for the sum of vector products. In Fig. 2/21 is shown a system of forces $\mathbf{F}_1$, $\mathbf{F}_2$, $\mathbf{F}_3$, ... concurrent at point A whose position vector from point O is **r**. The sum of the moments about O of the forces of the system is

$$\mathbf{r} \times \mathbf{F}_1 + \mathbf{r} \times \mathbf{F}_2 + \mathbf{r} \times \mathbf{F}_3 + \cdots = \mathbf{r} \times (\mathbf{F}_1 + \mathbf{F}_2 + \mathbf{F}_3 + \cdots) = \mathbf{r} \times \Sigma\mathbf{F}$$

or

$$\boxed{\Sigma\mathbf{M}_O = \mathbf{r} \times \mathbf{R}} \tag{2/13}$$

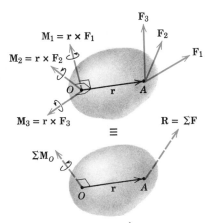

Figure 2/21

Thus the *sum of the moments of a system of concurrent forces about a given point equals the moment of their sum about the same point.* As mentioned in Art. 2/4 this principle finds repeated application in

Figure 2/22

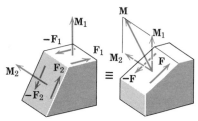

Figure 2/23

mechanics not only for the moments of force vectors but for the moments of other vectors as well.

Couple. The concept of the couple was introduced in Art. 2/5 and is easily extended to three dimensions. Figure 2/22 shows two equal and opposite forces **F** and −**F** acting on a body. The vector **r** joins *any* point *B* on the line of action of −**F** to *any* point *A* on the line of action of **F**. Points *A* and *B* are located by position vectors $\mathbf{r}_A$ and $\mathbf{r}_B$ from *any* point *O*. The combined moment of the two forces about *O* is

$$\mathbf{M} = \mathbf{r}_A \times \mathbf{F} + \mathbf{r}_B \times (-\mathbf{F}) = (\mathbf{r}_A - \mathbf{r}_B) \times \mathbf{F}$$

But $\mathbf{r}_A - \mathbf{r}_B = \mathbf{r}$, so that all reference to the moment center *O* disappears, and the *moment of the couple* becomes

$$\boxed{\mathbf{M} = \mathbf{r} \times \mathbf{F}} \qquad (2/14)$$

Thus the moment of a couple is the *same about all points*. We see that the magnitude of **M** is $M = Fd$ where d is the perpendicular distance between the lines of action of the two forces, as described in Art. 2/5.

The moment of a couple is a *free vector,* whereas the moment of a force about a point (which is also the moment about a defined axis through the point) is a *sliding vector* whose direction is along the axis through the point. As in the case of two dimensions the action of the couple is to produce a pure rotation on the body about an axis normal to the plane of the forces which constitute the couple.

Couple vectors obey all of the rules which govern vector quantities. Thus in Fig. 2/23 the couple vector $\mathbf{M}_1$ due to $\mathbf{F}_1$ and $-\mathbf{F}_1$ may be added as shown to the couple vector $\mathbf{M}_2$ due to $\mathbf{F}_2$ and $-\mathbf{F}_2$ to produce the couple **M** which, in turn, can be produced by **F** and −**F**.

In Art. 2/5 we learned to replace a force by its equivalent consisting of a force and a couple. We should also be prepared to carry out this replacement in three dimensions. The procedure is represented in Fig. 2/24 where the force **F** acting on a rigid body at point *A* is replaced by an equal force at point *B* and the couple $\mathbf{M} = \mathbf{r} \times \mathbf{F}$. By adding the equal and opposite forces **F** and −**F** at *B* we have the couple composed of −**F** and the original **F**. Thus we see that the couple vector is simply the moment of the original force about the point to which the force is being moved. Again we note that **r** is a vector from *B* to *any* point on the line of action of the original force passing through *A*.

Figure 2/24

Sample Problem   2/8

A tension **T** of magnitude 10 kN is applied to the cable attached to the top A of the rigid mast and secured to the ground at B. Determine the moment M_z of **T** about the z-axis passing through the base O of the mast.

Solution (a).   The required moment may be obtained by finding the component along the z-axis of the moment $\mathbf{M}_O$ of **T** about point O. The vector $\mathbf{M}_O$ is normal to the plane defined by **T** and point O as shown in the accompanying figure. In using Eq. 2/9 to find $\mathbf{M}_O$, the vector **r** is any vector from point O to the line of action of **T**. The simplest choice is the vector from O to A, which is written as $\mathbf{r} = 15\mathbf{j}$ m. The vector expression for **T** requires finding its direction cosines, which are $12/\overline{AB} = 0.566$, $-15/\overline{AB} = -0.707$,   and $9/\overline{AB} = 0.424$ where $\overline{AB} = \sqrt{9^2 + 15^2 + 12^2} = \sqrt{450} = 21.2$ m. Therefore we have

$$\mathbf{T} = 10(0.566\mathbf{i} - 0.707\mathbf{j} + 0.424\mathbf{k}) \text{ kN}$$

From Eq. 2/9,

$$[\mathbf{M}_O = \mathbf{r} \times \mathbf{F}] \quad \mathbf{M}_O = 15\mathbf{j} \times 10(0.566\mathbf{i} - 0.707\mathbf{j} + 0.424\mathbf{k})$$

$$= 150(-0.566\mathbf{k} + 0.424\mathbf{i}) \text{ kN} \cdot \text{m}$$

The magnitude M_z of the desired moment is the component of $\mathbf{M}_O$ in the z-direction or $M_z = \mathbf{M}_O \cdot \mathbf{k}$. Therefore

$$M_z = 150(-0.566\mathbf{k} + 0.424\mathbf{i}) \cdot \mathbf{k} = -84.9 \text{ kN} \cdot \text{m} \qquad Ans.$$

The minus sign indicates that the vector $\mathbf{M}_z$ is in the negative z-direction. Expressed as a vector, the moment is $\mathbf{M}_z = -84.9\mathbf{k}$ kN $\cdot$ m

Solution (b).   The force of magnitude T is resolved into components T_z and T_{xy} in the x-y plane, which is normal to the moment axis z. Since T_z is parallel to the z-axis, it can exert no moment about this axis. The moment M_z is, then, due only to T_{xy} and is $M_z = T_{xy} d$ where d is the perpendicular distance from T_{xy} to O. The cosine of the angle between T and T_{xy} is $\sqrt{15^2 + 12^2}/\sqrt{15^2 + 12^2 + 9^2} = 0.906$, and therefore

$$T_{xy} = 10(0.906) = 9.06 \text{ kN}$$

The moment arm d equals $\overline{OA}$ multiplied by the sine of the angle between T_{xy} and OA, or

$$d = 15 \frac{12}{\sqrt{12^2 + 15^2}} = 9.37 \text{ m}$$

Hence the moment of **T** about the z-axis is

$$M_z = 9.06(9.37) = 84.9 \text{ kN} \cdot \text{m} \qquad Ans.$$

and is clockwise when viewed in the x-y plane.

Solution (c).   The moment is also easily calculated by resolving T_{xy} into its components T_x and T_y. It is clear that T_y exerts no moment about the z-axis since it passes through it, so that the required moment is due to T_x alone. The direction cosine of **T** with respect to the x-axis is $12/\sqrt{9^2 + 12^2 + 15^2} = 0.566$ so that $T_x = 10(0.566) = 5.66$ kN. Thus

$$M_z = 5.66(15) = 84.9 \text{ kN} \cdot \text{m} \qquad Ans.$$

① We could also use the vector from O to B for **r** and obtain the same result, but using vector OA is simpler.

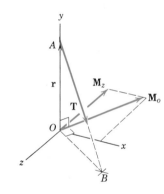

② It is always helpful to accompany your vector operations with a sketch of the vectors so as to retain a clear picture of the geometry of the problem.

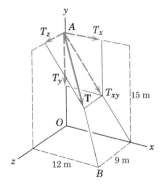

③ Sketch the x-y view of the problem and show d.

Sample Problem 2/9

Determine the magnitude and direction of the couple **M** that will replace the two given couples and still produce the same external effect on the block. Specify the two forces **F** and $-$**F**, applied in the two faces of the block parallel to the y-z plane, that may replace the four given forces. The 30-N forces act parallel to the y-z plane.

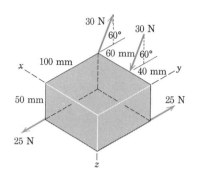

Solution. The couple due to the 30-N forces has the magnitude $M_1 = 30(0.06) = 1.80$ N·m. The direction of M_1 is normal to the plane defined by the two forces, and the sense, shown in the figure, is established by the right-hand convention. The couple due to the 25-N forces has the magnitude $M_2 = 25(0.10) = 2.50$ N·m with the direction and sense shown in the same figure. The two couple vectors combine to give the components

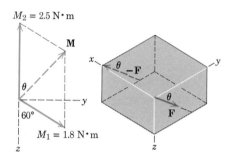

$$M_y = 1.80 \sin 60° = 1.559 \text{ N·m}$$

$$M_z = -2.50 + 1.80 \cos 60° = 1.600 \text{ N·m}$$

① Thus $\quad M = \sqrt{(1.559)^2 + (-1.600)^2} = 2.23$ N·m *Ans.*

with $\quad \theta = \tan^{-1} \dfrac{1.559}{1.600} = \tan^{-1} 0.974 = 44.3°$ *Ans.*

The forces **F** lie in a plane normal to the couple **M**, and their moment arm as seen from the right-hand figure is 100 mm. Thus each force has the magnitude

$$F = \frac{2.23}{0.10} = 22.3 \text{ N}$$ *Ans.*

and the direction $\theta = 44.3°$.

① Bear in mind that the couple vectors are *free* vectors and therefore have no unique lines of action.

Sample Problem 2/10

A force of 40 lb is applied at A to the handle of the control lever that is attached to the fixed shaft OB. In determining the effect of the force on the shaft at a cross section such as at O, the force may be replaced by an equivalent force at O and a couple. Describe this couple as a vector **M**.

Solution. The couple may be expressed in vector notation as **M** = **r** × **F** where **r** = $\overrightarrow{OA}$ = 8**j** + 5**k** in. and **F** = $-$40**i** lb. Thus

$$\mathbf{M} = (8\mathbf{j} + 5\mathbf{k}) \times (-40\mathbf{i})$$

$$= -200\mathbf{j} + 320\mathbf{k} \text{ lb-in.}$$ *Ans.*

from which the magnitude and direction of **M** may be written. Alternatively we see that moving the 40-lb force through a distance $d = \sqrt{5^2 + 8^2} = 9.43$ in. to a parallel position through O requires the addition of a couple **M** whose magnitude is

$$M = Fd = 40(9.43) = 377 \text{ lb-in.}$$ *Ans.*

The couple vector is perpendicular to the plane in which the force is shifted, and its sense is that of the moment of the given force about O. The direction of **M** in the y-z plane is given by

$$\theta = \tan^{-1}\tfrac{5}{8} = 32.0°$$ *Ans.*

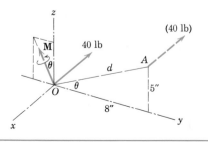

PROBLEMS

2/88 Determine the moment of force **F** about point *A*.
$$\text{Ans. } \mathbf{M}_A = F(b\mathbf{i} + a\mathbf{j})$$

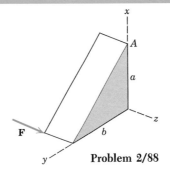

Problem 2/88

2/89 Determine the moment of the force **P** about point *A*.

Problem 2/89

2/90 A steel H-beam acts as a column and supports the two vertical loads shown. Replace these forces by a single equivalent force along the vertical centerline of the column and a couple **M**.
$$\text{Ans. } \mathbf{M} = (-0.450\mathbf{i} + 0.300\mathbf{j})10^6 \text{ lb-in.}$$

Problem 2/90

2/91 The directions of rotation of the input shaft *A* and output shaft *B* of the worm-gear reducer are indicated by the curved dotted arrows. An input torque (couple) of 80 N·m is applied to shaft *A* in the direction of rotation. The output shaft *B* supplies a torque of 320 N·m to the machine that it drives (not shown). The shaft of the driven machine exerts an equal and opposite reacting torque on the output shaft of the reducer. Determine the resultant **M** of the two couples that act on the reducer unit and calculate the direction cosine of **M** with respect to the *x*-axis.

Problem 2/91

150 N

250 mm

150 mm

250 mm

150 N

y

x

Problem 2/92

2/92 The two forces acting on the handles of the pipe wrenches constitute a couple **M.** Express the couple as a vector.   *Ans.* $\mathbf{M} = -75\mathbf{i} + 22.5\mathbf{j}$ N·m

250 lb

O

12"

60°

40°

4.2"

8"

O

Problem 2/93

2/93 Compute the magnitude of the moment of the 250-lb force about the axis O-O.

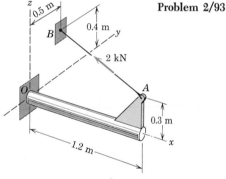

z

0.5 m

B

0.4 m

y

2 kN

A

O

0.3 m

1.2 m

x

Problem 2/94

2/94 Calculate the moment about the z-axis of the 2-kN tension in cable AB of Prob. 2/76, repeated here.   *Ans.* $M_z = 0.920$ kN·m

z

1.6 m

A

2 m

O

y

x

1.5 m

0.8 m

B **Problem 2/95**

2/95 The figure for Prob. 2/75 is repeated here where the cable AB exerts a force of 1.2 kN on the lever in the direction from A to B. Calculate the magnitude of the moment of this force about point O.

2/96 In picking up a load from position *A* a cable tension **T** of 21 kN is developed. Calculate the moment that **T** produces about the base *O* of the construction crane. *Ans.* $\mathbf{M}_O = -374\mathbf{i} + 93.5\mathbf{j} - 56.1\mathbf{k}$ kN·m

2/97 If the crane of Prob. 2/96 picks up a load at *B* rather than at *A* and develops an initial 21-kN tension *T* in its cable, determine the moment $\mathbf{M}_O$ of this force about the origin *O*.

Problem 2/96

2/98 Three couples are formed by the three pairs of equal and opposite forces. Determine the resultant **M** of the three couples.
$Ans.$ $\mathbf{M} = -20\mathbf{i} - 6.77\mathbf{j} - 37.2\mathbf{k}$ N·m

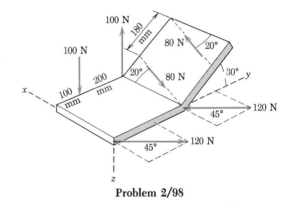

Problem 2/98

2/99 Calculate the magnitude of the moment about the *x*-axis of the 100-lb force acting on the bracket at *A*. Also determine the vector expression for the moment of the force about point *O* without using the cross product.

Problem 2/99

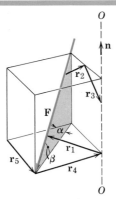

Problem 2/100

2/100 The force **F** exerts a certain moment **M** about the axis *O-O*. Determine which of the expressions cited correctly describes **M**. The vector **n** is a unit vector along *O-O*, and the box is rectangular.

$$\mathbf{r}_1 \times \mathbf{F} \cdot \mathbf{n} \qquad\qquad (\mathbf{r}_1 \times \mathbf{F} \cdot \mathbf{n})\mathbf{n}$$

$$(-\mathbf{r}_4 \times \mathbf{F} \cdot \mathbf{n})\mathbf{n} \qquad [\mathbf{F} \times (\mathbf{r}_2 + \mathbf{r}_3) \cdot \mathbf{n}]\mathbf{n}$$

$$(r_5 F \cos \beta \cos \alpha)\mathbf{n}$$

2/101 A 50-kN force with direction cosines proportional to $2, 6, 9$ passes through a point P whose x-, y-, z-coordinates in millimeters are $3, 2, -5$. Determine the moment **M** of the force about a point whose coordinates in millimeters are $2, 2, -3$.

Ans. $\mathbf{M} = \frac{50}{11}(12\mathbf{i} - 13\mathbf{j} + 6\mathbf{k}) \text{ N·m}$

2/102 A 100-lb force passes through the two points A and B in the sense from A to B. The x-, y-, z-coordinates of A and B expressed in inches are $-3, -1, 4$ and $3, 4, 5$, respectively. Calculate the moment **M** of the force about point C whose coordinates in inches are $2, -2, 1$.

2/103 The special-purpose milling cutter is subjected to the force of 1200 N and a couple of 240 N·m as shown. Determine the moment of this system about point O.   *Ans.* $\mathbf{M}_O = -260\mathbf{i} + 328\mathbf{j} + 88\mathbf{k} \text{ N·m}$

Problem 2/103

2/104 The rectangular steel plate is tilted about its horizontal edge *AB* and held in the position shown by a cable from corner *C* to point *O* on the ground. The tension in the cable is 20 kN. Calculate the magnitude of the moment of the cable tension about edge *AB*. Work the problem using the approach most appropriate for each of the two ways in which the same problem is specified. (Observe that the choice of method of solution often depends on how the data are given.)

(a)

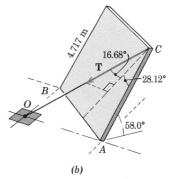

(b)

Problem 2/104

2/105 The figure for Prob. 2/84 is repeated here. The tension in the cable equals 200 lb. For the position *x* = 5.33 ft for pulley *C*, calculate the magnitude of the moment of the tension about the axis *AB*.

Ans. $M = 141.4$ lb-ft

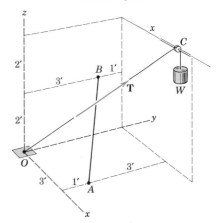

Problem 2/105

2/106 The figure of Prob. 2/86 is shown again here. If the magnitude of the moment of **F** about line *CD* is 50 N·m, determine the magnitude of **F**.

Problem 2/106

Problem 2/107

2/107 Replace the two forces that act on the 3-m cube by an equivalent single force $\mathbf{F}$ at A and a couple $\mathbf{M}$.

Ans. $\mathbf{F} = -1.72\mathbf{j} + 28.28\mathbf{k}$ kN
$\mathbf{M} = 5.15\mathbf{i} - 90\mathbf{k}$ kN·m

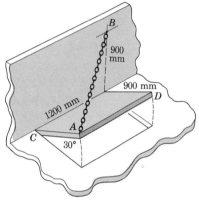

Problem 2/108

▶2/108 The access door of Prob. 2/87 is shown again here. If the tension in the chain AB is 100 N, determine the magnitude M of its moment about the hinge axis.   *Ans.* $M = 46.8$ N·m

Problem 2/109

▶2/109 The threading die is screwed onto the end of the fixed pipe which is bent through an angle of 20°. Replace the two forces by an equivalent force at O and a couple $\mathbf{M}$. Find $\mathbf{M}$ and calculate the magnitude M' of the moment which tends to screw the pipe into the fixed block about its angled axis through O.

Ans. $\mathbf{M} = 136\mathbf{i} - 679\mathbf{k}$ lb-in., $M' = 685$ lb-in.

2/9 RESULTANTS.

In Art. 2/6 we defined the resultant as the simplest force combination that can replace a given system of forces without altering the external effect on the rigid body upon which the forces act. We found the magnitude and direction of the resultant force for the two-dimensional force system by a vector summation of forces, Eq. 2/4, and we located the line of action of the resultant force by applying the principle of moments, Eq. 2/5. These same principles may be extended to three dimensions.

In the previous article we showed that a force could be moved to a parallel position by adding a corresponding couple. Thus for the system of forces $\mathbf{F}_1, \mathbf{F}_2, \mathbf{F}_3 \dots$ acting on a rigid body in Fig. 2/25a, we may move each of them in turn to the arbitrary point O provided we also introduce a couple for each force transferred. Thus, for example, we may move force $\mathbf{F}_1$ to O provided we introduce the couple $\mathbf{M}_1 = \mathbf{r}_1 \times \mathbf{F}_1$ where $\mathbf{r}_1$ is a vector from O to any point on the line of action of $\mathbf{F}_1$. When all forces are shifted to O in this manner, we have a system of concurrent forces at O and a system of couple vectors, as represented in the b-part of the figure. The concurrent forces may then be added vectorially to produce a resultant force $\mathbf{R}$, and the couples may also be added to produce a resultant couple $\mathbf{M}$, Fig. 2/25c. The general force system, then, is reduced to

(a)

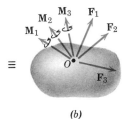

(b)

$$\boxed{\begin{aligned}\mathbf{R} &= \mathbf{F}_1 + \mathbf{F}_2 + \mathbf{F}_3 + \cdots = \Sigma\mathbf{F} \\ \mathbf{M} &= \mathbf{M}_1 + \mathbf{M}_2 + \mathbf{M}_3 + \cdots = \Sigma(\mathbf{r} \times \mathbf{F})\end{aligned}} \qquad (2/15)$$

The couple vectors are shown through point O, but since they are free vectors, they may be represented in any parallel positions. The magnitudes of the resultants and their components are

$$\boxed{\begin{aligned}R_x &= \Sigma F_x \qquad R_y = \Sigma F_y \qquad R_z = \Sigma F_z \\ R &= \sqrt{(\Sigma F_x)^2 + (\Sigma F_y)^2 + (\Sigma F_z)^2} \\ M_x &= \Sigma(\mathbf{r} \times \mathbf{F})_x \qquad M_y = \Sigma(\mathbf{r} \times \mathbf{F})_y \qquad M_z = \Sigma(\mathbf{r} \times \mathbf{F})_z \\ M &= \sqrt{M_x^2 + M_y^2 + M_z^2}\end{aligned}} \qquad (2/16)$$

(c)

Figure 2/25

The point O that is selected as the point of concurrency for the forces is arbitrary, and the magnitude and direction of $\mathbf{M}$ will depend on the particular point O selected. The magnitude and direction of $\mathbf{R}$, however, are the same no matter which point is selected. In general any system of forces may be replaced by its resultant force $\mathbf{R}$ and the resultant couple $\mathbf{M}$. In dynamics we usually select the mass center as the reference point, and the change in the linear motion of the body is determined by the resultant force and the change in the angular motion of the body is determined by the resultant couple. In statics the complete equilibrium of a body is specified when the resultant force $\mathbf{R}$ is zero and the resultant couple $\mathbf{M}$ is zero. Thus the determination of resultants is essential in both statics and dynamics.

The resultants for several special force systems will now be noted.

Concurrent forces. When forces are concurrent at a point, only the first of Eqs. 2/15 need to be used since there are no moments about the point of concurrency.

Parallel forces. For a system of parallel forces not all in the same plane the magnitude of the parallel resultant force is simply the algebraic sum of the given forces, and the position of its line of action is easily obtained by applying the two components of the moment principle about axes in a plane normal to the forces.

Coplanar forces. Article 2/6 was devoted to this force system.

Wrench resultant. When the resultant couple vector **M** is parallel to the resultant force **R**, as shown in Fig. 2/26, the resultant is said to be a *wrench*. By definition a wrench is positive if the couple and force vectors point in the same direction and negative if they point in opposite directions. A common example of a positive wrench is found with the application of a screwdriver.

Any general force system may be represented by a wrench applied along a unique line of action. This reduction is illustrated in Fig. 2/27, where the *a*-part of the figure represents for the general force system the resultant force **R** acting at some point *O* and the corresponding resultant couple **M**. Although **M** is a free vector, for convenience we represent it through *O*. In the *b*-part of the figure, **M** is resolved into components **M₁** along the direction of **R** and **M₂** normal to **R**. In the *c*-part of the figure the couple **M₂** is replaced by its equivalent of two forces **R** and −**R** separated a distance $d = M_2/R$ with −**R** applied at *O* to cancel the original **R**. This step leaves the resultant **R**, which acts along a new and unique line of action, and the parallel couple **M₁**, which is a free vector, as shown in the *d*-part of the figure. Thus the resultants of the original general force system have been transformed into a wrench (positive in this illustration) with its unique axis defined by the new position of **R**. We see from Fig. 2/27 that the axis of the wrench resultant lies in a plane through *O* normal to the plane defined by **R** and **M**. The wrench is the simplest form in which the resultant of a general force system may be expressed. This form of the resultant, however, has limited application, since it is usually more convenient to use as the reference point some point *O* such as the mass center of the body or other convenient origin of coordinates not on the wrench axis.

Positive wrench

Negative wrench

Figure 2/26

(a)   (b)   (c)   (d)

Figure 2/27

Sample Problem 2/11

Replace the two forces and the negative wrench by a single force **R** applied at A and the corresponding couple **M**.

Solution. The resultant force has the components

$[R_x = \Sigma F_x]$ $\quad R_x = 500 \sin 40° + 700 \sin 60° = 928$ N

$[R_y = \Sigma F_y]$ $\quad R_y = 600 + 500 \cos 40° \cos 45° = 871$ N

$[R_z = \Sigma F_z]$ $\quad R_z = 700 \cos 60° + 500 \cos 40° \sin 45° = 621$ N

Thus $\qquad\qquad$ **R** $= 928\mathbf{i} + 871\mathbf{j} + 621\mathbf{k}$ N

and $\qquad R = \sqrt{(928)^2 + (871)^2 + (621)^2} = 1416$ N $\qquad$ *Ans.*

The couple to be added as a result of moving the 500-N force is

① $[\mathbf{M} = \mathbf{r} \times \mathbf{F}]$ $\quad \mathbf{M}_{500} = (0.08\mathbf{i} + 0.12\mathbf{j} + 0.05\mathbf{k}) \times 500(\mathbf{i} \sin 40°$
$\qquad\qquad\qquad\qquad\qquad + \mathbf{j} \cos 40° \cos 45° + \mathbf{k} \cos 40° \sin 45°)$

where **r** is the vector from A to B.

The term-by-term, or determinant, expansion gives

$$\mathbf{M}_{500} = 18.95\mathbf{i} - 5.59\mathbf{j} - 16.90\mathbf{k} \text{ N} \cdot \text{m}$$

The moment of the 600-N force about A is written by inspection of its x- and z-components, which give

$$\mathbf{M}_{600} = (600)(0.060)\mathbf{i} + (600)(0.040)\mathbf{k}$$
$$= 36.0\mathbf{i} + 24.0\mathbf{k} \text{ N} \cdot \text{m}$$

② The moment of the 700-N force about A is easily obtained from the moments of the x- and z-components of the force. The result becomes

$$\mathbf{M}_{700} = (700 \cos 60°)(0.030)\mathbf{i} - [(700 \sin 60°)(0.060)$$
$$+ (700 \cos 60°)(0.100)]\mathbf{j} - (700 \sin 60°)(0.030)\mathbf{k}$$
$$= 10.5\mathbf{i} - 71.4\mathbf{j} - 18.19\mathbf{k} \text{ N} \cdot \text{m}$$

③ Also, the couple of the given wrench may be written

$$\mathbf{M}' = 25.0(-\mathbf{i} \sin 40° - \mathbf{j} \cos 40° \cos 45° - \mathbf{k} \cos 40° \sin 45°)$$
$$= -16.07\mathbf{i} - 13.54\mathbf{j} - 13.54\mathbf{k} \text{ N} \cdot \text{m}$$

Therefore, the resultant couple upon adding together the **i**-, **j**-, and **k**-terms of the four **M**'s is

④ $$\mathbf{M} = 49.4\mathbf{i} - 90.5\mathbf{j} - 24.6\mathbf{k} \text{ N} \cdot \text{m}$$

and $\quad M = \sqrt{(49.4)^2 + (90.5)^2 + (24.6)^2} = 106.0$ N $\cdot$ m $\qquad$ *Ans.*

① *Suggestion:* Check the cross-product results by evaluating the moments about A of the components of the 500-N force directly from the sketch.

② For the 600-N and 700-N forces it is easier to obtain the components of their moments about the coordinate directions through A by inspection of the figure than it is to set up the cross-product relations.

③ The 25-N $\cdot$ m couple vector of the *wrench* points in the direction opposite to that of the 500-N force, and we must resolve it into its x-, y-, and z-components to be added to the other couple-vector components.

④ Although the resultant couple vector **M** in the sketch of the resultants is shown through A, we recognize that a couple vector is a free vector and therefore has no specified line of action.

Sample Problem   2/12

Determine the wrench resultant of the three forces acting on the bracket. Calculate the coordinates of the point P in the x-y plane through which the resultant force of the wrench acts. Also find the magnitude of the couple $\mathbf{M}$ of the wrench.

① *Solution.*   The direction cosines of the couple $\mathbf{M}$ of the wrench must be the same as those of the resultant force $\mathbf{R}$, assuming that the wrench is positive. The resultant force is

$$\mathbf{R} = 20\mathbf{i} + 40\mathbf{j} + 40\mathbf{k} \text{ lb} \qquad R = \sqrt{(20)^2 + (40)^2 + (40)^2} = 60 \text{ lb}$$

and its direction cosines are

$$\cos \theta_x = 20/60 = 1/3, \ \cos \theta_y = 40/60 = 2/3, \ \cos \theta_z = 40/60 = 2/3$$

The moment of the wrench couple must equal the sum of the moments of the given forces about point P through which $\mathbf{R}$ passes. The moments about P of the three forces are

$$(\mathbf{M})_{R_x} = 20y\mathbf{k} \text{ lb-in.}$$

$$(\mathbf{M})_{R_y} = -40(3)\mathbf{i} - 40x\mathbf{k} \text{ lb-in.}$$

$$(\mathbf{M})_{R_z} = 40(4 - y)\mathbf{i} - 40(5 - x)\mathbf{j} \text{ lb-in.}$$

and the total moment is

$$\mathbf{M} = (40 - 40y)\mathbf{i} + (-200 + 40x)\mathbf{j} + (-40x + 20y)\mathbf{k} \text{ lb-in.}$$

The direction cosines of $\mathbf{M}$ are

$$\cos \theta_x = (40 - 40y)/M$$

$$\cos \theta_y = (-200 + 40x)/M$$

$$\cos \theta_z = (-40x + 20y)/M$$

where M is the magnitude of $\mathbf{M}$. Equating the direction cosines of $\mathbf{R}$ and $\mathbf{M}$ gives

$$40 - 40y = \frac{M}{3}$$

$$-200 + 40x = \frac{2M}{3}$$

$$-40x + 20y = \frac{2M}{3}$$

Solution of the three equations gives

$$M = -120 \text{ lb-in.}, \quad x = 3 \text{ in.}, \quad y = 2 \text{ in.} \qquad Ans.$$

We see that M turned out to be negative, which means that the couple vector is pointing in the direction opposite to $\mathbf{R}$, which makes the wrench negative.

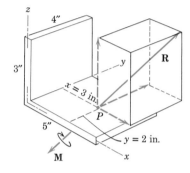

① We shall assume initially that the wrench is positive. If $\mathbf{M}$ turns out to be negative, then the direction of the couple vector is opposite to that of the resultant force.

PROBLEMS

2/110 The concrete slab supports the six vertical loads shown. Determine the resultant of these forces and the *x*- and *y*-coordinates of a point through which it acts.   *Ans.* $x = 7.30$ m, $y = 15.83$ m

Problem 2/110

2/111 Determine the *x*- and *y*-coordinates of a point through which the resultant of the parallel forces passes.

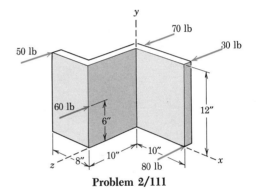

Problem 2/111

2/112 Determine the angle θ so that the net downward force applied to the fixed eyebolt is 750 lb. Determine the corresponding magnitude of the resultant of the three forces and the direction cosines of the resultant.

     Ans. $\theta = 33.3°$, $R = 1021$ lb, $\cos \theta_x = 0.525$,
         $\cos \theta_y = 0.430$, $\cos \theta_z = -0.734$

Problem 2/112

Problem 2/113

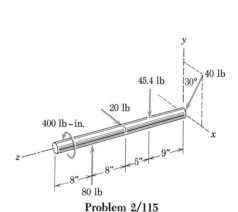

Problem 2/115

Problem 2/116

2/113 The four forces are concurrent at the origin O of coordinates. If the x-component of their resultant **R** is -50 N and the z-component is 100 N, determine F, θ, and R.

2/114 Use Fig. 2/25c and show that the most general force system may be represented by two nonintersecting forces.

2/115 Determine the resultant of the four forces and couple acting on the shaft.
  Ans. Couple $\mathbf{M} = -1351\mathbf{i} + 280\mathbf{j} + 400\mathbf{k}$ lb-in.

2/116 Represent the resultant of the force system acting on the pipe assembly by a single force **R** at A and a couple **M**.

2/117 The four forces are acting along the edges of the 0.8-m cube as shown. Represent the resultant of these forces by a force **R** through point *A* and a couple **M**.

   Ans. **R** = 200(**j** + **k**) N
   M = 560**i** − 320**j** + 400**k** N·m

Problem 2/117

2/118 Replace the two forces of equal magnitude *F* by a single force **R** passing through point *A* and a couple **M**.

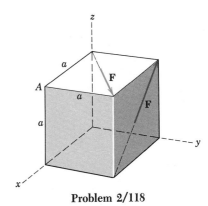

Problem 2/118

2/119 The combined action of the three forces on the base at *O* may be obtained by establishing their resultant through *O*. Determine the magnitudes of **R** and the accompanying couple **M**.

   Ans. *R* = 1093 lb, *M* = 9734 lb-ft

Problem 2/119

Problem 2/120

Problem 2/121

2/120 The motor mounted on the bracket is acted upon by its 160-N weight, and its shaft resists the 120-N thrust and 25-N·m couple applied to it. Determine the resultant of the force system shown in terms of a force **R** at A and a couple **M**.

▶**2/121** The resultant of the two forces and couple may be represented by a wrench. Determine the vector expression for the moment **M** of the wrench and find the coordinates of the point P in the x-z plane through which the resultant force of the wrench passes.

Ans. $\mathbf{M} = 10(\mathbf{i} + \mathbf{j})\ \text{N·m}, x = 100\ \text{mm}, z = 100\ \text{mm}$

▶**2/122** Replace the force system acting on the pipe assembly of Prob. 2/116 by a wrench. Find the magnitude of the moment **M** of the wrench and the coordinates of the point P in the x-y plane through which the resultant force of the wrench passes.

Ans. $M = 95.6\ \text{N·m},\ x = 271\ \text{mm},\ y = -486\ \text{mm}$

▶**2/123** The resultant of a general force system may be expressed as a wrench along a unique line of action. For the force system of Prob. 2/120 determine the coordinates of the point P which is the intersection of the line of action of the wrench with the x-y plane.

Ans. $x = 56.2\ \text{mm}, y = 100\ \text{mm}$

EQUILIBRIUM

3

3/1 INTRODUCTION. The subject of statics deals primarily with the description of the conditions of force that are both necessary and sufficient to maintain the state of equilibrium of engineering structures. This chapter on equilibrium, therefore, constitutes the most central part of statics and should be mastered thoroughly. We shall make continuous use of the concepts developed in Chapter 2 involving forces, moments, couples, and resultants as we apply the principles of equilibrium. The procedures which we shall develop in Chapter 3 constitute a comprehensive introduction to the approach used in the solution of countless problems in mechanics and in other engineering areas as well. The approach which we shall develop is basic to the successful mastery of statics, and the student is urged to read and study the following articles with special effort and attention to detail.

A body is in equilibrium when the resultant of *all* forces acting on it is zero. Thus the resultant force $\mathbf{R}$ and the resultant couple $\mathbf{M}$ are both zero, and we have the equilibrium equations

$$\mathbf{R} = \Sigma\mathbf{F} = 0 \qquad \mathbf{M} = \Sigma\mathbf{M} = 0 \tag{3/1}$$

These requirements are both necessary and sufficient conditions for equilibrium.

We shall follow the arrangement used in Chapter 2 and discuss in Section A the equilibrium of two-dimensional force systems and in Section B the equilibrium of three-dimensional force systems.

SECTION A. EQUILIBRIUM IN TWO DIMENSIONS

3/2 MECHANICAL SYSTEM ISOLATION. Before we apply Eqs. 3/1 it is essential that we define unambiguously the particular body or mechanical system to be analyzed and represent clearly and completely *all* forces which act *on* the body. The omission of a force or the inclusion of a force which does not act on the body in question will give erroneous results.

A mechanical system is defined as a body or group of bodies that can be isolated from all other bodies. Such a system may be a single body or a combination of connected bodies. The bodies may be rigid or nonrigid. The system may also be a defined fluid mass, liquid or gas, or the system may be a combination of fluids and solids. In statics we direct our attention primarily to a description of the forces that act on rigid bodies at rest, although consideration is also given to the statics of fluids. Once we reach a decision about which body or combination of bodies is to be analyzed, then this body or combination treated as a single body is *isolated* from all surrounding bodies. This isolation is accomplished by means of the *free-body diagram,* which is a diagrammatic representation of the isolated body or combination of bodies considered as a single body, showing all forces applied to it by other bodies that are imagined to be removed. Only after such a diagram has been carefully drawn should the equilibrium equations be written. Because of its critical importance we emphasize here that

the free-body diagram is the most important single step in the solution of problems in mechanics.

Before we attempt to draw free-body diagrams, the mechanical characteristics of force application must be recognized. In Art. 2/2 the basic characteristics of force were described with primary attention focused on the vector properties of force. We noted that forces are applied both by direct physical contact and by remote action and that forces may be either internal or external to the body under consideration. We further observed that the application of external forces is accompanied by reactive forces and that both applied and reactive forces may be either concentrated or distributed. Additionally the principle of transmissibility was introduced which permits the treatment of force as a sliding vector insofar as its external effects on a rigid body are concerned. We will now use these characteristics of force in developing the analytical model of an isolated mechanical system to which the equations of equilibrium will then be applied.

Figure 3/1 shows the common types of force application on mechanical systems for analysis in two dimensions. In each example the force exerted *on* the body to be isolated *by* the body to be removed is indicated. Newton's third law, which notes the existence of an equal and opposite reaction to every action, must be carefully observed.

In example 1 the action of a flexible cable, belt, rope, or chain on the body to which it is attached is depicted. Because of its flexibility a rope or cable is unable to offer any resistance to bending, shear, or compression and therefore exerts a tension force in a direction tangent to the cable at its point of attachment. The force exerted *by* the cable *on* the body to which it is attached is always *away* from the body. When the tension T is large compared with the weight of the

MECHANICAL ACTION OF FORCES IN TWO–DIMENSIONAL ANALYSIS

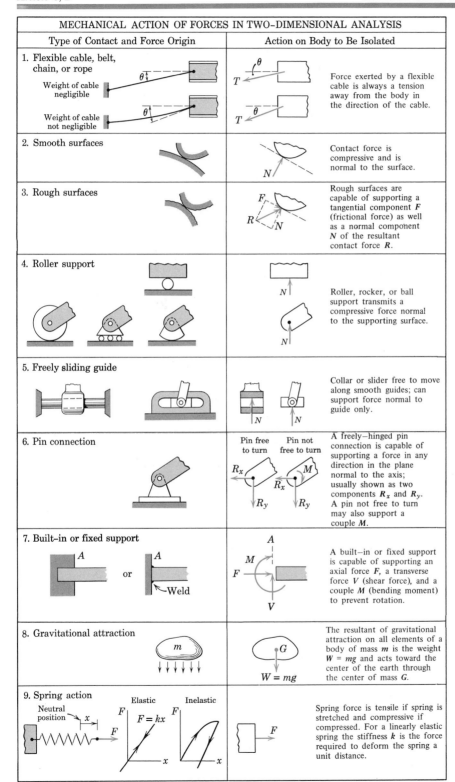

Type of Contact and Force Origin	Action on Body to Be Isolated
1. Flexible cable, belt, chain, or rope	Force exerted by a flexible cable is always a tension away from the body in the direction of the cable.
2. Smooth surfaces	Contact force is compressive and is normal to the surface.
3. Rough surfaces	Rough surfaces are capable of supporting a tangential component F (frictional force) as well as a normal component N of the resultant contact force R.
4. Roller support	Roller, rocker, or ball support transmits a compressive force normal to the supporting surface.
5. Freely sliding guide	Collar or slider free to move along smooth guides; can support force normal to guide only.
6. Pin connection	A freely–hinged pin connection is capable of supporting a force in any direction in the plane normal to the axis; usually shown as two components R_x and R_y. A pin not free to turn may also support a couple M.
7. Built–in or fixed support	A built–in or fixed support is capable of supporting an axial force F, a transverse force V (shear force), and a couple M (bending moment) to prevent rotation.
8. Gravitational attraction	The resultant of gravitational attraction on all elements of a body of mass m is the weight $W = mg$ and acts toward the center of the earth through the center of mass G.
9. Spring action	Spring force is tensile if spring is stretched and compressive if compressed. For a linearly elastic spring the stiffness k is the force required to deform the spring a unit distance.

Figure 3/1

cable, we may assume that the cable forms a straight line. When the cable weight is not negligible compared with its tension, the sag of the cable becomes important, and the tension in the cable changes direction and magnitude along its length. At its attachment the cable exerts a force tangent to itself.

When the smooth surfaces of two bodies are in contact, as in example 2, the force exerted by one on the other is *normal* to the tangency of the surfaces and is compressive. Although no actual surfaces are perfectly smooth, we are justified in making this assumption for practical purposes in many instances.

When mating surfaces of contacting bodies are rough, example 3, the force of contact may not necessarily be normal to the tangent to the surfaces but may be resolved into a *tangential* or frictional component F and a *normal component N*.

Example 4 illustrates a number of forms of mechanical support which effectively eliminate tangential friction forces, and here the net reaction is normal to the supporting surface.

Example 5 shows the action of a smooth guide on the body it supports. Resistance parallel to the guide is absent.

Example 6 illustrates the action of a pin connection. Such a connection is able to support force in any direction normal to the axis of the pin. We usually represent this action in terms of two rectangular components. If the joint is free to turn about the pin, only the force R can be supported. If the joint is not free to turn, a resisting couple M may also be supported.

Example 7 shows the resultants of the rather complex distribution of force over the cross section of a slender bar or beam at a built-in or fixed support.

One of the most common forces is that due to gravitational attraction, example 8. This force affects all elements of mass of a body and is, therefore, distributed throughout it. The resultant of the gravitational forces on all elements is the weight $W = mg$ of the body, which passes through the center of mass G and is directed toward the center of the earth for earthbound structures. The position of G is frequently obvious from the geometry of the body, particularly where conditions of symmetry exist. When the position is not readily apparent, the location of G must be calculated or determined by experiment. Similar remarks apply to the remote action of magnetic and electric forces. These forces of remote action have the same overall effect on a rigid body as forces of equal magnitude and direction applied by direct external contact.

Example 9 illustrates the action of a linear elastic spring and of a nonlinear spring. The force exerted by a linear spring, in tension or compression, is given by $F = kx$, where k is the stiffness of the spring and x is its deformation measured from the neutral or undeformed position. The linearity of the force-deformation relation tells us that the magnitudes of the tensile and compressive forces in a given spring are equal if the magnitudes of the respective tension and compression deformations are equal.

The student is urged to study these nine conditions and to identify them in the problem work so that the correct free-body diagrams may be drawn. The representations in Fig. 3/1 are *not* free-body diagrams but are merely elements in the construction of free-body diagrams.

The full procedure for drawing a free-body diagram which accomplishes the isolation of the body or system under consideration will now be described.

Construction of free-body diagrams.   The following steps are involved.

Step 1. A clear decision is made concerning which body or combination of bodies is to be isolated. The body chosen will usually involve one or more of the desired unknown quantities.

Step 2. The body or combination chosen is next isolated by a diagram that represents its *complete external boundary.*

This boundary defines the isolation of the body from *all* other contacting or attracting bodies, which are considered removed. This step is often the most crucial of all. We should always be certain that we have *completely isolated* the body before proceeding with the next step.

Step 3. All forces that act *on* the isolated body as applied *by* the removed contacting and attracting bodies are next represented in their proper positions on the diagram of the isolated body. A systematic traverse of the entire boundary will disclose all contact forces. Weights, where appreciable, must be included. Known forces should be represented by vector arrows with their proper magnitude, direction, and sense indicated. Unknown forces should be represented by vector arrows with the unknown magnitude or direction indicated by symbol. If the sense of the vector is also unknown, it may be arbitrarily assumed. The calculations will reveal a positive quantity if the correct sense was assumed and a negative magnitude if the incorrect sense was assumed. It is necessary to be *consistent* with the assigned characteristics of unknown forces throughout all of the calculations.

Step 4. The choice of coordinate axes should be indicated directly on the diagram. Pertinent dimensions may also be represented for convenience. Note, however, that the free-body diagram serves the purpose of focusing accurate attention on the action of the external forces, and therefore the diagram should not be cluttered with excessive extraneous information. Force arrows should be clearly distinguished from any other arrows which may appear so that confusion will not result. For this purpose a colored pencil may be used.

When the foregoing four steps are completed, a correct free-body diagram will result, and the way will be clear for a straightforward and successful application of the governing equations, both in statics and in dynamics.

Many students are often tempted to omit from the free-body diagram certain forces that may not appear at first glance to be needed in the calculations. When we yield to this temptation we invite serious error. It is only through *complete* isolation and a systematic representation of *all* external forces that a reliable accounting of the effects of all applied and reactive forces can be made. Very often a force that at first glance may not appear to influence a desired result does indeed have an influence. Hence the only safe procedure is to make certain that all forces whose magnitudes are not negligible appear on the free-body diagram.

The free-body diagram has been explained in some detail because of its great importance in mechanics. The free-body method ensures an accurate definition of a mechanical system and focuses attention on the exact meaning and application of the force laws of statics and dynamics. Indeed the free-body method is so important that students are strongly urged to reread this section several times in conjunction with their study of the sample free-body diagrams shown in Fig. 3/2 and the sample problems which appear at the end of the next article.

Figure 3/2 gives four examples of mechanisms and structures together with their correct free-body diagrams. Dimensions and magnitudes are omitted for clarity. In each case we treat the entire system as a single body, so that the internal forces are not shown. The characteristics of the various types of contact forces illustrated in Fig. 3/1 are included in the four examples as they apply.

In example 1 the truss is composed of structural elements that, taken all together, constitute a rigid framework. Thus we may remove the entire truss from its supporting foundation and treat it as a single rigid body. In addition to the applied external load P, the free-body diagram must include the reactions on the truss at A and B. The rocker at B can support a vertical force only, and this force is transmitted to the structure at B (example 4 of Fig. 3/1). The pin connection at A (example 6 of Fig. 3/1) is capable of supplying both a horizontal and a vertical component of force to the truss. In this relatively simple example it is clear that the vertical component A_y must be directed down to prevent the truss from rotating clockwise about B. Also, the horizontal component A_x will be to the left to keep the truss from moving to the right under the influence of the horizontal component of P. If the total weight of the truss members is appreciable compared with P and the forces at A and B, then the weights of the members must be included on the free-body diagram as external forces.

SAMPLE FREE-BODY DIAGRAMS

Mechanical System	Free-Body Diagram of Isolated Body

1. Plane truss

   Weight of truss assumed negligible compared with P

2. Cantilever beam

   Mass m

3. Beam

   Smooth surface contact at A. Mass m

4. Rigid system of interconnected bodies analyzed as a single unit

   Weight of mechanism neglected

Figure 3/2

In example 2 the cantilever beam is secured to the wall and subjected to the three applied loads. When we isolate that part of the beam to the right of the section at A, we must include the reactive forces applied to the beam by the wall. The resultants of these reactive forces are shown acting on the section of the beam (example 7 of Fig. 3/1). A vertical force V to counteract the excess of downward applied force is shown, and a tension T to balance the excess of applied force to the right must also be included. Then to prevent the beam from rotating about A a counterclockwise couple M is also required. The weight mg of the beam must also be represented through the mass center (example 8 of Fig. 3/1). The free-body diagram is now complete and shows the beam in equilibrium under the action of six forces and one couple.

In example 3 the weight $W = mg$ is shown acting through the center of mass of the beam, which is assumed known (example 8 of Fig. 3/1). The force exerted by the corner A on the beam is normal to the smooth surface of the beam (example 2 of Fig. 3/1). If the contacting surfaces at the corner were not smooth, a tangential frictional component of force could be developed. In addition to the applied force P and couple M, there is the pin connection at B, which exerts both an x- and a y-component of force on the beam. The positive senses of these components are assigned arbitrarily.

In example 4 the free-body diagram of the entire isolated mechanism discloses three unknown quantities for equilibrium with the given loads mg and P. Any one of many internal configurations for securing the cable leading from the mass m would be possible without affecting the external response of the mechanism as a whole, and this fact is brought out by the free-body diagram.

The positive senses of B_x and B_y in example 3 and B_y in example 4 are assumed on the free-body diagrams, and the correctness of the assumptions would be proved or disproved according to whether the algebraic signs of the terms were plus or minus when the calculations were carried out in the actual problems.

The isolation of the mechanical system under consideration will be recognized as a crucial step in the formulation of the mathematical model. The student is again urged to devote special attention to this step. Before direct use is made of the free-body diagram in the application of the principles of force equilibrium in the next article, some initial practice with the drawing of free-body diagrams is helpful. The problems that follow are designed to provide this practice.

PROBLEMS

3/1 In each of the five following examples, the body to be isolated is shown in the left-hand diagram, and an *incomplete* free-body diagram (FBD) of the isolated body is shown on the right. Add whatever forces are necessary in each case to form a complete free-body diagram. The weights of the bodies are negligible unless otherwise indicated. Dimensions and numerical values are omitted for simplicity.

	Body	Incomplete *FBD*
1. Bell crank supporting mass *m* with pin support at *A*		
2. Control lever applying torque to shaft at *O*		
3. Boom *OA*, of negligible mass compared with mass *m*. Boom hinged at *O* and supported by hoisting cable at *B*.		
4. Uniform crate of mass *m* leaning against smooth vertical wall and supported on a rough horizontal surface		
5. Loaded bracket supported by pin connection at *A* and fixed pin in smooth slot at *B*		

Problem 3/1

3/2 In each of the five following examples, the body to be isolated is shown in the left-hand
diagram, and either a *wrong* or an *incomplete* free-body diagram (FBD) is shown on the
right. Make whatever changes or additions are necessary in each case to form a correct
and complete free-body diagram. The weights of the bodies are negligible unless
otherwise indicated. Dimensions and numerical values are omitted for simplicity.

	Body	Wrong or Incomplete *FBD*
1. Lawn roller of mass m being pushed up incline θ		
2. Pry bar lifting body A having smooth horizontal surface. Bar rests on horizontal rough surface		
3. Uniform pole of mass m being hoisted into position by winch. Horizontal supporting surface notched to prevent slipping of pole		
4. Supporting angle bracket for frame. Pin joints		
5. Bent rod welded to support at A and subjected to two forces and couple		

Problem 3/2

3/3 Draw a complete and correct free-body diagram of each of the bodies designated in the statements. The weights of the bodies are significant only if the mass is stated. All forces, known and unknown, should be labeled. (*Note:* The sense of some reaction components cannot always be determined without numerical calculation.)

1. Uniform horizontal bar of mass m suspended by vertical cable at A and supported by rough inclined surface at B	5. Uniform grooved wheel of mass m supported by a rough surface and by action of horizontal cable
2. Wheel of mass m on verge of being rolled over curbing by pull **P**	6. Bar, initially horizontal but deflected under load **L**. Pinned to rigid support at each end
3. Loaded truss supported by pin joint at A and by cable at B	7. Uniform heavy plate of mass m supported in vertical plane by cable C and hinge A
4. Uniform bar of mass m and roller of mass m_0 taken together. Subjected to couple M and supported as shown. Roller is free to turn	8. Entire frame, pulleys, and contacting cable to be isolated as a single unit

Problem 3/3

3/3 EQUILIBRIUM CONDITIONS. In Art. 3/1 we defined equilibrium as the condition in which the resultant of all forces acting on a body is zero. Stated in another way, a body is in equilibrium if all forces and moments applied to it are in balance. These requirements are contained in the vector equations of equilibrium, Eqs. 3/1, which in two dimensions may be written in scalar form as

$$\Sigma F_x = 0 \qquad \Sigma F_y = 0 \qquad \Sigma M_O = 0 \qquad (3/2)$$

The third equation represents the zero sum of the moments of all forces about any point O on or off the body. Equations 3/2 are the necessary and sufficient conditions for complete equilibrium in two dimensions. They are necessary conditions because, if not satisfied, there can be no force or moment balance. They are sufficient since once satisfied there can be no unbalance, and equilibrium is assured.

The equations relating force and acceleration for rigid-body motion are developed in Vol. 2 *Dynamics* from Newton's second law of motion. These equations show that the acceleration of the mass center of a body is proportional to the resultant force $\Sigma \mathbf{F}$ acting on the body. Consequently, if a body moves with constant velocity (zero acceleration), the resultant force on it must be zero, and the body may be treated as in a state of equilibrium.

For complete equilibrium in two dimensions all three of Eqs. 3/2 must hold. However, these conditions are independent requirements, and one may hold without another. Take, for example, a body which slides along a horizontal surface with increasing velocity under the action of applied forces. The force-equilibrium equations will be satisfied in the vertical direction where the acceleration is zero but not in the horizontal direction. Also, a body, such as a flywheel, which rotates about its fixed mass center with increasing angular speed is not in rotational equilibrium, but the two force-equilibrium equations will be satisfied.

(a) Categories of Equilibrium. Applications of Eqs. 3/2 fall naturally into a number of categories that are easily identified. These categories of force systems acting on bodies in two-dimensional equilibrium are summarized in Fig. 3/3 and are explained further as follows:

Case 1, equilibrium of collinear forces, clearly requires only the one force equation in the direction of the forces (x-direction), since all other equations are automatically satisfied.

Case 2, equilibrium of forces that lie in a plane (x-y plane) and are concurrent at a point O, requires the two force equations only, since the moment sum about O, that is, about a z-axis through O, is necessarily zero.

Case 3, equilibrium of parallel forces in a plane, requires the one

CATEGORIES OF EQUILIBRIUM IN TWO DIMENSIONS		
Force System	Free–Body Diagram	Independent Equations
1. Collinear	$\mathbf{F}_3$ $\mathbf{F}_2$ $\mathbf{F}_1$	$\Sigma F_x = 0$
2. Concurrent at a point	$\mathbf{F}_1$ $\mathbf{F}_2$ O $\mathbf{F}_4$ $\mathbf{F}_3$	$\Sigma F_x = 0$ $\Sigma F_y = 0$
3. Parallel	$\mathbf{F}_2$ $\mathbf{F}_1$ $\mathbf{F}_3$ $\mathbf{F}_4$	$\Sigma F_x = 0$ $\Sigma M_z = 0$
4. General	$\mathbf{F}_1$ $\mathbf{F}_2$ M $\mathbf{F}_3$ $\mathbf{F}_4$	$\Sigma F_x = 0$ $\Sigma M_z = 0$ $\Sigma F_y = 0$

Figure 3/3

Two–force members
(*a*)
Figure 3/4

force equation in the direction of the forces (*x*-direction) and one moment equation about an axis (*z*-axis) normal to the plane of the forces.

Case 4, equilibrium of a general system of forces in a plane (*x-y*), requires the two force equations in the plane and one moment equation about an axis (*z*-axis) normal to the plane.

There are two equilibrium situations that occur frequently to which the student should be alerted. The first situation is the equilibrium of a body under the action of two forces only. Two examples are shown in Fig. 3/4, and we see that for such a *two-force member* the forces must be equal, opposite, and collinear. The shape of the member should not obscure this simple requirement. In the illustrations cited we consider the weights of the members to be negligible compared with the applied forces.

The second situation is the equilibrium of a body under the action of three forces, Fig. 3/5*a*. We see that the lines of action of the three forces must be *concurrent*. If they were not concurrent, then one of the forces would exert a resultant moment about the point of concurrency of the other two, which would violate the requirement of zero moment about every point. The only exception occurs when

(*a*) Three–force member

(*b*) Closed polygon
satisfies $\Sigma \mathbf{F} = 0$
Figure 3/5

the three forces are parallel. In this case we may consider the point of concurrency to be at infinity. The principle of the concurrency of three forces in equilibrium is of considerable use in carrying out a graphical solution of the force equations. In this case the polygon of forces is drawn and made to close, as shown in Fig. 3/5*b*. Frequently, a body in equilibrium under the action of more than three forces may be reduced to a *three-force member* by a combination of two or more of the known forces.

(*b*) *Alternative Equilibrium Equations.* There are two additional ways in which we may express the general conditions for the equilibrium of forces in two dimensions. For the body shown in Fig. 3/6*a*, if $\Sigma M_A = 0$, then the resultant **R**, if it still exists, cannot be a

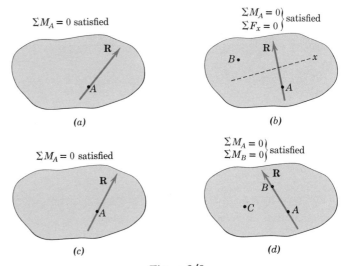

Figure 3/6

couple but must be a force **R** passing through A. If now the equation $\Sigma F_x = 0$ holds, where the x-direction is perfectly arbitrary, it follows from Fig. 3/6*b* that the resultant force **R**, if it still exists, not only must pass through A, but also must be perpendicular to the x-direction as shown. Now, if $\Sigma M_B = 0$, where B is any point such that the line AB is not perpendicular to the x-direction, we see that **R** must be zero, and hence the body is in equilibrium. Therefore an alternative set of equilibrium equations is

$$\Sigma F_x = 0 \qquad \Sigma M_A = 0 \qquad \Sigma M_B = 0$$

where the two points A and B must not lie on a line perpendicular to the x-direction.

A third formulation of the conditions of equilibrium may be

made for a coplanar force system. Again, if $\Sigma M_A = 0$ for any body such as that shown in Fig. 3/6c, the resultant, if any, must be a force **R** through A. In addition if $\Sigma M_B = 0$, the resultant, if one still exists, must pass through B as shown in Fig. 3/6d. Such a force cannot exist, however, if $\Sigma M_C = 0$, where C is not collinear with A and B. Hence we may write the equations of equilibrium as

$$\Sigma M_A = 0 \qquad \Sigma M_B = 0 \qquad \Sigma M_C = 0$$

where A, B, and C are any three points not on the same straight line.

When equilibrium equations are written which are not independent, redundant information is obtained, and solution of the equations will yield $0 = 0$. For example, for a general problem in two dimensions with three unknowns, three moment equations written about three points which lie on the same straight line are not independent. Such equations will contain duplicated information, and solution of two of them can at best determine two of the unknowns with the third equation merely verifying the identity $0 = 0$.

(*c*) *Constraints and Statical Determinacy.* The equilibrium equations developed in this article, once satisfied, are both necessary and sufficient conditions to establish the equilibrium of a body. However, they do not necessarily provide all the information that is required to calculate all the unknown forces that may act on a body in equilibrium. The question of adequacy lies in the characteristics of the constraints to possible movement of the body provided by its supports. By constraint we mean the restriction of movement. In example 4 of Fig. 3/1 the roller, ball, and rocker provide constraint normal to the surface of contact but none tangent to the surface. Hence a tangential force cannot be supported. For the collar and slider of example 5 constraint exists only normal to the guide. In example 6 the fixed-pin connection provides constraint in both directions but offers no resistance to rotation about the pin unless the pin is not free to turn. The fixed support of example 7, however, offers constraint against rotation as well as lateral movement.

If the rocker that supports the truss of example 1 in Fig. 3/2 were replaced by a pin joint, as at A, there would be one additional constraint beyond that required to support an equilibrium configuration without collapse. The three scalar conditions of equilibrium, Eqs. 3/2, would not provide sufficient information to determine all four unknowns, since A_x and B_x could not be separated. These two components of force would be dependent on the deformation of the members of the truss as influenced by their corresponding stiffness properties. The horizontal reactions A_x and B_x would also be dependent on any initial deformation required to fit the dimensions of the structure to those of the foundation between A and B. Again referring to Fig. 3/2, if the pin B in example 3 were not free to turn, the support could transmit a couple to the beam through the pin. Therefore there would be four unknown supporting reactions acting on the beam, namely, the force at A, the two components of force at B, and the couple at B. Consequently the three independent scalar equations of equilibrium would not provide enough information to compute all four unknowns.

A body, or rigid combination of elements considered as a single body, that possesses more external supports or constraints than are necessary to maintain an equilibrium position is called *statically indeterminate*. Supports that can be removed without destroying the equilibrium condition of the body are said to be *redundant*. The number of redundant supporting elements present corresponds to the degree of statical indeterminacy and equals the total number of unknown external forces minus the number of available independent equations of equilibrium. On the other hand, bodies that are supported by the minimum number of constraints necessary to ensure an equilibrium configuration are called *statically determinate,* and for such bodies the equilibrium equations are sufficient to determine the unknown external forces.

The problems on equilibrium included in this article and throughout *Statics* are generally restricted to statically determinate bodies where the constraints are just sufficient to ensure a stable position and where the unknown supporting forces can be completely determined by the available independent equations of equilibrium. The student is alerted at this point by this brief discussion, however, to the fact that we must be aware of the nature of the constraints before we attempt to solve an equilibrium problem. A body will be recognized as statically indeterminate when there are more unknown external reactions than there are available independent equilibrium equations for the force system involved. It is always well to count the number of unknown forces on a given body and to be certain that an equal number of independent equations may be written; otherwise effort may be wasted in attempting an impossible solution with the aid of the equilibrium equations only. Unknowns may be forces, couples, distances, or angles.

In discussing the relationship between constraints and equilibrium, we should look further at the question of the adequacy of constraints. The existence of three constraints for a two-dimensional problem does not always guarantee a stable configuration. Figure 3/7 shows four different types of constraints. In the *a*-part of the figure point A of the rigid body is fixed by the two links and cannot move, and the third link prevents any rotation about A. Thus this body is completely fixed with three adequate (proper) constraints. In the *b*-part of the figure the third link is positioned so that the force transmitted by it passes through point A where the other two constraint forces act. Thus this configuration of constraints can offer no initial resistance to rotation about A, which would occur when exter-

(a) Complete fixity   (b) Incomplete fixity   (c) Incomplete fixity   (d) Excessive fixity
   Adequate constraints   Partial constraints   Partial constraints   Redundant constraint

Figure 3/7

nal loads were applied to the body. We conclude, therefore, that this body is incompletely fixed under partial constraints. The configuration in the c-part of the figure gives us a similar condition of incomplete fixity since the three parallel links could offer no initial resistance to a small vertical movement of the body as a result of external loads applied to it in this direction. The constraints in these two examples are often termed *improper*. In the d-part of Fig. 3/7 we have a condition of complete fixity with link 4 acting as an unnecessary fourth constraint to maintain a fixed position. Link 4, then, is a *redundant* constraint, and the body is statically indeterminate.

As in the four examples of Fig. 3/7 it is generally possible by direct observation to conclude whether the constraints on a body in two-dimensional equilibrium are adequate (proper), partial (improper), or redundant. As indicated previously, the vast majority of problems in this book are statically determinate with adequate (proper) constraints.

The sample problems at the end of the article illustrate the application of free-body diagrams and the equations of equilibrium to typical statics problems. When equilibrium equations are applied, one of the most useful steps is an expeditious choice of reference axes and moment centers. Generally, the best choice of a moment center is one through which as many unknown forces pass as possible. Simultaneous solutions of equilibrium equations are frequently necessary but can be minimized or avoided by a careful choice of reference axes and moment centers.

In the problem work it is always advisable to precede each major application of a principle of mechanics by a symbolic statement of the principle or governing equation which is involved. In the sample problems these statements are set forth in brackets to the left of the calculations and serve as a reminder of the justification for each major step. Also, the recommendations set forth in Art. 1/8 will be valuable to the student particularly at this stage as he begins to form habits of approach to the solution of engineering problems.

Sample Problem 3/1

Determine the magnitudes of the forces **C** and **T** which, along with the other three forces shown, act on the members of the bridge-truss joint.

Solution. The given sketch constitutes the free-body diagram of the isolated section of the joint in question and shows the five forces which are in equilibrium.

Solution I (scalar algebra). For the *x-y* axes as shown we have

$[\Sigma F_x = 0]$ $8 + T\cos 40° + C\sin 20° - 16 = 0$

$$0.766\,T + 0.342\,C = 8 \qquad (a)$$

$[\Sigma F_y = 0]$ $T\sin 40° - C\cos 20° - 3 = 0$

$$0.643\,T - 0.940\,C = 3 \qquad (b)$$

Simultaneous solution of Eqs. (*a*) and (*b*) produces

$$T = 9.09 \text{ kN} \qquad C = 3.03 \text{ kN} \qquad \textit{Ans.}$$

① *Solution II (scalar algebra).* To avoid a simultaneous solution we may use axes *x'-y'* with the first summation in the *y'*-direction to eliminate reference to *T*. Thus

$[\Sigma F_{y'} = 0]$ $-C\cos 20° - 3\cos 40° - 8\sin 40° + 16\sin 40° = 0$

$$C = 3.03 \text{ kN} \qquad \textit{Ans.}$$

$[\Sigma F_{x'} = 0]$ $T + 8\cos 40° - 16\cos 40° - 3\sin 40° - 3.03\sin 20° = 0$

$$T = 9.09 \text{ kN} \qquad \textit{Ans.}$$

Solution III (vector algebra). With unit vectors **i** and **j** in the *x*- and *y*-directions the zero summation of forces for equilibrium yields the vector equation

$[\Sigma \mathbf{F} = \mathbf{0}]$ $8\mathbf{i} + (T\cos 40°)\mathbf{i} + (T\sin 40°)\mathbf{j} - 3\mathbf{j} + (C\sin 20°)\mathbf{i}$

$$- (C\cos 20°)\mathbf{j} - 16\mathbf{i} = \mathbf{0}$$

Equating the coefficients of the **i**- and **j**-terms to zero gives

$$8 + T\cos 40° + C\sin 20° - 16 = 0$$
$$T\sin 40° - 3 - C\cos 20° = 0$$

which are the same, of course, as Eqs. (*a*) and (*b*) which we solved above.

Solution IV (geometric). The polygon representing the zero vector sum of the five forces is shown. Equations (*a*) and (*b*) are seen immediately to give the projections of the vectors onto the *x*- and *y*-directions. Similarly, projections onto the *x'*- and *y'*-directions give the alternative equations in Solution II.

② A graphical solution is easily obtained. The known vectors are laid off head-to-tail to some convenient scale, and the directions of **T** and **C** are then drawn to close the polygon. The resulting intersection at point *P* completes the solution, thus enabling us to measure the magnitudes of **T** and **C** directly from the drawing to whatever degree of accuracy we incorporate into the construction.

① The selection of reference axes to facilitate computation is always an important consideration. Alternatively in this example we could take a set of axes along and normal to the direction of **C** and employ a force summation normal to **C** to eliminate it.

② The known vectors may be added in any order desired, but they must be added before the unknown vectors.

Sample Problem 3/2

Determine the magnitude T of the tension in the supporting cable and the magnitude of the force on the pin at A for the jib crane shown. The beam AB is a standard 0.5-m I-beam with a mass of 95 kg per meter of length.

0.25 m

0.5 m

0.12 m

1.5 m

10 kN

5 m

Algebraic solution. The system is symmetrical about the vertical x-y plane through the center of the beam, so the problem may be analyzed as the equilibrium of a coplanar force system. The free-body diagram of the beam is shown in the figure with the pin reaction at A represented in terms of its two rectangular components. The weight of the beam is $95(10^{-3})(5)9.81 = 4.66$ kN and acts through its center. Note that there are three unknowns A_x, A_y, and T which may be found from the three equations of equilibrium. We begin with a moment equation about A which eliminates two of the three unknowns from the equation. In applying the moment equation about A it is simpler to consider the moments of ① the x- and y-components of $\mathbf{T}$ than it is to compute the perpendicular distance from $\mathbf{T}$ to A. Hence with the counterclockwise sense as positive we write

Free–Body Diagram

② $[\Sigma M_A = 0]$ $(T \cos 25°)0.25 + (T \sin 25°)(5 - 0.12)$

$$- 10(5 - 1.5 - 0.12) - 4.66(2.5 - 0.12) = 0$$

from which

$$T = 19.61 \text{ kN} \qquad \textit{Ans.}$$

Equating the sum of forces in the x- and y-directions to zero gives

$[\Sigma F_x = 0]$ $A_x - 19.61 \cos 25° = 0,$ $A_x = 17.77$ kN

$[\Sigma F_y = 0]$ $A_y + 19.61 \sin 25° - 4.66 - 10 = 0,$ $A_y = 6.37$ kN

③ $[A = \sqrt{A_x{}^2 + A_y{}^2}]$ $A = \sqrt{(17.77)^2 + (6.37)^2},$ $A = 18.88$ kN   *Ans.*

Graphical solution. The principle that three forces in equilibrium must be concurrent is utilized for a graphical solution by combining the two known vertical forces of 4.66 and 10 kN into a single 14.66-kN force located as shown on the modified free-body diagram of the beam in the lower figure. The position of this resultant load may easily be determined graphically or algebraically. The intersection of the 14.66-kN force with the line of action of the unknown tension $\mathbf{T}$ defines the point of concurrency O through which the pin reaction $\mathbf{A}$ must pass. The unknown magnitudes of $\mathbf{T}$ and $\mathbf{A}$ may now be found by adding the forces head-to-tail to form the closed equilibrium polygon of forces, thus satisfying their zero vector sum. After the known vertical load is laid off to a convenient scale, as shown in the lower part of the figure, a line representing the given direction·of the tension $\mathbf{T}$ is drawn through the tip of the 14.7-kN vector. Likewise a line representing the direction of the pin reaction $\mathbf{A}$, determined from the concurrency established with the free-body diagram, is drawn through the tail of the 14.7-kN vector. The intersection of the lines representing vectors $\mathbf{T}$ and $\mathbf{A}$ establishes the magnitudes T and A which are necessary to make the vector sum of the forces equal to zero. These magnitudes may be scaled directly from the diagram. The x- and y-components of $\mathbf{A}$ may be constructed on the force polygon if desired.

① The justification for this step is, of course, Varignon's theorem, the principle of moments, explained in Art. 2/4. Be prepared to take full advantage of this principle frequently.

② The calculation of moments in two-dimensional problems is generally handled more simply by scalar algebra than by the vector cross-product $\mathbf{r} \times \mathbf{F}$. In three dimensions, as we shall see later, the reverse is often the case.

③ The direction of the force at A could be easily calculated if desired. However, in designing the pin A or in checking its strength it is the magnitude only of the force which matters.

Graphical Solution

Problem 3/4

Problem 3/5

Problem 3/6

Problem 3/7

PROBLEMS

3/4 The homogeneous cylinder has a mass of 40 kg and rests on smooth surfaces *A* and *B* which are inclined 30° and 60°, respectively, from the horizontal. Determine the contact forces at *A* and *B*.

Ans. A = 340 N, *B* = 196 N

3/5 Determine the magnitude *P* of the horizontal force which the gardener must exert on the light handle of the 200-lb lawn roller to hold it in place on the 10° slope. Also determine the force exerted by the ground on the roller. The center of gravity of the roller is at its center *O*.

3/6 To facilitate shifting the position of a lifting hook when it is not under load, the sliding hanger shown is used. The projections at *A* and *B* engage the flanges of a box beam when a load is supported, and the hook projects through a horizontal slot in the beam. Compute the forces at *A* and *B* when the hook supports a 300-kg mass.

Ans. A = 4.91 kN, *B* = 1.96 kN

3/7 If the screw *B* of the wood clamp is tightened so that the two blocks are under a compression of 500 N, determine the force in screw *A*. (*Note:* The force supported by each screw may be taken in the direction of the screw.)

3/8 A spring scale indicates a tension T in the right-hand cable of the pulley-cable system which supports the body of mass m. Express m in terms of T. Neglect the mass of the pulleys.   *Ans. $m = 8T/g$*

Problem 3/8

3/9 The uniform 18-ft beam weighs 500 lb and is loaded in the vertical plane by the parallel forces shown. Calculate the reactions at the supports A and B.

Problem 3/9

3/10 Calculate the pull P on the rope which the man must exert in order to suspend the 200-kg crate in the deflected position shown. Choose your reference axis so as to solve for P in one equation without involving the tension in the upper rope.

$Ans.\ P = 871\ N$

3/11 Solve for the pull P and the tension T in the upper rope of Prob. 3/10 graphically.

Problem 3/10

3/12 Compute the magnitudes T_1 and T_2 of the tensions in the two cables attached to the 100-kg crate.

$Ans.\ T_1 = 1067\ N,\ T_2 = 1730\ N$

Problem 3/12

Problem 3/13

3/13 Determine the force P that the 80-kg worker must exert on the rope in order to support himself in the bosun's chair. What force R does the man exert on the seat of the chair?

Ans. $P = 157\,\text{N}$, $R = 628\,\text{N}$

Problem 3/14

3/14 To test the deflection of the uniform 200-lb beam the 120-lb boy exerts a pull of 40 lb on the rope rigged as shown. Compute the force supported by the pin at the hinge O.   *Ans.* $F_O = 820$ lb

Problem 3/15

3/15 The resistance of the bracket to bending is tested under the 2-kN load. Compute the force on the roller at A and the total force supported by the pin at O.

3/16 The T-frame weighs 500 lb with mass center at G. Calculate the total force supported by the pin at O after the 600-lb force is applied to the cable.

Ans. $O = 465$ lb

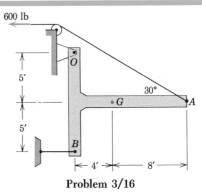

Problem 3/16

3/17 A jet airplane having a mass of 8 Mg is flying horizontally at a constant speed of 1000 km/h under a thrust of 16 kN from its turbojet engines. If the pilot increases the fuel rate to give a thrust of 20 kN and noses the plane upward to maintain a constant 1000-km/h air speed, determine the angle θ made by the new line of flight with the horizontal. Note that the air resistance in the line of flight at the particular altitude involved is a function of air speed.

Ans. $\theta = 2.92°$

3/18 If the mass of the boom is negligible compared with that of the load L, find the force F on the ball joint at A and show that the magnitude F is constant for all values of θ. Determine the limiting value of T as θ approaches 90°.

Problem 3/18

3/19 The hinged member is used to activate a latching device for securing a large trailer body to its undercarriage. If a tension $T = 80$ lb is required in the horizontal control rod to trip the plunger against which the member acts at C, estimate the force supported by the pin at A by means of a free-hand vector solution.

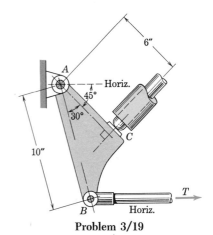

Problem 3/19

3/20 Calculate the value of the couple M required to roll the 40-kg wheel up the incline. Also determine the contact force R at A. The surface of the incline is sufficiently rough to prevent slipping.

Ans. $M = 47.1$ N·m, $R = 393$ N

Problem 3/20

Problem 3/21

3/21 Replace the couple M of Prob. 3/20 by the horizontal force P and compute its value required to roll the 40-kg wheel up the incline. Also determine the contact force R at A. No slipping occurs.

Ans. $R = 406$ N

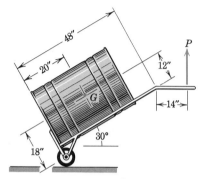

Problem 3/22

3/22 The oil drum weighs 620 lb when full and has a mass center at G. Calculate the vertical force P required to maintain equilibrium of the drum and dolly in the position shown. The weight of the dolly may be neglected compared with that of the drum.

Problem 3/23

3/23 The uniform cantilever beam has a mass of 50 kg per meter of length and supports the couple and three forces shown. Isolate the beam to the right of section A and calculate the moment (couple) M and shear (vertical force) V exerted by the wall support on the beam at this section.

Ans. $V = 3.44$ kN up

$M = 8.33$ kN·m counterclockwise

Problem 3/24

3/24 The uniform 15-m pole has a mass of 150 kg and is supported by its smooth ends against the vertical walls and by the tension T in the vertical cable. Compute the reactions at A and B.

Problem 3/25

3/25 The hook wrench or pin spanner is used to turn shafts and collars. If a moment of 60 lb-ft is required to turn the 8-in.-diameter collar about its center O under the action of the applied force P, determine the contact force R on the smooth surface at A. Engagement of the pin at B may be considered to occur at the periphery of the collar.

Ans. $R = 236$ lb

3/26 A drum of 600-mm radius with mass center G at its geometric center has a total mass of 1600 kg and rests on a cradle consisting of two long rollers of 240-mm diameter. The rollers rest on a smooth horizontal surface and are prevented from separating by horizontal links, one on each end of the rollers as shown. Calculate the tension T in each link and the reaction R between the drum and each roller.   *Ans.* $T = 3.79$ kN, $R = 10.91$ kN

Problem 3/26

3/27 The rigid truss is subjected to the four loads shown and is supported by the hinge at A and the link BC. Neglect the weight of the truss and compute the magnitude of the force supported by the pin at A. Why is the direction of the force at A parallel to BC with the given loading?

Problem 3/27

3/28 Magnetic tape under a tension of 10.0 N at D passes around the guide pulleys and through the erasing head at C at constant speed. As a result of a small amount of friction in the bearings of the pulleys, the tape at E is under a tension of 11.0 N. Determine the tension T in the supporting spring at B. The plate is horizontal and is mounted on a precision needle bearing at A.   *Ans.* $T = 10.6$ N

Problem 3/28

3/29 Calculate the magnitude of the force supported by the pin at A under the action of the 250-lb-in. couple applied to the light bracket.

Problem 3/29

Problem 3/30

3/30 The applied couple in Prob. 3/29 is replaced by the 60-lb horizontal force shown here. Calculate the magnitude of the force supported by the pin at *A* which holds the light bracket.   *Ans.* *A* = 147.5 lb

Problem 3/31

3/31 The uniform 50-kg bar with end rollers is held in the equilibrium position by the horizontal cord *CD* as shown. Determine the tension *T* in the cord by writing only one equation of equilibrium. Then obtain the forces at *A* and *B* by inspection.

Problem 3/32

3/32 When the 60-lb force is applied to the tire wrench as shown, the wheel is prevented from turning because of the friction force between it and the ground. By using only one equation of equilibrium determine the horizontal component O_x of the reaction exerted on the wheel by the fixed bearing at *O*. Analyze the wrench and wheel as a single body.
   Ans. O_x = 102 lb

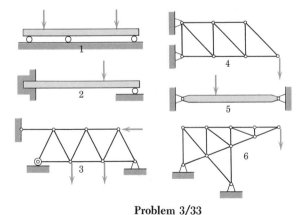

Problem 3/33

3/33 Each of the structures is statically indeterminate. Describe at least one modification in the supports for each case which would make the structure statically determinate.

3/34 The figure shows a series of rectangular plates and their constraints, all confined to the plane of representation. The plates could be subjected to various known loads applied in the plane of the plate. Identify the plates that belong to each of the following categories.

(A) Complete fixity with minimum number of adequate constraints
(B) Partial fixity with inadequate constraints
(C) Complete fixity with redundant constraints
(D) Partial fixity with redundant constraint

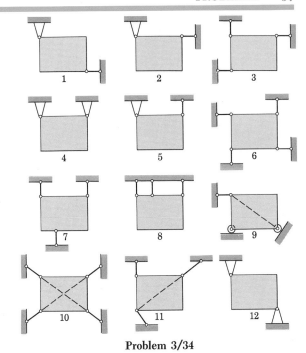

Problem 3/34

3/35 The uniform bar with end rollers weighs 60 lb and is supported by the horizontal and vertical surfaces and by the wire AC. Calculate the tension T in the wire and the reactions against the rollers at A and B.
Ans. $T = 60.2$ lb, $A = 15$ lb, $B = 40$ lb

Problem 3/35

3/36 The uniform concrete slab shown in edge view has a mass of 25 Mg and is being hoisted slowly into a vertical position by the tension P in the hoisting cable. For the position where $\theta = 60°$ calculate the tension T in the horizontal anchor cable by using only one equation of equilibrium.

Problem 3/36

Problem 3/37

3/37 An experimental boat is equipped with four hydro-foils, two on each side as shown. The boat has a total mass m with center of mass at G. Thrust is provided with the air screw. The ratio of lift to drag for each foil is n. Lift is the vertical force supported by each foil, and drag is the horizontal resistance to motion through the water. For a given thrust T of the propeller write the expression for the drag D on each of the two forward foils.

$$Ans. \ D = \frac{1}{4n}\left(mg + T\frac{d+h}{b}\right)$$

Problem 3/38

3/38 The supporting slippers, which hold the missile to the launcher, slide in T-slots in the guiding rail. In a static test of a missile the slipper at A is clamped securely to the rail whereas the slipper at B is not clamped. If the missile has a mass of 1500 kg with center of mass at G and if the static thrust T is 20 kN, determine the force supported by the pin that connects the missile to the slipper at B.

Problem 3/39

3/39 The bar and its end rollers have a mass center at G and are placed on the inclined surfaces shown. For a given value of θ determine the one angle β for which the bar will be in an unstable equilibrium position. Also indicate two additional values of β for stable positions.   *Ans.* $\beta = \tan^{-1}(\frac{1}{2}\tan\theta)$

Problem 3/40

3/40 The dump truck is used to lift a cut section of trunk from the stump of a large tree. The section of trunk is measured and calculated to weigh 1200 lb. The body of the rotatable dump weighs 600 lb, and its center of mass G_1 is directly over the rear wheels in the position for lifting. Calculate the necessary torque M applied to the dump through its shaft at O to make the lift. Also calculate the corresponding forces under the front and rear pairs of wheels of the truck. The truck weighs 6200 lb exclusive of the dump, and its center of mass is at G_2.

3/41 The 300-mm-radius wheel has a mass of 60 kg with center of mass at G and rests on a rough horizontal surface. Under the action of the couple $M = 50$ N·m applied at A the wheel bears against the roller at B, which prevents it from rolling further. Calculate the magnitude of the total force exerted on the wheel by the horizontal surface at C.

$Ans.\ F_C = 343$ N

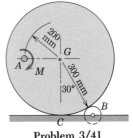

Problem 3/41

3/42 The specially built trailer is used to transport and erect a missile to its vertical launch position. The trailer body and missile have a combined mass of 6.20 Mg with center of mass at G. The unit is tilted into position by two hydraulic cylinders, one on each side of the trailer. Compute the compressive force C in each piston rod of the cylinders for the position where the axis AB of the cylinder is perpendicular to the longitudinal axis of the trailer and missile.

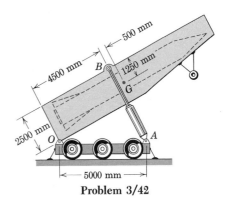

Problem 3/42

3/43 Determine the magnitude F of the total force supported by the hinge at O for the missile transport described in Prob. 3/42. A graphical solution is suggested. $Ans.\ F \approx 26$ kN

3/44 The portable floor crane in the automotive shop is lifting a 420-lb engine. For the position shown compute the magnitude of the force supported by the pin at C and the oil pressure p against the 3.20-in.-diameter piston of the hydraulic-cylinder unit AB.

Problem 3/44

Problem 3/45

Problem 3/46

Problem 3/47

3/45 Determine the dimension l if the mass m maintains a specified tension T in the belt for the position shown. Neglect the mass of the arm and central pulley compared with m. Also find the magnitude of the force R supported by the pin at O.

$$Ans. \ R = \sqrt{3T^2 + (mg)^2}$$

3/46 The uniform 400-kg drum is mounted on a line of rollers at A and a line of rollers at B. An 80-kg man moves slowly a distance of 700 mm from the vertical centerline before the drum begins to rotate. All rollers are perfectly free to rotate except one of them at B which must overcome appreciable friction in its bearing. Calculate the friction force F exerted by that one roller tangent to the drum and find the magnitude R of the force exerted by all rollers at A on the drum for this condition.

3/47 The lunar shelter is fabricated from aluminum and has a mass of 240 kg. The boom OC of the lunar crane has a mass of 48 kg, and its center of mass is at its midlength. Determine the force F in the hydraulic cylinder AB of the crane in the position shown as it is placing the shelter in position on the lunar surface. Recall that gravitational attraction on the moon is $\frac{1}{6}$ that on the earth.   *Ans. $F = 2340 \ N$*

3/48 The crane is hoisting a 4.20-Mg bulldozer. The mass center of the 2-Mg boom *OA* is at its midlength. Calculate the tension *T* in the cable where it attaches at *B* and the magnitude of the force supported by the hinge at *O* for equilibrium in the 60° position. Neglect the width of the boom.

Ans. T = 61.5 kN, *O* = 99.2 kN

Problem 3/48

3/49 The weight of the rigid truss *ABCDE* is small compared with that of the 12-Mg mass which it supports. Compute the force in the horizontal link at *B* and the magnitude of the force supported by the pin joint at *A*.

Problem 3/49

3/50 The member *OBC* and sheave at *C* together weigh 500 lb, with a combined center of mass at *G*. Calculate the magnitude of the force supported by the pin connection at *O* when the 300-lb load is applied. The collar at *A* can provide support in the horizontal direction only. *Ans. O* = 1345 lb

Problem 3/50

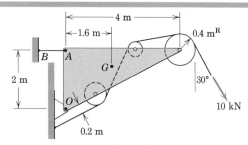

Problem 3/51

3/51 The triangular frame with pulleys has a total mass of 200 kg with mass center at G. Calculate the magnitude of the force supported by the pin at O when the 10-kN force is applied to the cable. (*Hint:* In using the moment equation take advantage of the replacement of a force by a force and a couple.)

Ans. O = 34.3 kN

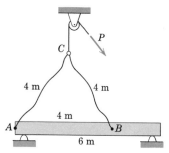

Problem 3/52

▶3/52 The uniform beam has an overall length of 6 m and a mass of 300 kg. The force P applied to the hoisting cable is slowly increased to raise the ring C, the two 4-m ropes AC and BC, and the beam. Compute the tensions in the ropes at A and B when the beam is clear of its supports and the force P is equal to the weight of the beam.

Ans. T_A = 816 N, T_B = 2449 N

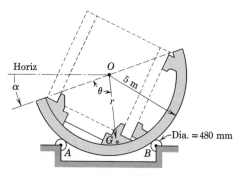

Problem 3/53

▶3/53 A special jig for turning large concrete pipe sections (shown dotted) consists of an 80-Mg sector mounted on a line of rollers at A and a line of rollers at B. One of the rollers at B is a gear which meshes with a ring of gear teeth on the sector so as to turn the sector about its geometric center O. When $\alpha = 0$, a counterclockwise torque of 2460 N·m must be applied to the gear at B to keep the assembly from rotating. When $\alpha = 30°$, a clockwise torque of 4680 N·m is required to prevent rotation. Locate the mass center G of the jig by calculating $\bar{r}$ and θ.

Ans. $\bar{r}$ = 367 mm, θ = 79.8°

SECTION B. EQUILIBRIUM IN THREE DIMENSIONS

3/4 EQUILIBRIUM CONDITIONS. We now extend our principles and methods developed for two-dimensional equilibrium to the case of three dimensional equilibrium. In Art. 3/1 the general conditions for the equilibrium of a body were stated in Eqs. 3/1 which require that the resultant force and resultant couple on a body in equilibrium must be zero. These two vector equations of equilibrium and their scalar components may be written as

$$\Sigma \mathbf{F} = \mathbf{0} \quad \text{or} \quad \begin{cases} \Sigma F_x = 0 \\ \Sigma F_y = 0 \\ \Sigma F_z = 0 \end{cases}$$

$$\Sigma \mathbf{M} = \mathbf{0} \quad \text{or} \quad \begin{cases} \Sigma M_x = 0 \\ \Sigma M_y = 0 \\ \Sigma M_z = 0 \end{cases} \tag{3/3}$$

The first three scalar equations state that there is no resultant force acting on a body in equilibrium in any of the three coordinate directions. The second three scalar equations express the further equilibrium requirement that there is no resultant moment acting on the body about any of the coordinate axes or about axes parallel to the coordinate axes. These six equations are both necessary and sufficient conditons for complete equilibrium. The reference axes may be chosen arbitrarily as a matter of convenience, the only restriction being that a right-handed coordinate system must be used with vector notation.

The six scalar relationships of Eqs. 3/3 are independent conditions since any of them may be valid without the others. For example, for a car which accelerates on a straight and level road in the x-direction, Newton's second law tells us that the resultant force on the car equals its mass times its acceleration. Hence, $\Sigma F_x \neq 0$ but the remaining force-equilibrium equations are satisfied since all other acceleration components are zero. Similarly, if the flywheel of the engine of the accelerating car is rotating with increasing angular speed about the x-axis, it is not in rotational equilibrium about this axis. Hence $\Sigma M_x \neq 0$ along with $\Sigma F_x \neq 0$, but the remaining four equilibrium equations would be satisfied for mass-center axes.

In applying the vector form of Eqs. 3/3 we first express each of the forces in terms of the coordinate unit vectors $\mathbf{i}$, $\mathbf{j}$, and $\mathbf{k}$. For the first equation, $\Sigma \mathbf{F} = \mathbf{0}$, the vector sum will be zero only if the coefficients of $\mathbf{i}$, $\mathbf{j}$, and $\mathbf{k}$ in the expression are, respectively, zero. These three sums when each set equal to zero yield precisely the three scalar equations of equilibrium, $\Sigma F_x = 0$, $\Sigma F_y = 0$, $\Sigma F_z = 0$.

For the second equation, $\Sigma \mathbf{M} = \mathbf{0}$, where the moment sum may be taken about any convenient point O, we express the moment of each force as the cross product $\mathbf{r} \times \mathbf{F}$ where $\mathbf{r}$ is the position vector from O to any point on the line of action of the force $\mathbf{F}$. Thus $\Sigma \mathbf{M} = \Sigma \mathbf{r} \times \mathbf{F} = \mathbf{0}$. The coefficients of $\mathbf{i}$, $\mathbf{j}$, and $\mathbf{k}$ in the resulting moment equation when set equal to zero, respectively, produce exactly the three scalar moment equations $\Sigma M_x = 0$, $\Sigma M_y = 0$, and $\Sigma M_z = 0$.

(*a*) *Free-Body Diagrams.* The summations in Eqs. 3/3 include the effects of *all* forces on the body under consideration. We learned in the previous article that the free-body diagram is the only reliable method for disclosing all forces and moments which should be included in our equilibrium equations. In three dimensions the free-body diagram serves the same essential purpose as it does in two dimensions and should *always* be drawn. We have our choice either of drawing a pictorial view of the isolated body with all external forces represented or of drawing the orthogonal projections of the free-body diagram. Both representations will be illustrated in the sample problems at the end of the article.

The correct representation of forces on the free-body diagram requires a knowledge of the characteristics of contacting surfaces. These characteristics were set forth in Fig. 3/1 for two-dimensional problems, and their extension to three-dimensional problems is represented in Fig. 3/8 for the most common situations of force transmission. The representations in both Figs. 3/1 and 3/8 will be used in three-dimensional analysis.

(*b*) *Categories of Equilibrium.* Application of Eqs. 3/3 falls into four categories which we can easily identify with the aid of Fig. 3/9 (pg. 106).

Case 1, equilibrium of forces that are concurrent at a point O, requires all three force equations but no moment equations since the moment of the forces about any axis through O is automatically zero.

Case 2, equilibrium of forces that are concurrent with a line, requires all equations except the moment equation about that line, which is automatically satisfied.

Case 3, equilibrium of parallel forces, requires only one force equation in the direction of the forces (*x*-direction as shown) and two moment equations about axes (*y* and *z*) that are normal to the direction of the forces.

Case 4, equilibrium of a general system of forces, requires all three force equations and all three moment equations.

The observations contained in these statements are generally quite evident when a given problem is being solved.

(*c*) *Constraints and Statical Determinacy.* The six scalar relations of Eqs. 3/3, although necessary and sufficient conditions to

MECHANICAL ACTION OF FORCES IN THREE–DIMENSIONAL ANALYSIS

Type of Contact and Force Origin	Action on Body to be Isolated
1. Member in contact with smooth surface, or ball–supported member	Force must be normal to the surface and directed toward the member.
2. Member in contact with rough surface	The possibility exists for a force F tangent to the surface (friction force) to act on the member, as well as a normal force N.
3. Roller or wheel support with lateral constraint	A lateral force P exerted by the guide on the wheel can exist, in addition to the normal force N.
4. Ball–and–socket joint	A ball–and–socket joint free to pivot about the center of the ball can support a force R with all three components.
5. Fixed connection (embedded or welded)	In addition to three components of force, a fixed connection can support a couple M represented by its three components.
6. Thrust–bearing support	Thrust bearing is capable of supporting axial force R_y as well as radial forces R_x and R_y. Unless bearing is pivoted about x– and z–axes, it can support couples M_x and M_z.

Figure 3/8

CATEGORIES OF EQUILIBRIUM IN THREE DIMENSIONS		
Force System	Free–Body Diagram	Independent Equations
1. Concurrent at a point		$\Sigma F_x = 0$ $\Sigma F_y = 0$ $\Sigma F_z = 0$
2. Concurrent with a line		$\Sigma F_x = 0 \qquad \Sigma M_y = 0$ $\Sigma F_y = 0 \qquad \Sigma M_z = 0$ $\Sigma F_z = 0$
3. Parallel		$\Sigma F_x = 0 \qquad \Sigma M_y = 0$ $\qquad\qquad \Sigma M_z = 0$
4. General		$\Sigma F_x = 0 \qquad \Sigma M_x = 0$ $\Sigma F_y = 0 \qquad \Sigma M_y = 0$ $\Sigma F_z = 0 \qquad \Sigma M_z = 0$

Figure 3/9

establish equilibrium, do not necessarily provide all of the information required to calculate the unknown forces acting in a three-dimensional equilibrium situation. Again, as we found to be the case with two dimensions, the question of adequacy of information lies in the characteristics of the constraints provided by the supports. An analytical criterion for determining the adequacy of constraints is available but is beyond the scope of this treatment.[*] In Fig. 3/10, however, we cite four examples of constraint conditions to alert the reader to the problem. In the *a*-part of the figure is shown a rigid body whose corner A is completely fixed by the links 1, 2, and 3. Links 4, 5, and 6 prevent rotations about the axes of links 1, 2, and 3, respectively, so

[*] See the author's *Statics, 2nd Edition* or *SI Version*, Art. 16.

that the body is completely fixed and the constraints are said to be adequate. The *b*-part of Fig. 3/10 shows the same number of constraints, but we see that they provide no resistance to a moment which might be applied about axis *AE*. Here the body is incompletely fixed and only partially constrained. Similarly, in Fig. 3/10c the constraints provide no resistance to an unbalanced force in the *y*-direction, so here also is a case of incomplete fixity with partial constraints. In Fig. 3/10d, if a seventh constraining link were imposed on a system of six constraints placed properly for complete fixity, more supports would be provided than would be necessary to establish the equilibrium position, and link 7 would be *redundant*. The body would then be *statically indeterminate* with such a seventh link in place. With only a few exceptions the supporting constraints for rigid bodies in equilibrium in this book are adequate and the bodies are statically determinate.

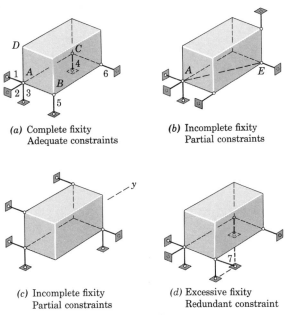

(a) Complete fixity
Adequate constraints

(b) Incomplete fixity
Partial constraints

(c) Incomplete fixity
Partial constraints

(d) Excessive fixity
Redundant constraint

Figure 3/10

Sample Problem 3/3

The uniform 7-m steel shaft has a mass of 200 kg and is supported by a ball-and-socket joint at A in the horizontal floor. The ball end B rests against the smooth vertical walls as shown. Compute the forces exerted by the walls and the floor on the ends of the shaft.

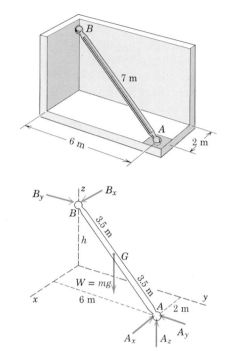

Solution. The free-body diagram of the shaft is first drawn where the contact forces acting on the shaft at B are shown normal to the wall surfaces. In addition to the weight $W = mg = 200(9.81) = 1962$ N, the force exerted by the floor on the ball joint at A is represented by its x-, y-, and z-components. These components are shown in their correct physical sense as should be evident from the requirement that A is held in place. The vertical position of B is found from $7 = \sqrt{2^2 + 6^2 + h^2}$, $h = 3$ m. Right-handed coordinate axes are conveniently assigned as shown.

Vector solution. We will use A as a moment center to eliminate reference to the forces at A. The position vectors needed to compute the moments about A are

$$\mathbf{r}_{AG} = -1\mathbf{i} - 3\mathbf{j} + 1.5\mathbf{k} \text{ m} \qquad \text{and} \qquad \mathbf{r}_{AB} = -2\mathbf{i} - 6\mathbf{j} + 3\mathbf{k} \text{ m}$$

where the mass center G is located halfway between A and B for the uniform shaft.

The vector moment equation gives

$$[\Sigma \mathbf{M}_A = 0] \qquad \mathbf{r}_{AB} \times (\mathbf{B}_x + \mathbf{B}_y) + \mathbf{r}_{AG} \times \mathbf{W} = 0$$

$$(-2\mathbf{i} - 6\mathbf{j} + 3\mathbf{k}) \times (i B_x + j B_y) + (-\mathbf{i} - 3\mathbf{j} + 1.5\mathbf{k}) \times (-1962\mathbf{k}) = 0$$

$$\begin{vmatrix} \mathbf{i} & \mathbf{j} & \mathbf{k} \\ -2 & -6 & 3 \\ B_x & B_y & 0 \end{vmatrix} + \begin{vmatrix} \mathbf{i} & \mathbf{j} & \mathbf{k} \\ -1 & -3 & 1.5 \\ 0 & 0 & -1962 \end{vmatrix} = 0$$

$$(-3B_y + 5886)\mathbf{i} + (3B_x - 1962)\mathbf{j} + (-2B_y + 6B_x)\mathbf{k} = 0$$

Equating the coefficients of $\mathbf{i}$, $\mathbf{j}$, and $\mathbf{k}$ to zero and solving give

$$B_x = 654 \text{ N} \qquad \text{and} \qquad B_y = 1962 \text{ N} \qquad\qquad \textit{Ans.}$$

The forces at A are easily determined by

$$[\Sigma \mathbf{F} = 0] \qquad (654 - A_x)\mathbf{i} + (1962 - A_y)\mathbf{j} + (-1962 + A_z)\mathbf{k} = 0$$

and
$$A_x = 654 \text{ N} \qquad A_y = 1962 \text{ N} \qquad A_z = 1962 \text{ N}$$

Finally
$$A = \sqrt{A_x{}^2 + A_y{}^2 + A_z{}^2}$$

$$= \sqrt{(654)^2 + (1962)^2 + (1962)^2} = 2851 \text{ N} \qquad\qquad \textit{Ans.}$$

Scalar solution. If we evaluate the scalar moment equations about axes through A parallel, respectively, to the x- and y-axes, we may write

$$[\Sigma M_{A_x} = 0] \qquad 1962(3) - 3B_y = 0 \qquad B_y = 1962 \text{ N}$$

$$[\Sigma M_{A_y} = 0] \qquad -1962(1) + 3B_x = 0 \qquad B_x = 654 \text{ N}$$

The force equations give, simply,

$$[\Sigma F_x = 0] \qquad -A_x + 654 = 0 \qquad A_x = 654 \text{ N}$$

$$[\Sigma F_y = 0] \qquad -A_y + 1962 = 0 \qquad A_y = 1962 \text{ N}$$

$$[\Sigma F_z = 0] \qquad A_z - 1962 = 0 \qquad A_z = 1962 \text{ N}$$

① We could, of course, assign all of the unknown components of force in the positive mathematical sense in which case A_x and A_y would turn out to be negative upon computation. The free-body diagram describes the physical situation, so it is generally preferable to show the forces in their correct physical senses wherever possible.

② Note that the third equation $-2B_y + 6B_x = 0$ merely checks the results of the first two equations. This result could be anticipated from the fact that an equilibrium system of forces concurrent with a line requires only two moment equations (Case 2 under *Categories of equilibrium*).

③ We observe that a moment sum about an axis through A parallel to the z-axis merely gives us $6B_x - 2B_y = 0$, which serves only as a check as noted previously. Alternatively we could have first obtained A_z from $\Sigma F_z = 0$ and then taken our moment equations about axes through B to obtain A_x and A_y.

Sample Problem 3/4

A 200-N force is applied to the handle of the hoist in the direction shown. The bearing A supports the thrust (force in the direction of the shaft axis) while bearing B supports only radial load (load normal to the shaft axis). Determine the mass m which can be supported and the total radial force exerted on the shaft by each bearing.

Solution. The system is clearly three-dimensional with no lines or planes of symmetry, and therefore the problem must be analyzed as a general space system of forces. A scalar solution is used here to illustrate this approach, although a solution using vector notation would also be satisfactory. The free-body diagram of the shaft, lever, and drum considered a single body could be shown by a space view if desired but is represented here by its three orthogonal projections.

The 200-N force is resolved into its three components, and each of the three views shows two of these components. The correct directions of A_x and B_x may be seen by inspection by observing that the line of action of the resultant of the two 70.7-N forces passes between A and B. The correct sense of the forces A_y and B_y cannot be determined until the magnitudes of the moments are obtained, so they are arbitrarily assigned. The x-y projection of the bearing forces is shown in terms of the sums of the unknown x- and y-components. The addition of A_z and the weight $W = mg$ completes the free-body diagrams. It should be noted that the three views represent three two-dimensional problems related by the corresponding components of the forces.

From the x-y projection

$$[\Sigma M_O = 0] \qquad 100(9.81m) - 250(173.2) = 0, \qquad m = 44.1 \text{ kg} \qquad Ans.$$

From the x-z projection

$$[\Sigma M_A = 0] \qquad 150\,B_x + 175(70.7) - 250(70.7) = 0, \qquad B_x = 35.4 \text{ N}$$

$$[\Sigma F_x = 0] \qquad A_x + 35.4 - 70.7 = 0, \qquad A_x = 35.3 \text{ N}$$

The y-z view gives

$$[\Sigma M_A = 0] \qquad 150\,B_y + 175(173.2) - 250(44.1)(9.81) = 0, \qquad B_y = 520 \text{ N}$$

$$[\Sigma F_y = 0] \qquad A_y + 520 - 173.2 - (44.1)(9.81) = 0, \qquad A_y = 86.8 \text{ N}$$

$$[\Sigma F_z = 0] \qquad A_z = 70.7 \text{ N}$$

The total radial forces on the bearings become

$$[A_r = \sqrt{A_x{}^2 + A_y{}^2}] \qquad A_r = \sqrt{(35.3)^2 + (86.8)^2} = 93.5 \text{ N} \qquad Ans.$$

$$[B = \sqrt{B_x{}^2 + B_y{}^2}] \qquad B = \sqrt{(35.4)^2 + (520)^2} = 521 \text{ N} \qquad Ans.$$

① If the standard three views of orthographic projection are not entirely familiar, then review and practice them. Visualize the three views as the images of the body projected onto the front, top, and end surfaces of a clear plastic box placed over and aligned with the body.

② We could have started with the x-z projection rather than with the x-y projection.

③ The y-z view could have followed immediately after the x-y view since the determination of A_y and B_y may be made after mg is found.

Sample Problem 3/5

The welded tubular frame is secured to the horizontal x-y plane by a ball-and-socket joint at A and receives support from the loose-fitting ring at B. Under the action of the 2-kN load, rotation about a line from A to B is prevented by the cable CD, and the frame is stable in the position shown. Neglect the weight of the frame compared with the applied load and determine the tension T in the cable, the reaction at the ring, and the reaction components at A.

Solution. The system is clearly three-dimensional with no lines or planes of symmetry, and therefore the problem must be analyzed as a general space system of forces. The free-body diagram is drawn, where the ring reaction is shown in terms of its two components. All unknowns except **T** may be eliminated by a moment sum about the line AB. The

① direction of AB is specified by the unit vector $\mathbf{n} = \dfrac{1}{\sqrt{6^2 + 4.5^2}}(4.5\mathbf{j} + 6\mathbf{k})$

$= \frac{1}{5}(3\mathbf{j} + 4\mathbf{k})$. The moment of **T** about AB is the component in the direction of AB of the vector moment about the point A and equals

② $\mathbf{r}_1 \times \mathbf{T} \cdot \mathbf{n}$. Similarly the moment of the applied load F about AB is $\mathbf{r}_2 \times \mathbf{F} \cdot \mathbf{n}$. With $\overrightarrow{CD} = \sqrt{46.25}$ m the vector expressions for **T**, **F**, $\mathbf{r}_1$, and $\mathbf{r}_2$ are

$$\mathbf{T} = \frac{T}{\sqrt{46.25}}(2\mathbf{i} + 2.5\mathbf{j} - 6\mathbf{k}), \qquad \mathbf{F} = 2\mathbf{j} \text{ kN}$$

$$\mathbf{r}_1 = -\mathbf{i} + 2.5\mathbf{j} \text{ m}, \qquad \mathbf{r}_2 = 2.5\mathbf{i} + 6\mathbf{k} \text{ m}$$

The moment equation now becomes

$$[\Sigma M_{AB} = 0] \quad (-\mathbf{i} + 2.5\mathbf{j}) \times \frac{T}{\sqrt{46.25}}(2\mathbf{i} + 2.5\mathbf{j} - 6\mathbf{k}) \cdot \tfrac{1}{5}(3\mathbf{j} + 4\mathbf{k})$$

$$+ (2.5\mathbf{i} + 6\mathbf{k}) \times (2\mathbf{j}) \cdot \tfrac{1}{5}(3\mathbf{j} + 4\mathbf{k}) = 0$$

Completion of the vector operations gives

$$-\frac{48T}{\sqrt{46.25}} + 20 = 0, \qquad T = 2.83 \text{ kN} \qquad Ans.$$

and the components of T become,

$$T_x = 0.833 \text{ kN} \qquad T_y = 1.042 \text{ kN} \qquad T_z = -2.50 \text{ kN}$$

We may find the remaining unknowns by moment and force summations as follows.

③

$[\Sigma M_z = 0] \quad 2(2.5) - 4.5B_x - 1.042(3) = 0 \qquad B_x = 0.417 \text{ kN} \qquad Ans.$

$[\Sigma M_x = 0] \quad 4.5B_z - 2(6) - 1.042(6) = 0 \qquad B_z = 4.06 \text{ kN} \qquad Ans.$

$[\Sigma F_x = 0] \quad A_x + 0.417 + 0.833 = 0 \qquad A_x = -1.250 \text{ kN} \qquad Ans.$

$[\Sigma F_y = 0] \quad A_y + 2 + 1.042 = 0 \qquad A_y = -3.04 \text{ kN} \qquad Ans.$

$[\Sigma F_z = 0] \quad A_z + 4.06 - 2.50 = 0 \qquad A_z = -1.56 \text{ kN} \qquad Ans.$

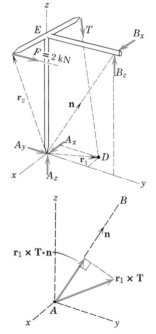

① Recall that the vector **r** in the expression **r** × **F** for the moment of a force is a vector from the moment center to *any* point on the line of action of the force. Instead of $\mathbf{r}_1$ an equally simple choice would be the vector $\overrightarrow{AC}$.

② The advantage of using vector notation in this problem is the freedom to take moments directly about any axis. In this problem this freedom permits the choice of an axis that eliminates five of the unknowns.

③ The negative signs with the A-components indicate that they are in the opposite direction to those shown on the diagram.

PROBLEMS

3/54 As a check on the balance of the aircraft each of the three wheels is run onto a scale, and the force readings are $A = 22.0$ kN, $B = 22.4$ kN, $C = 3.43$ kN. Calculate the x-y coordinates of the mass center of the plane through which the total weight of the airplane acts.

$Ans.\ \bar{x} = -0.0201$ m, $\bar{y} = 0.215$ m

Problem 3/54

3/55 The two steel I-beams, each weighing 28 lb per foot of length, are welded together at right angles and lifted by the vertical cables so that the beams remain in a horizontal plane. Compute the tension in each of the cables A, B, and C.

Problem 3/55

3/56 The mass center of the 30-kg door is in the center of the panel. If the weight of the door is supported entirely by the lower hinge A, calculate the magnitude of the total force supported by the hinge at B.

$Ans.\ B = 190.2$ N

Problem 3/56

3/57 The industrial door is a uniform rectangular panel weighing 1200 lb and rolls along the fixed rail D on its hanger-mounted wheels A and B. The door is maintained in a vertical plane by the floor-mounted guide roller C which bears against the bottom edge. For the position shown compute the horizontal side thrust on each of the wheels A and B.

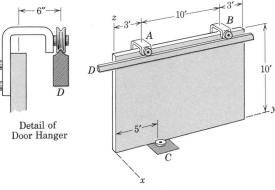

Detail of
Door Hanger

Problem 3/57

Problem 3/58

End View
of V-groove

θ — Horizontal

Problem 3/59

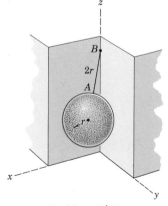

Problem 3/60

3/58 The large rectangular sign has a mass of 800 kg with center of mass in the center of the rectangle. The sign is prevented from swinging by the light chains *OC* and *OD*. The two chains together are slightly longer than the distance from *C* to *D* so that only one of them is taut at any time. Calculate the magnitude of the total force supported by the hinge at *A* when a horizontal wind exerts a force of 2 kN normal to the sign and acting at its center.

Ans. A = 4200 N

3/59 The smooth homogeneous sphere rests in the 120° groove and bears against the end plate which is normal to the direction of the groove. Determine the angle θ, measured from the horizontal, for which the reaction on each side of the groove equals the force supported by the end plate.

3/60 A smooth homogeneous sphere of mass *m* and radius *r* is suspended by a wire *AB* of length 2*r* from point *B* on the line of intersection of the two vertical walls at right angles to one another. Determine the reaction *R* of each wall against the sphere.

Ans. R = $mg/\sqrt{7}$

3/61 A high-voltage power line is suspended from a transmission tower by the projecting framework shown. If the tension in the power line is 3 kN, calculate the tension T in link AD and the compression C in links AB and AC.

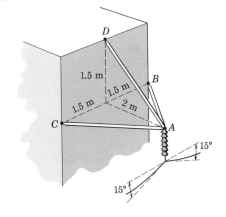

3/62 The center of mass of the 120-kg workbench is on a vertical line through the center of the square top. A 400-N downward force is applied to the handle of the pipe wrench used to twist the pipe in its flange. If legs A and C are slightly shorter than legs B and D, calculate the forces supported by the three legs which contact the floor. Assume contact takes place under the outside corners of the legs and neglect the mass of the wrench, pipe, and vise compared with the mass of the table.

$Ans.\ B = 522$ N, $C = 367$ N, $D = 689$ N

3/63 The uniform ventilator door has a mass of 200 kg and is hinged at the corners A and B of its upper edge. The door is held open in a horizontal position by the wire from C to point D in the vertical wall. Compute the tension in the wire and the forces normal to the hinge axis supported by the hinge pins at A and B.

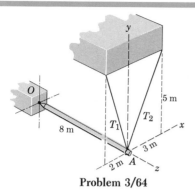

Problem 3/64

3/64 The uniform horizontal boom has a mass of 160 kg and is supported by the two cables in the vertical x-y plane and by the ball-and-socket joint at O. Calculate the tensions T_1 and T_2 in the cables.

Ans. $T_1 = 507$ N, $T_2 = 366$ N

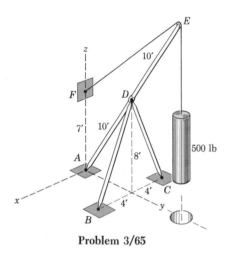

Problem 3/65

3/65 The three poles are erected as shown and support the 500-lb cylinder prior to lowering it into the hole. The cable attached to the cylinder passes over a small pulley at E and is secured at point F. The connections at the ends of the poles may be treated as ball-and-socket joints, and the weight of the poles is small compared with the loads they support. Calculate the compression P in each of the equal legs BD and CD and find the magnitude of the total force at A.

Problem 3/66

3/66 The uniform horizontal boom has a mass of 240 kg and is supported by the two cables anchored at B and C and by the ball-and-socket joint at O. Calculate the tension T in cable AC.   Ans. $T = 1310$ N

3/67 A uniform steel ring 60 in. in diameter and weighing 600 lb is lifted by the three cables, each 50 in. long, attached at points A, B, and C as shown. Compute the tension in each cable.

Problem 3/67

3/68 The square steel plate has a mass of 1800 kg with mass center at its center G. Calculate the tension in each of the three cables with which the plate is lifted while remaining horizontal.

Ans. $T_A = T_B = 5.41$ kN, $T_C = 9.87$ kN

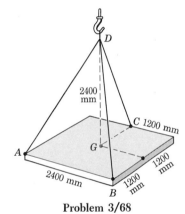

Problem 3/68

3/69 Determine the force in each member of the tripod. Each of the three members is secured at its ends by ball-and-socket connections and is capable of supporting tension or compression. The weights of the members may be neglected.

Ans. $A = 2.041$ kN tension
$B = 0.861$ kN compression
$C = 1.269$ kN compression

Problem 3/69

Problem 3/70

Problem 3/71

Problem 3/72

3/70 The rigid unit of post, bracket, and motor weighs 60 lb with its mass center at G located 12 in. from the vertical centerline of the post. The post is welded to the fixed base at A. The motor, which drives a machine through a flexible shaft, turns in the direction indicated and delivers a torque of 1500 lb-in. In addition, a 50-lb force is applied to the bracket as shown. Determine the vector expressions for the total force $\mathbf{R}$ and moment $\mathbf{M}$ applied to the post at A by the supporting base. (*Caution:* Be careful to assign the torque (couple) which acts on the motor shaft in its correct sense consistent with Newton's third law.)

3/71 A torque of 20 N·m is applied to the flexible shaft as shown. After the initial angular twist has taken place, one revolution of the input end will be accompanied by one revolution of the output end, and the output torque will equal the input torque. The shaft housing is mounted in a self-aligning support at B which is capable of exerting a single force normal to the shaft axis. Support A, on the other hand, is a loose-fitting sleeve and can support the shaft housing in either of the two ways shown in the separate sectional views. Insofar as equilibrium is concerned the shaft housing may be treated as a rigid body for the position in which it is bent. Determine the forces exerted by the supports at B, C, and D on the housing. Neglect the weight of the shaft. (*Caution:* Observe Newton's third law at the output end regarding the torque applied *to* the shaft.)   *Ans.* $\mathbf{B} = \mathbf{D} = 50\mathbf{k}$ N, $\mathbf{C} = -100\mathbf{k}$ N

3/72 The uniform door weighs 50 lb and is hinged at A and B with hinge A carrying the entire weight. A 40-lb horizontal pull parallel to the suface of the door is applied to the handle of a claw hammer to remove a nail. The door is prevented from opening further by a stop at C which exerts a force at the lower edge of the door normal to its surface. Determine the magnitude of the horizontal force supported by the hinge at B.

3/73 Gear *C* drives the V-belt pulley *D* at a constant speed. For the belt tensions shown calculate the gear-tooth force *P* and the magnitudes of the total forces supported by the bearings at *A* and *B*.
Ans. $P = 70.9$ N, $A = 83.3$ N, $B = 208$ N

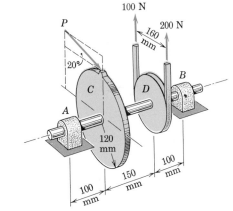

Problem 3/73

3/74 The crankshaft for a two-cylinder compressor is mounted in bearings at *A* and *B*. For the position shown the forces exerted by the connecting rods on the crank pins are $F_1 = 16$ kN and $F_2 = 8$ kN in the directions indicated. For an equilibrium condition of the crankshaft calculate the *x*- and *y*-components of the total forces supported by the bearings at *A* and *B* and the torque *M* applied to the shaft.

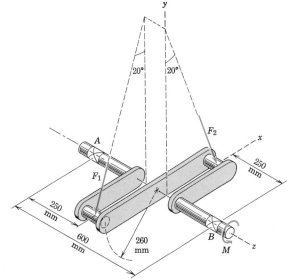

Problem 3/74

3/75 If the ventilator door of Prob. 3/63 is held in position by a wire from *C* to *E* rather than from *C* to *D*, compute the tension in the wire and the forces normal to the hinge axis supported by the hinge pins.   *Ans.* $T = 1907$ N, $A = 523$ N, $B = 1256$ N

3/76 If the weight of the mast is negligible compared with the applied 30-kN load, determine the two cable tensions T_1 and T_2 and the force *A* acting at the ball joint at *A*.

Problem 3/76

Problem 3/77

3/77 The uniform 30- by 40-in. trap door weighs 200 lb and is propped open by the light strut AB at the angle $\theta = \tan^{-1}(4/3)$. Calculate the compression F_B in the strut and the force supported by the hinge D normal to the hinge axis. Assume that the hinges act at the extreme ends of the lower edge.

Ans. $F_B = 70$ lb, $D_n = 101$ lb

Problem 3/78

3/78 One of the vertical walls supporting end B of the 200-kg uniform shaft of Sample Problem 3/3 is turned through a 30° angle as shown here. End A is still supported by the ball-and-socket connection in the horizontal x-y plane. Calculate the magnitudes of the forces $\mathbf{P}$ and $\mathbf{R}$ exerted on the ball end B of the shaft by the vertical walls C and D, respectively.

Ans. $P = 1584$ N, $R = 755$ N

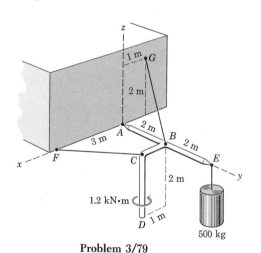

Problem 3/79

3/79 The light rigid frame is secured to the vertical wall by a ball-and-socket joint at A and by cables BG and CF. In addition to supporting the 500-kg load, the frame supports a 1.2-kN·m couple applied to the vertical leg CD. Compute the y-component of the force supported by the ball-and-socket joint at A. (*Suggestion:* Use one vector equation which eliminates reference to all unknowns other than A_y.)

3/80 The 9-m steel boom has a mass of 600 kg with center of mass at midlength. It is supported by a ball-and-socket joint at A and the two cables under tensions T_1 and T_2. The cable which supports the 2000-kg load leads through a sheave at B and is secured to the vertical x-y plane at F. Calculate the magnitude of the tension T_1. (*Hint:* Write a moment equation which eliminates all unknowns except T_1.)

Ans. $T_1 = 19.76$ kN

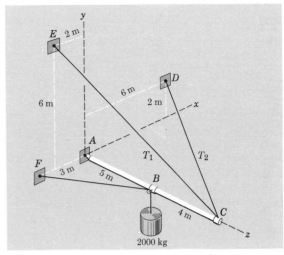

Problem 3/80

3/81 The welded tubular frame of negligible mass is supported by a ball-and-socket joint at A, by a smooth ring at B, and by the wire from the strut at C to the fixed point D. The frame is loaded by the 300-lb ball that is welded to the horizontal strut. Calculate the force exerted on the frame by the fixed ring at B. (*Hint:* The problem may be solved without involving the force at A.)

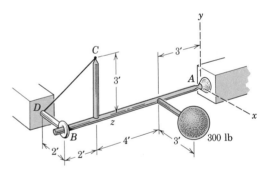

Problem 3/81

▶3/82 The rigid member ABC is attached to the vertical x-y surface by a ball-and-socket joint at A and is supported by the cables BE and CD. The mass of the member may be neglected compared with that of the 5-Mg load which it supports. There is one position of D along the horizontal slot through which the cable must be passed and secured for the member to maintain the position shown. Find x.

Ans. $x = 3.75$ m

Problem 3/82

Problem 3/83

▶3/83 A rectangular sign over a store has a mass of 100 kg, with the center of mass in the center of the rectangle. The support against the wall at point C may be treated as a ball-and-socket joint. At corner D support is provided in the y-direction only. Calculate the tensions T_1 and T_2 in the supporting wires, the total force supported at C, and the lateral force R supported at D.

$$Ans.\ T_1 = 347\ \text{N} \qquad T_2 = 431\ \text{N}$$
$$R = 63.1\ \text{N} \qquad C = 768\ \text{N}$$

3/5 PROBLEM FORMULATION AND REVIEW.

In Chapter 3 we have applied our knowledge of the properties of forces, moments, and couples studied in Chapter 2 to solve problems in the equilibrium of rigid bodies. Each such body in complete equilibrium is characterized by the requirement that the vector resultant of all forces acting on it is zero ($\Sigma \mathbf{F} = \mathbf{0}$) and the vector resultant of all moments on the body about a point (or axis) is also zero ($\Sigma \mathbf{M} = \mathbf{0}$). We are guided in all of our solutions by these two requirements which are easily comprehended physically.

As is often the case it is not the theory but its application which presents the difficulty. The crucial steps in applying our principles of equilibrium should be quite familiar by now. They are:

1. Make an unequivocal decision as to which body in equilibrium is to be analyzed.
2. Isolate the body in question from all contacting bodies by drawing its *free-body diagram* upon which *all* forces that act on the isolated body from external sources are represented.
3. Observe the principle of action and reaction (Newton's third law) when assigning the sense of each force.
4. Label all forces, known and unknown.
5. Choose and label reference axes, always using a right-handed set for three-dimensional analysis.

These five steps should become automatic, and following them consistently before beginning the equilibrium calculation is the very best assurance of a correct solution.

When solving an equilibrium problem we should first check to see that the body is statically determinate. If there are more supports than are necessary to hold the body in place, the body is statically indeterminate, and the equations of equilibrium by themselves will not enable us to solve for all of the external reactions. In applying the equations of equilibrium, we choose scalar algebra, vector algebra, or graphical analysis according to both preference and experience, with vector algebra being particularly useful for many three-dimensional problems.

One of our most useful procedures is to simplify the algebra of a solution by the choice of a moment axis that eliminates as many unknowns as possible or by the choice of a direction for a force summation that avoids reference to certain unknowns. A few moments of thought to take advantage of these simplifications can save appreciable time and effort.

The principles and methods which are covered in Chapters 2 and 3 constitute the most basic part of statics. They lay the foundation for what follows not only in statics but in dynamics as well.

REVIEW PROBLEMS

0.5 m

1 m

5 m

A

O

30°

Problem 3/84

3/84 The uniform 5-m bar with a mass of 100 kg is hinged at O and prevented from rotating in the vertical plane beyond the 30° position by the fixed roller at A. Calculate the magnitude of the total force supported by the pin at O.   *Ans. O = 1769 N*

P

300 mm

50 mm

Problem 3/85

3/85 The jaws of the open-ended wrench are smooth case-hardened steel and, for the given thickness of the wrench, can withstand up to 7.5 kN of concentrated force against the corner of a hexagonal hardened bolt without surface damage. Calculate the maximum force P at a lever arm of 300 mm that can safely be applied to the wrench. Assume that there is a very slight clearance between the head of the bolt and the jaws of the wrench.
                                             Ans. P = 0.689 kN

M

r_0

θ

r

G

b

Problem 3/86

3/86 A wheel of mass m and radius r with its mass center G at the geometric center rests in a small depression of width b. Determine the minimum couple M applied to the wheel at the location shown in order to roll the wheel out of the depression. Assume no slipping occurs. What is the influence of r_0, r, and θ?

M

C

30°

A

300 mm

Problem 3/87

3/87 The 100-kg wheel rests on a rough surface and bears against the roller A when the couple M is applied to it. If $M = 60$ N·m with the wheel not slipping, compute the reaction on the roller A.
                                             Ans. F_A = 231 N

3/88 A large symmetrical drum for drying sand is operated by the geared motor drive shown. If the mass of
the sand is 750 kg and an average gear-tooth force of
2.6 kN is supplied by the motor pinion A to the
drum gear normal to the contacting surfaces at B,
calculate the average offset $\bar{x}$ of the center of mass
G of the sand from the vertical centerline. Neglect
all friction in the supporting rollers.

Ans. $\bar{x} = 199$ mm

Detail of contact
at B

Problem 3/88

3/89 The roller-band device consists of two rollers, each
of radius r, encircled by a flexible band of negligible
thickness and subjected to the two tensions T. Write
the expression for the contact force R between the
band and the flat supporting surfaces at A and B.
The action is in the horizontal plane, so that the
weights of the rollers and band are not involved.

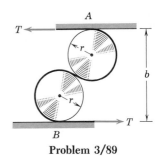

Problem 3/89

3/90 The device shown in section can support the load L
at various heights by resetting the pawl C in another
tooth at the desired height on the fixed vertical
column D. Determine the distance b at which the
load should be positioned in order for the two rollers
A and B to support equal forces. The weight of the
device is negligible compared with L.

Ans. $b = 10.33$ in.

Problem 3/90

Problem 3/91

3/91 The elevating structure for a rocket test stand and the rocket within it have a combined mass of 635 Mg with center of mass at G. For the position in which $x = 5$ m determine the equilibrium value of the force on the hinge axis A. Solve graphically.

Problem 3/92

3/92 The rubber-tired tractor shown has a mass of 13.5 Mg with center of mass at G and is used for pushing or pulling heavy loads. Determine the load P which the tractor can pull at a constant speed of 5 km/h up the 15-percent grade if the driving force exerted by the ground on each of its four wheels is 80 percent of the normal force under that wheel. Also find the total normal reaction N_B under the rear pair of wheels at B.

Ans. $P = 85$ kN, $N_B = 125$ kN

Problem 3/93

3/93 The power unit of the post-hole digger supplies a torque of 4000 lb-in. to the auger. The arm B is free to slide in the supporting sleeve C but is not free to rotate about the horizontal axis of C. If the unit is free to swivel about the vertical axis of the mount D, determine the force exerted against the right rear wheel by the block A (or A') which prevents the unbraked truck from rolling. (*Hint:* View the system from above.) *Ans.* $A' = 41.7$ lb

3/94 The uniform steel plate of mass m is triangular in shape and is supported in the horizontal plane by three vertical wires from its vertices. Show that the tension in each wire is always $mg/3$ regardless of the shape of the triangle.

Problem 3/94

3/95 A vertical force P on the foot pedal of the bell crank is required to produce a tension T of 400 N in the vertical control rod. Determine the corresponding bearing reactions at A and B.

Ans. $A = 184$ N, $B = 424$ N

Problem 3/95

3/96 One of the three landing pads for the Mars Viking lander is shown in the figure with its approximate dimensions. The mass of the lander is 600 kg. Compute the force in each leg when the lander is resting on a horizontal surface on Mars. (Assume equal support by the pads and consult Table C2 in Appendix C as needed.)

Ans. $F_{CD} = 1046$ N compression
$F_{AC} = F_{CB} = 240$ N tension

Dimensions in Millimeters
Problem 3/96

Problem 3/97

3/97 Three identical steel balls each of mass m are placed in the cylindrical ring which rests on a horizontal surface and whose height is slightly greater than the radius of the balls. The diameter of the ring is such that the balls are virtually touching one another. A fourth identical ball is then placed on top of the three balls. Determine the force P exerted by the ring on each of the three lower balls.

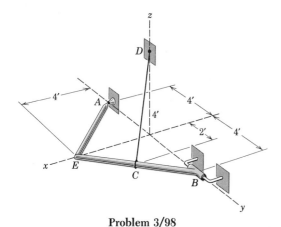

Problem 3/98

3/98 Each of the two legs of the welded frame weighs 100 lb. A wire from C to D prevents the frame from rotating out of the horizontal plane about an axis through its bearing at B and its ball-and-socket joint at A. Calculate the tension T in the wire and the magnitude of the total force supported by the connection at A.   *Ans.* $T = 245$ lb, $A = 122.5$ lb

Problem 3/99

3/99 The mast has a mass of 300 kg and is supported by a ball-and-socket joint at A. Calculate the tension T_1 if the horizontal 10-kN force is applied at F.
        Ans. $T_1 = 11.34$ kN

STRUCTURES

4

4/1 INTRODUCTION. In Chapter 3 we focused our attention on the equilibrium of a single rigid body or upon a system of connected members which, when taken as a whole, could be treated as a single rigid body. In these problems we first drew a free-body diagram of this single body showing all forces external to the isolated body and then we applied the force and moment equations of equilibrium. In Chapter 4 our attention is directed toward the determination of the forces internal to a structure, that is, forces of action and reaction between the connected members. An engineering structure is any connected system of members built to support or transfer forces and to withstand safely the loads applied to it. In the force analysis of structures it is necessary to dismember the structure and to analyze separate free-body diagrams of individual members or combinations of members in order to determine the forces internal to the structure. This analysis calls for very careful observance of Newton's third law, which states that each action is accompanied by an equal and opposite reaction.

In Chapter 4 we shall analyze the internal forces acting in several types of structures, namely, trusses, frames, and machines. In this treatment we shall consider only *statically determinate* structures, that is, structures which do not have more supporting constraints than are necessary to maintain an equilibrium configuration. Thus, as we have already seen, the equations of equilibrium are adequate to determine all unknown reactions.

The student who has mastered the basic procedure developed in Chapter 3 of defining unambiguously the body under consideration by constructing a correct free-body diagram will have little difficulty with the analysis of statically determinate structures. The analysis of trusses, frames and machines, and beams under concentrated loads constitutes a straightforward application of the material developed in the previous two chapters.

4/2 PLANE TRUSSES. A framework composed of members joined at their ends to form a rigid structure is known as a truss. Bridges, roof supports, derricks, and other such structures are common examples of trusses. Structural members used are I-beams, channels, angles, bars, and special shapes which are fastened together at their ends by welding, riveted connections, or large bolts or pins. When the members of the truss lie essentially in a single plane, the truss is known as a *plane truss*. Plane trusses, such as those used for bridges, are commonly designed in pairs with one truss panel placed on each side of the bridge and connected together by cross beams that support the roadway and transfer the applied loads to the truss members. Several examples of commonly used trusses that can be analyzed as plane trusses are shown in Fig. 4/1.

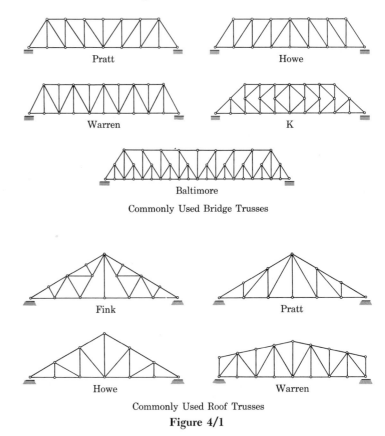

Pratt   Howe

Warren   K

Baltimore

Commonly Used Bridge Trusses

Fink   Pratt

Howe   Warren

Commonly Used Roof Trusses

Figure 4/1

(a)

(b)

(c)

Figure 4/2

The basic element of a plane truss is the triangle. Three bars joined by pins at their ends, Fig. 4/2a, constitute a rigid frame. On the other hand, four or more bars pin-jointed to form a polygon of as many sides constitute a nonrigid frame. We can make the nonrigid frame in Fig. 4/2b stable or rigid by adding a diagonal bar joining *A* and *D* or *B* and *C* and thereby forming two triangles. The structure may be extended by adding additional units of two end-connected bars, such as *DE* and *CE* or *AF* and *DF*, Fig. 4/2c, which are pinned to two fixed joints, and in this way the entire structure will remain rigid. The term rigid is used in the sense of noncollapsible and also in the sense that deformation of the members due to induced internal strains is negligible.

Structures that are built from a basic triangle in the manner described are known as *simple trusses*. When more members are present than are needed to prevent collapse, the truss is statically indeterminate. A statically indeterminate truss cannot be analyzed by the equations of equilibrium alone. Additional members or supports that are not necessary for maintaining the equilibrium position are called *redundant*.

The design of a truss involves the determination of the forces in the various members and the selection of appropriate sizes and structural shapes to withstand the forces. Several assumptions are made in the force analysis of simple trusses. First, we assume all members to be *two-force members*. A two-force member is one in equilibrium under the action of two forces only, as defined in general terms with Fig. 3/4 in Art. 3/3. For trusses each member is a straight link joining the two points of application of force. The two forces are applied at the ends of the member and are necessarily equal, opposite, and *collinear* for equilibrium. We see that the member may be in tension or compression, as shown in Fig. 4/3. When we represent the equilibrium of a portion of a two-force member, the tension T or compression C acting on the cut section is the same for all sections. We assume here that the weight of the member is small compared with the force it supports. If it is not, or if the small effect of the weight is to be accounted for, the weight W of the member, if uniform, may be assumed to be replaced by two forces, each $W/2$, acting at each end of the member. These forces, in effect, are treated as loads externally applied to the pin connections. Accounting for the weight of a member in this way gives the correct result for the average tension or compression along the member but will not account for the effect of bending of the member.

When welded or riveted connections are used to join structural members, the assumption of a pin-jointed connection is usually satisfactory if the centerlines of the members are concurrent at the joint as in Fig. 4/4.

We also assume in the analysis of simple trusses that all external forces are applied at the pin connections. This condition is satisfied in most trusses. In bridge trusses the deck is usually laid on cross beams that are supported at the joints.

Provision for expansion and contraction due to temperature changes and for deformations resulting from applied loads is usually made at one of the supports for large trusses. A roller, rocker, or some kind of slip joint is provided. Trusses and frames wherein such provision is not made are statically indeterminate, as explained in Art. 3/3.

Two methods for the force analysis of simple trusses will be

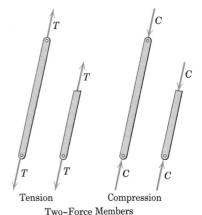

Tension   Compression

Two–Force Members

Figure 4/3

Figure 4/4

(a)

(b)

Figure 4/5

Figure 4/6

given, and reference will be made to the simple truss shown in Fig. 4/5a for each of the two methods. The free-body diagram of the truss as a whole is shown in Fig. 4/5b. The external reactions are usually determined by computation from the equilibrium equations applied to the truss as a whole before proceeding with the force analysis of the remainder of the truss.

4/3 METHOD OF JOINTS. This method for finding the forces in the members of a simple truss consists of satisfying the conditions of equilibrium for the forces acting on the connecting pin of each joint. The method therefore deals with the equilibrium of concurrent forces, and only two independent equilibrium equations are involved. We begin the analysis with any joint where at least one known load exists and where not more than two unknown forces are present. Solution may be started with the pin at the left end, and its free-body diagram is shown in Fig. 4/6. With the joints indicated by letters, we may designate the force in each member by the two letters defining the ends of the member. The proper directions of the forces should be evident for this simple case by inspection. The free-body diagrams of portions of members AF and AB are also shown to indicate clearly the mechanism of the action and reaction. The member AB actually makes contact on the left side of the pin, although the force AB is drawn from the right side and is shown acting away from the pin. Thus, if we consistently draw the force arrows on the *same* side of the pin as the member, then tension (such as AB) will always be indicated by an arrow *away* from the pin, and compression (such as AF) will always be indicated by an arrow *toward* the pin. The magnitude of AF is obtained from the equation $\Sigma F_y = 0$ and AB is then found from $\Sigma F_x = 0$.

Joint F must be analyzed next, since it now contains only two unknowns, EF and BF. Joints B, C, E, and D are subsequently analyzed in that order. The free-body diagram of each joint and its corresponding force polygon which represents graphically the two equilibrium conditions $\Sigma F_x = 0$ and $\Sigma F_y = 0$ are shown in Fig. 4/7. The numbers indicate the order in which the joints are analyzed. We note that, when joint D is finally reached, the computed reaction R_2 must be in equilibrium with the forces in members CD and ED, determined previously from the two neighboring joints. This requirement will provide a check on the correctness of our work. We should also note that isolation of joint C quickly discloses the fact that the force in CE is zero when the equation $\Sigma F_y = 0$ is applied. The force in this member would not be zero, of course, if an external vertical load were applied at C.

It is often convenient to indicate the tension T and compression C of the various members directly on the original truss diagram by drawing arrows away from the pins for tension and toward the pins

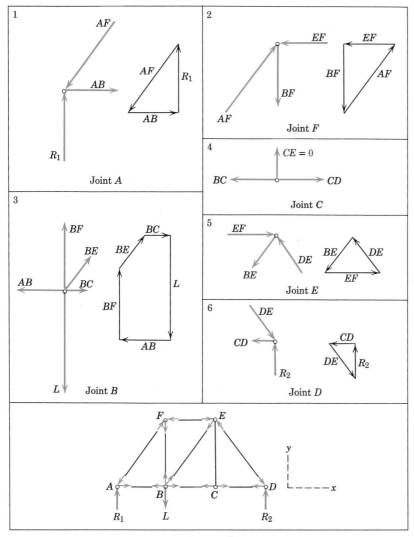

Figure 4/7

for compression. This designation is illustrated at the bottom of Fig. 4/7.

In some instances it is not possible to assign initially the correct direction of one or both of the unknown forces acting on a given pin. In this event we may make an arbitrary assignment. A negative value from the computation indicates that the assumed direction is incorrect.

If a simple truss has more external supports than are necessary to ensure a stable equilibrium configuration, the truss as a whole is statically indeterminate, and the extra supports constitute *external redundancy*. If the truss has more internal members than are necessary to prevent collapse, then the extra members constitute *internal redundancy* and the truss is statically indeterminate. For a truss that is statically determinate externally, there is a definite relation between the number of its members and the number of its joints necessary for internal stability without redundancy. Since we can specify the equilibrium of each joint by two scalar force equations, there are in all $2j$ such equations for a simple truss with j joints. For the entire truss composed of m two-force members with a maximum of three unknown support reactions, there are in all $m + 3$ unknowns. Thus, for a simple plane truss composed of triangular elements, the equation $m + 3 = 2j$ will be satisfied if the truss is statically determinate internally.

This relation is a necessary condition for stability but it is not a sufficient condition, since one or more of the m members can be arranged in such a way as not to contribute to a stable configuration of the entire truss. If $m + 3 > 2j$, there are more members than there are independent equations, and the truss is statically indeterminate internally with redundant members present. If $m + 3 < 2j$, there is a deficiency of internal members, and the truss is unstable and will collapse under load.

The force polygon for each joint, shown in Fig. 4/7, may be constructed graphically to obtain the unknown forces in the members as an alternative to or as a check on the algebraic calculations using the force equations of equilibrium. If a consistent sequence around each joint, clockwise, for example, has been used for the addition of the forces, we may superpose these force polygons on one another to form a composite graphical figure known as the *Maxwell diagram*.° The force and its sense may be obtained directly from the diagram. The student who is interested in structures may wish to experiment with this construction and to consult other books dealing more completely with structural analysis for a more detailed description of the Maxwell diagram.

Special conditions. We draw attention to several special conditions which occur frequently in the analysis of simple trusses. When two collinear members are under compression, as indicated in Fig. 4/8*a*, it is necessary to add a third member to maintain alignment of the two members and prevent buckling. We see very quickly from a force summation in the y-direction that the force F_3 in the third member must be zero and from the x-direction that $F_1 = F_2$. This conclusion holds regardless of the angle θ and, of course, holds if the collinear members are in tension. If an external force with a compo-

---

° The method was published by James Clerk Maxwell in 1864.

nent in the y-direction were applied to the joint, then, of course, F_3 would no longer be zero.

When two noncollinear members are joined as shown in Fig. 4/8b, then in the absence of an externally applied load at this joint, the forces in both members must be zero as we see from the two force summations.

When two pairs of collinear members are joined as shown in Fig. 4/8c, the forces in each pair must be equal and opposite. This conclusion follows from the force summations indicated in the figure.

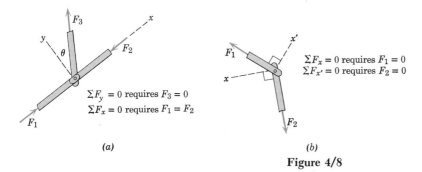

$\Sigma F_y = 0$ requires $F_3 = 0$
$\Sigma F_x = 0$ requires $F_1 = F_2$

(a)

$\Sigma F_x = 0$ requires $F_1 = 0$
$\Sigma F_{x'} = 0$ requires $F_2 = 0$

(b)

$\Sigma F_x = 0$ requires $F_1 = F_2$
$\Sigma F_{x'} = 0$ requires $F_3 = F_4$

(c)

Figure 4/8

Truss panels are frequently cross-braced as shown in Fig. 4/9a. Such a panel is statically indeterminate if each brace is capable of supporting either tension or compression. However, when the braces are flexible members incapable of supporting compression, as are cables, then only the tension member acts and the other member is disregarded. It is usually evident from the asymmetry of the loading how the panel will deflect. If the deflection is as indicated in Fig. 4/9b, then member AB should be retained and CD disregarded. When this choice cannot be made by inspection, we may make an arbitrary selection of the member to be retained. If the assumed tension turns out to be positive upon calculation, then the choice was correct. If the assumed tension force turns out to be negative, then the opposite member must be retained and the calculation redone.

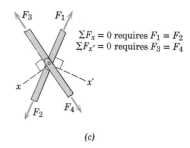

(a)   (b)

Figure 4/9

The simultaneous solution of the equations for two unknown forces at a joint may be avoided by a careful choice of reference axes. Thus for the joint indicated schematically in Fig. 4/10 where L is known and F_1 and F_2 are unknown, a force summation in the x-direction eliminates reference to F_1 and a force summation in the x'-direction eliminates reference to F_2. When the angles involved are not easily found, then a simultaneous solution of the equations using one set of reference directions for both unknowns may be preferable.

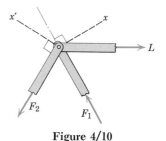

Figure 4/10

Sample Problem 4/1

Compute the force in each member of the loaded cantilever truss by the method of joints.

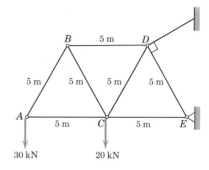

Solution. If it were not desired to calculate the external reactions at D and E, the analysis for a cantilever truss could begin with the joint at the loaded end. However, this truss will be analyzed completely, so the first step will be to compute the external forces at D and E from the free-body diagram of the truss as a whole. The equations of equilibrium give

$[\Sigma M_E = 0]$ $\quad 5T - 20(5) - 30(10) = 0$ $\qquad T = 80.0$ kN

$[\Sigma F_x = 0]$ $\quad 80.0 \cos 30° - E_x = 0$ $\qquad E_x = 69.3$ kN

$[\Sigma F_y = 0]$ $\quad 80.0 \sin 30° + E_y - 20 - 30 = 0$ $\qquad E_y = 10.0$ kN

Next we draw free-body diagrams showing the forces acting on each of the connecting pins. The correctness of the assigned directions of the forces is verified when each joint is considered in sequence. There should be no question about the correct direction of the forces on joint A. Equilibrium requires

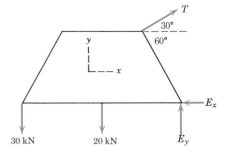

$[\Sigma F_y = 0]$ $\quad 0.866AB - 30 = 0$ $\quad AB = 34.64$ kN T $\qquad$ *Ans.*

$[\Sigma F_x = 0]$ $\quad AC - 0.5(34.64) = 0$ $\quad AC = 17.32$ kN C $\qquad$ *Ans.*

where T stands for tension and C stands for compression.

Joint B must be analyzed next, since there are more than two unknown forces on joint C. The force BC must provide an upward component, in which case BD must balance the force to the left. Again the forces are obtained from

$[\Sigma F_y = 0]$ $\ 0.866BC - 0.866(34.64) = 0$ $\quad BC = 34.64$ kN C $\quad$ *Ans.*

$[\Sigma F_x = 0]$ $\quad BD - 2(0.5)(34.64) = 0$ $\quad BD = 34.64$ kN T $\quad$ *Ans.*

Joint C now contains only two unknowns, and these are found in the same way as before:

$[\Sigma F_y = 0]$ $\quad 0.866CD - 0.866(34.64) - 20 = 0$

$\qquad CD = 57.74$ kN T $\qquad$ *Ans.*

$[\Sigma F_x = 0]$ $\quad CE - 17.32 - 0.5(34.64) - 0.5(57.74) = 0$

$\qquad CE = 63.51$ kN C $\qquad$ *Ans.*

Finally, from joint E there results

$[\Sigma F_y = 0]$ $\qquad 0.866DE = 10.00$ $\quad DE = 11.55$ kN $\qquad$ *Ans.*

and the equation $\Sigma F_x = 0$ checks.

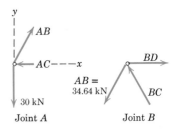

① Note that we draw the force arrow on the same side of the joint as the member which exerts the force. In this way tension (arrow away from the joint) is distinguished from compression (arrow toward the joint).

PROBLEMS

(Solve the following problems by the method of joints. Neglect the weights of the members compared with the forces they support unless otherwise indicated.)

4/1 Calculate the force in each member of the loaded
truss.   Ans. $AB = 3.52$ kN T
$BC = 4.62$ kN C
$AC = 3.27$ kN T

Problem 4/1

4/2 Calculate the force in each member of the truss.
(1 kip = 1000 lb.)

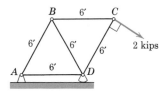

Problem 4/2

4/3 Calculate the force in member CF of the loaded
truss.   Ans. $CF = 3.33$ kN C

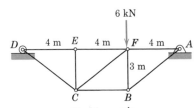

Problem 4/3

4/4 Calculate the force in each member of the loaded
truss. All triangles are isosceles.

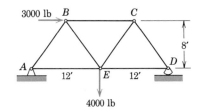

Problem 4/4

4/5 Calculate the forces in members CG and CF for the
truss shown.
Ans. $CG = 2.24$ kN T
$CF = 1.00$ kN C

Problem 4/5

Problem 4/7

Problem 4/8

Problem 4/9

Problem 4/10

4/6 If the 2-kN force acting on the truss of Prob. 4/5 were removed, identify by inspection those members in which the forces are zero. On the other hand, if the 2-kN force were applied at G instead of B, would there be any zero-force members?

4/7 Each member of the truss is a uniform 20-ft bar weighing 400 lb. Calculate the average tension or compression in each member due to the weights of the members.

$$Ans.\ AE = CD = 2000/\sqrt{3}\ \text{lb } C$$
$$AB = BC = 1000/\sqrt{3}\ \text{lb } T$$
$$BE = BD = 800/\sqrt{3}\ \text{lb } T$$
$$DE = 1400/\sqrt{3}\ \text{lb } C$$

4/8 Calculate the forces in members FG, EG, and GD for the loaded cantilever truss.

4/9 Calculate the forces in members JB and BH for the loaded truss. $$Ans.\ JB = 56.6\ \text{kN } C$$
$$BH = 47.1\ \text{kN } C$$

4/10 Determine the forces in members BI, CI, and HI for the loaded truss. All angles are 30°, 60°, or 90°.

4/11 A snow load transfers the forces shown to the upper joints of a Pratt roof truss. Neglect any horizontal reactions at the supports and compute the forces in members *BH*, *BC*, and *CH*.

Problem 4/11

4/12 Calculate the forces induced in members *GH* and *ED* for the crane truss when it lifts an 1800-kg car.

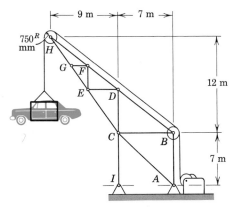

Problem 4/12

4/13 The signboard truss is designed to support a horizontal wind load of 800 lb. A separate analysis shows that $\frac{5}{8}$ of this force is transmitted to the center connection at *C* and the rest is equally divided between *D* and *B*. Calculate the forces in members *BE* and *BC*.   *Ans. BE* = 559 lb *T, BC* = 300 lb *T*

Problem 4/13

4/14 Calculate the forces in members *CF, BF, BG,* and *FG* for the simple crane truss.

Problem 4/14

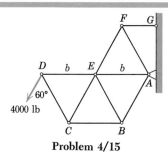

Problem 4/15

4/15 Calculate the forces in all members of the loaded truss supported by the horizontal link *FG* and the hinge at *A*. All interior angles are 60°.

Ans. AB = CB = DC = 4000 lb C

BE = CE = DE = 4000 lb T, *AE =* 0

EF = 8000 lb T, *AF =* 8000 lb C

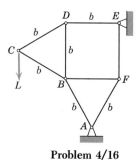

Problem 4/16

4/16 Show that the truss is statically determinate and determine the forces in members *BD* and *BF* in terms of the applied load *L*.

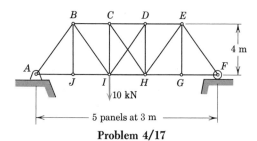

Problem 4/17

4/17 Calculate the forces in members *AB, BJ, BI,* and *CI*. Members *CH* and *DI* are cables that are capable of supporting tension only.

Ans. AB = 7.5 kN C, *BJ = CI =* 0, *BI =* 7.5 kN T

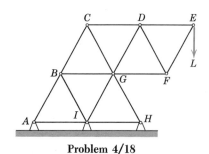

Problem 4/18

4/18 By inspection designate those members of the truss that cause the structure to be statically indeterminate.

4/19 Each of the loaded trusses has supporting constraints which are statically indeterminate. List all
members of each truss whose forces are not affected
by the indeterminacy of the supports and that may
be computed directly by using only the equations of
equilibrium. Assume that the loading and dimensions of the trusses are known.

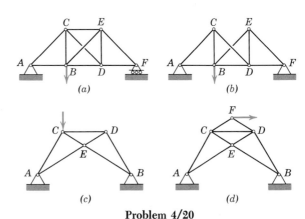

Problem 4/19

4/20 Verify the fact that each of the trusses contains one
or more elements of redundancy, and propose two
separate changes, either one of which would remove
the redundancy and produce complete statical
determinacy. All members can support compression
as well as tension.

Problem 4/20

4/21 The movable gantry is used to erect and prepare a
500-Mg rocket for firing. The primary structure of
the gantry is approximated by the symmetrical
plane truss shown, which is statically indeterminate.
As the gantry is positioning a 60-Mg section of the
rocket suspended from A, strain gage measurements
indicate a compressive force of 50 kN in member
AB and a tensile force of 120 kN in member CD due
to the 60-Mg load. Calculate the corresponding
forces in members BF and EF.

Ans. $BF = 188.4$ kN C, $EF = 120$ kN T

Problem 4/21

Problem 4/22

Problem 4/23

▶4/22 The tower for a transmission line is modeled by the truss shown. The crossed members in the center sections of the truss may be assumed capable of supporting tension only. For the loads of 1.8 kN applied in the vertical plane, compute the forces induced in members *AB*, *DB*, and *CD*.

Ans. *AB* = 3.89 kN *C*, *DB* = 0, *CD* = 0.93 kN *C*

▶4/23 Find the forces in members *EF*, *KL*, and *GL* for the Fink truss shown. (*Hint:* Note that the forces in *BP*, *PC*, *DN*, etc., are zero.)

Ans. *EF* = 202 kN *C*
KL = 100 kN *T*
GL = 50.0 kN *T*

4/4 METHOD OF SECTIONS. In the previous article on the analysis of plane trusses by the method of joints we took advantage of only two of the three equilibrium equations, since the procedures involve concurrent forces at each joint. We may take advantage of the third or moment equation of equilibrium by selecting an entire section of the truss for the free body in equilibrium under the action of a nonconcurrent system of forces. This *method of sections* has the basic advantage that the force in almost any desired member may be found directly from an analysis of a section which has cut that member. Thus it is not necessary to proceed with the calculation from joint to joint until the member in question has been reached. In choosing a section of the truss we note that, in general, not more than three members whose forces are unknown may be cut, since there are only three available equilibrium relations which are independent.

The method of sections will now be illustrated for the truss in Fig. 4/5, which was used in the explanation of the previous method. The truss is shown again in Fig. 4/11*a* for ready reference. The external reactions are first computed as before, considering the truss as a whole. Now let us determine the force in the member *BE* for example. An imaginary section, indicated by the dotted line, is passed through the truss, cutting it into two parts, Fig. 4/11*b*. This section has cut three members whose forces are initially unknown. In order for the portion of the truss on each side of the section to remain in equilibrium it is necessary to apply to each cut member the force that was exerted on it by the member cut away. These forces, either tensile or compressive, will always be in the directions of the respective members for simple trusses composed of two-force members. The left-hand section is in equilibrium under the action of the applied load *L*, the end reaction R_1, and the three forces exerted on the cut members by the right-hand section which has been removed. We may usually draw the forces with their proper senses by a visual approximation of the equilibrium requirements. Thus in balancing the moments about point *B* for the left-hand section, the force *EF* is clearly to the left, which makes it compressive, since it acts toward the cut section of member *EF*. The load *L* is greater than the reaction R_1, so

(a)

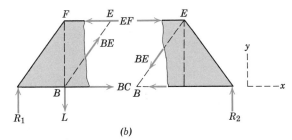

(b)

Figure 4/11

that the force *BE* must be up and to the right to supply the needed upward component for vertical equilibrium. Force *BE* is therefore tensile, since it acts away from the cut section. With the approximate magnitudes of R_1 and L in mind we see that the balance of moments about point E requires that *BC* be to the right. A casual glance at the truss should lead to the same conclusion when it is realized that the lower horizontal member will stretch under the tension caused by bending. The equation of moments about joint B eliminates three forces from the relation, and *EF* may be determined directly. The force *BE* is calculated from the equilibrium equation for the y-direction. Finally, we determine *BC* by balancing moments about point E. In this way each of the three unknowns has been determined independently of the other two.

The right-hand section of the truss, Fig. 4/11*b*, is in equilibrium under the action of R_2 and the same three forces in the cut members applied in the directions opposite to those for the left section. The proper sense for the horizontal forces may easily be seen from the balance of moments about points B and E.

We may use either section of a truss for the calculations, but the one involving the smaller number of forces will usually yield the simpler solution.

It is essential to understand that in the method of sections an entire portion of the truss is considered a single body in equilibrium. Thus the forces in members internal to the section are not involved in the analysis of the section as a whole. In order to clarify the free body and the forces acting externally on it, the section is preferably passed through the members and not the joints.

The moment equations are used to great advantage in the method of sections, and a moment center, either on or off the section, through which as many forces pass as possible should be chosen. It is not always possible to assign an unknown force in the proper sense when the free-body diagram of a section is initially drawn. With an arbitrary assignment made, a positive answer will verify the assumed sense and a negative result will indicate that the force is in the sense opposite to that assumed. Any system of notation desired may be used, although usually it is found convenient to letter the joints and designate a member and its force by the two letters defining the ends of the member.

An alternative notation preferred by some is to assign all unknown forces arbitrarily as positive in the tension direction (away from the section) and let the algebraic sign of the answer distinguish between tension and compression. Thus a plus sign would signify tension and a minus sign compression. On the other hand the advantage of assigning forces in their correct sense on the free-body diagram of a section wherever possible is that it emphasizes the physical action of the forces more directly and is preferred in this treatment.

Sample Problem 4/2

Calculate the forces induced in members KL, CL, and CB by the 20-ton load on the cantilever truss.

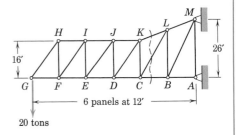

6 panels at 12′

20 tons

Solution. Although the vertical components of the reactions at A and M are statically indeterminate with the two fixed supports, all members other than AM are statically determinate. We may pass a section directly through members KL, CL, and CB and analyze the portion of the truss to the left of this section as a statically determinate rigid body.

The free-body diagram of the portion of the truss to the left of the section is shown. A moment sum about L quickly verifies the assignment of CB as compression, and a moment sum about C quickly discloses that KL is in tension. The direction of CL is not quite so obvious until we observe that KL and CB intersect at a point P to the right of G. A moment sum about P eliminates reference to KL and CB and shows that CL must be compressive to balance the moment of the 20-ton force about P. With these considerations in mind the solution becomes straightforward, as we now see how to solve for each of the three unknowns independently of the other two.

Summing moments about L requires the moment arm $BL = 16 + (26 - 16)/2 = 21$ ft. Thus

$$[\Sigma M_L = 0] \qquad 20(5)(12) - CB(21) = 0 \qquad CB = 57.1 \text{ tons } C \qquad Ans.$$

Next we take moments about C which requires a calculation of $\cos \theta$. From the given dimensions we see $\theta = \tan^{-1}(5/12)$ so that $\cos \theta = 12/13$. Therefore

$$[\Sigma M_C = 0] \qquad 20(4)(12) - \tfrac{12}{13}KL(16) = 0 \qquad KL = 65.0 \text{ tons } T \qquad Ans.$$

Finally we may find CL by a moment sum about P whose distance from C is given by $PC/16 = 24/(26 - 16)$ or $PC = 38.4$ ft. We also need β which is given by $\beta = \tan^{-1}(\overline{CB}/\overline{BL}) = \tan^{-1}(12/21) = 29.7°$ and $\cos \beta = 0.868$. We now have

$$[\Sigma M_P = 0] \qquad 20(48 - 38.4) - CL(0.868)(38.4) = 0$$

$$CL = 5.76 \text{ tons } C \qquad Ans.$$

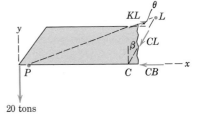

20 tons

① We note that analysis by the method of joints would necessitate working with eight joints in order to calculate the three forces in question. Thus the method of sections offers a considerable advantage in this case.

② We could have started with moments about C or P just as well.

③ We could also have determined CL by a force summation in either the x- or y-direction.

Sample Problem 4/3

Calculate the force in member DJ of the Howe roof truss illustrated. Neglect any horizontal components of force at the supports.

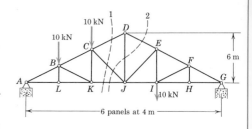

Solution. It is not possible to pass a section through DJ without cutting four members whose forces are unknown. Although three of these cut by section 2 are concurrent at J and therefore the moment equation about J could be used to obtain DE, the force in DJ cannot be obtained from the remaining two equilibrium principles. It is necessary to consider first the adjacent section 1 before considering section 2.

The free-body diagram for section 1 is drawn and includes the reaction of 18.3 kN at A, which is previously calculated from the equilibrium of the truss as a whole. In assigning the proper directions for the forces acting on the three cut members we see that a balance of moments about A eliminates the effects of CD and JK and clearly requires that CJ be up and to the left. A balance of moments about C eliminates the effect of the three forces concurrent at C and indicates that JK must be to the right to supply sufficient counterclockwise moment. Again it should be fairly obvious that the lower chord is under tension because of the bending tendency of the truss. Although it should also be apparent that the top chord is under compression, for purposes of illustration the force in CD will be arbitrarily assigned as tension.

By the analysis of section 1, CJ is obtained from

$$[\Sigma M_A = 0] \qquad 0.707CJ(12) - 10(4) - 10(8) = 0 \qquad CJ = 14.1 \text{ kN } C$$

In this equation the moment of CJ is calculated by considering its horizontal and vertical components acting at point J. Equilibrium of moments about J requires

$$[\Sigma M_J = 0] \qquad 0.894CD(6) + 18.3(12) - 10(4) - 10(8) = 0$$

$$CD = -18.6 \text{ kN}$$

The moment of CD about J is calculated here by considering its two components as acting through D. The minus sign indicates that CD was assigned in the wrong direction.

Hence $\qquad\qquad\qquad\qquad CD = 18.6 \text{ kN } C \qquad\qquad\qquad\qquad$ *Ans.*

From the free-body diagram of section 2, which now includes the known value of CJ, a balance of moments about G is seen to eliminate DE and JK. Thus

$$[\Sigma M_G = 0] \qquad 12DJ + 10(16) + 10(20) - 18.3(24) - 14.1(0.707)(12) = 0$$

$$DJ = 16.6 \text{ kN } T \qquad\qquad\qquad\qquad Ans.$$

Again the moment of CJ is determined from its components considered to be acting at J. The answer for DJ is positive, so that the assumed tensile direction is correct. An analysis of the joint D alone also verifies this conclusion.

In choosing a section it is always important to match the number of unknowns with the number of independent equilibrium equations which may be applied.

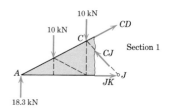

① There is no harm in assigning one or more of the forces in the wrong direction as long as the calculations are consistent with the assumption. A negative answer will show the need for reversing the direction of the force.

② If desired, the direction of CD may be changed on the free-body diagram and the algebraic sign of CD reversed in the calculations, or else the work may be left as it stands with a note stating the proper direction.

③ Observe that a section through members CD, DJ, and DE could be taken that would cut only three unknown members. However, since the forces in these three members are all concurrent at D, a moment equation about D would yield no information about them. The remaining two force equations would not be sufficient to solve for the three unknowns.

PROBLEMS

(Solve the following problems by the method of
sections. Neglect the weight of the members compared with the forces they support.)

4/24 Calculate the forces in members *AB*, *BF*, and *EF* in
the loaded truss.
Ans. AB = 8 kN *T*, *BF* = 2 kN *C*, *EF* = $4\sqrt{5}$ kN *C*

Problem 4/24

4/25 Calculate the forces in members *CD*, *BC*, and *CG* in
the loaded truss composed of equilateral triangles.

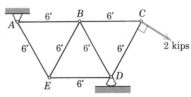

Problem 4/25

4/26 Determine the force in member *CF* in terms of the
applied load *L*. All interior angles are 60°.
Ans. CF = $2L/\sqrt{3}$, *C*

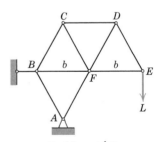

Problem 4/26

4/27 Calculate the forces in members *CD*, *BC*, and *CG* in
the loaded truss composed of equilateral triangles.

Problem 4/27

Problem 4/28

Problem 4/29

Problem 4/30

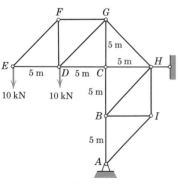

Problem 4/31

4/28 Calculate the force in members *AB*, *BG*, and *GF*. Solve for each force from an equilibrium equation which contains that force as the only unknown.

> *Ans.* $AB = 7.2$ kN T, $BG = 3$ kN C,
> $GF = 7.8$ kN C

4/29 Assume that the cross braces in the bridge truss are flexible members incapable of supporting compression. Calculate the force in member *DE* for the loading condition shown.

4/30 Calculate the forces in members *DI*, *DE*, and *EI* for the loaded truss shown.

> *Ans.* $DI = 18$ kN C, $DE = 25.5$ kN C,
> $EI = 0$

4/31 Calculate the forces in members *CH*, *CB*, and *GH* for the cantilevered truss. Solve for each force from a moment equation which contains that force as the only unknown.

4/32 Compute the forces in members *BC*, *CI*, and *HI* for the truss of Prob. 4/10 repeated here. Solve for each force from an equilibrium equation which contains that force as the only unknown.

Ans. $BC = 4.33$ kN C, $CI = 2.12$ kN T
$HI = 2.69$ kN T

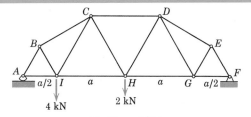

Problem 4/32

4/33 The roof truss is composed of 30°–60° right triangles and is loaded as shown. Compute the forces in members *BH* and *HG*.

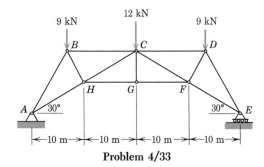

Problem 4/33

4/34 A crane is modeled by the simple truss shown. Compute the forces in members *DE*, *DG*, and *HG* under the load of the 2-ton tractor.

Ans. $DE = 2$ tons T, $DG = 4.24$ tons T
$HG = 5$ tons C

Problem 4/34

4/35 Compute the forces in members *CH*, *CD*, and *HI* for the crane truss of Prob. 4/34.

Problem 4/36

Problem 4/37

Problem 4/38

Problem 4/39

4/36 Solve for the forces in members *BG* and *BF* of the signboard truss of Prob. 4/13 repeated here. The resultant of the 800-lb wind load passes through *C*.
Ans. $BG = 400\sqrt{5}$ lb *C*, $BF = 800$ lb *T*

4/37 Calculate the forces in members *FC* and *FB* due to the 100-kN load on the crane truss.

4/38 Each of the members *BE* and *FC* is capable of supporting compression as well as tension. Compute the forces in members *CF*, *BE*, and *EF*.
Ans. $BE = 1.667$ kN *C*, $FC = 3.33$ kN *C*
$EF = 4$ kN *T*

4/39 The truss shown is composed of 45° right triangles. The crossed members in the center two panels are slender tie rods incapable of supporting compression. Retain the two rods which are under tension and compute the magnitudes of their tensions. Also find the force in member *MN*.

4/40 The crane truss is secured to the fixed supports at A and K, and its winch W is locked in position while supporting the 2000-lb tank. Identify any statically indeterminate members and calculate the force in member HG.   *Ans. HG = 8121 lb C*

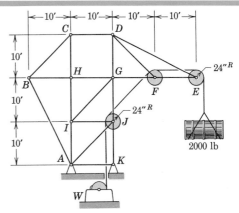

Problem 4/40

4/41 The transmission-line truss of Prob. 4/22 is shown again here. Assume that the crossed members are capable of supporting tension only and compute the force in member FC under the action of the loading shown.

Problem 4/41

4/42 Find the force in member JQ for the Baltimore truss where all angles are 30°, 60°, 90°, or 120°.
                                      Ans. JQ = 57.7 kN C

Problem 4/42

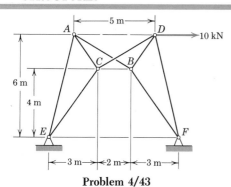

Problem 4/43

▶4/43 The hinged frames *ACE* and *DFB* are connected by two hinged bars, *AB* and *CD*, which cross without being connected. Compute the force in *AB*.

Ans. $AB = 3.78$ kN C

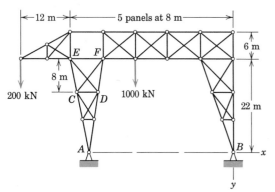

Problem 4/44

▶4/44 In the traveling bridge crane shown all crossed members are slender tie rods incapable of supporting compression. Determine the forces in members *DF* and *EF* and find the horizontal reaction on the truss at *A*. Show that if *CF* = 0, *DE* = 0 also.

Ans. $DF = 768$ kN C

$EF = 364$ kN C

$A_x = 101$ kN to the right

4/5 SPACE TRUSSES.   A space truss is the three-dimensional counterpart of the plane truss described in the two previous articles. The idealized space truss consists of rigid links connected at their ends by ball-and-socket joints. We saw that a triangle of pin-connected bars forms the basic noncollapsible unit for the plane truss. A space truss, on the other hand, requires six bars joined at their ends to form the edges of a tetrahedron for the basic noncollapsible unit. In Fig. 4/12*a* the two bars *AD* and *BD* joined at *D* require a third support *CD* to keep the triangle *ADB* from rotating about *AB*. In Fig. 4/12*b* the supporting base is replaced by three more bars *AB*, *BC*, and *AC* to form a tetrahedron not dependent on the foundation for its own rigidity. We may form a new rigid unit to the structure with three additional concurrent bars whose ends are attached to three fixed joints on the existing structure. Thus in Fig. 4/12*c* the bars *AF*, *BF*, and *CF* are attached to the foundation and therefore fix point *F* in space. Likewise point *H* is fixed in space by the bars *AH*, *DH*, and *CH*. The three additional bars *CG*, *FG*, and *HG* are attached to the three fixed points *C*, *F*, and *H* and therefore fix *G* in space. Point *E* is similarly established. We see now that the structure is entirely rigid. The two applied loads shown will result in forces in all of the members.

Ideally there must be point support, such as represented by a ball-and-socket joint, for the connections of a space truss so that there will be no bending in the members. Again, as in riveted and welded connections for plane trusses, if the centerlines of joined members intersect at a point, we may justify the assumption of two-force members under simple tension and compression.

(a)

(b)

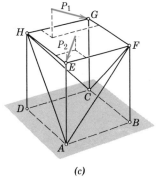

(c)

Figure 4/12

For a space truss which is supported externally in such a way that it is statically determinate as an entire unit, a relationship exists between the number of its joints and the number of its members necessary for internal stability without redundancy. Since the equilibrium of each joint is specified by three scalar force equations, there are in all $3j$ such equations for a simple space truss with j joints. For the entire truss composed of m members there are m unknowns plus six unknown support reactions in the general case of a statically determinate space structure. Thus, for a simple space truss composed of tetrahedral elements, the equation $m + 6 = 3j$ will be satisfied if the truss is statically determinate internally. Again, as in the case of the plane truss, this relation is a necessary condition for stability, but it is not a sufficient condition, since one or more of the m members can be arranged in such a way as not to contribute to a stable configuration of the entire truss. If $m + 6 > 3j$, there are more members than there are independent equations, and the truss is statically indeterminate internally with redundant members present. If $m + 6 < 3j$, there is a deficiency of internal members, and the truss is unstable and subject to collapse under load. The foregoing relationship between the number of joints and the number of members for a space truss is very helpful in the preliminary design for such a truss, since the configuration is not nearly so obvious as in the case of a plane truss, where the geometry for statical determinancy is generally quite apparent.

The method of joints developed in Art. 4/3 for plane trusses may be extended directly to space trusses by satisfying the complete vector equation

$$\Sigma \mathbf{F} = \mathbf{0}$$

for each joint. It is necessary to start at some joint where at least one known force acts and not more than three unknown forces are present. Adjacent joints upon which not more than three unknown forces act may be analyzed in turn.

The method of sections developed in the previous article may also be applied to space trusses. The two vector equations

$$\Sigma \mathbf{F} = \mathbf{0} \quad \text{and} \quad \Sigma \mathbf{M} = \mathbf{0}$$

must be satisfied for any section of the truss, where the zero moment sum will hold for all moment axes. Since the two vector equations are equivalent to six scalar equations, we conclude that a section should in general not be passed through more than six members whose forces are unknown. The method of sections for space trusses is not widely used, however, because a moment axis can seldom be found which eliminates all but one unknown as in the case of plane trusses.

Vector notation for expressing the terms in the force and moment equations for space trusses is of considerable advantage and is used in the sample problem that follows.

Sample Problem 4/4

The space truss consists of the rigid tetrahedron $ABCD$ anchored by a ball-and-socket connection at A and prevented from any rotation about the x-, y-, or z-axes by the respective links 1, 2, and 3. The load L is applied to joint E, which is rigidly fixed to the tetrahedron by the three additional links. Solve for the forces in the members at joint E and indicate the procedure for the solution of the forces in the remaining members of the truss.

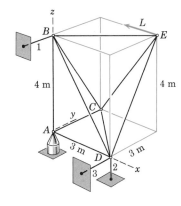

Solution. We note first that the truss is supported with six properly placed constraints, which are the three at A and the links 1, 2, and 3. Also, with $m = 9$ members and $j = 5$ joints the condition $m + 6 = 3j$ for a sufficiency of members to provide a noncollapsible structure is satisfied.

① The external reactions at A, B, and D can be calculated easily as a first step, although their values will be determined from the solution of all forces on each of the joints in succession.

② We must start with a joint upon which at least one known force and not more than three unknown forces act, which in this case is joint E. The free-body diagram of joint E is shown with all force vectors arbitrarily but consistently indicated in their positive tension directions (away from the joint). The vector expressions for the three unknown forces are

$$\mathbf{F}_{EB} = \frac{F_{EB}}{\sqrt{2}}(-\mathbf{i} - \mathbf{j}), \quad \mathbf{F}_{EC} = \frac{F_{EC}}{5}(-3\mathbf{i} - 4\mathbf{k}), \quad \mathbf{F}_{ED} = \frac{F_{ED}}{5}(-3\mathbf{j} - 4\mathbf{k})$$

① *Suggestion:* Draw a free-body diagram of the truss as a whole and verify that the external forces acting on the truss are $\mathbf{A}_x = L\mathbf{i}$, $\mathbf{A}_y = L\mathbf{j}$, $\mathbf{A}_z = (4L/3)\mathbf{k}$, $\mathbf{B}_y = 0$, $\mathbf{D}_y = -L\mathbf{j}$, $\mathbf{D}_z = -(4L/3)\mathbf{k}$.

Equilibrium of joint E requires

$$[\Sigma\mathbf{F} = 0] \qquad \mathbf{L} + \mathbf{F}_{EB} + \mathbf{F}_{EC} + \mathbf{F}_{ED} = 0 \qquad \text{or}$$

$$-L\mathbf{i} + \frac{F_{EB}}{\sqrt{2}}(-\mathbf{i} - \mathbf{j}) + \frac{F_{EC}}{5}(-3\mathbf{i} - 4\mathbf{k}) + \frac{F_{ED}}{5}(-3\mathbf{j} - 4\mathbf{k}) = 0$$

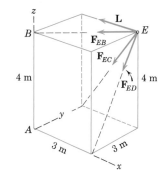

Rearranging terms gives

$$\left(-L - \frac{F_{EB}}{\sqrt{2}} - \frac{3F_{EC}}{5}\right)\mathbf{i} + \left(-\frac{F_{EB}}{\sqrt{2}} - \frac{3F_{ED}}{5}\right)\mathbf{j} + \left(-\frac{4F_{EC}}{5} - \frac{4F_{ED}}{5}\right)\mathbf{k} = 0$$

Equating the coefficients of the $\mathbf{i}$-, $\mathbf{j}$-, and $\mathbf{k}$-terms to zero gives the three equations

$$\frac{F_{EB}}{\sqrt{2}} + \frac{3F_{EC}}{5} = -L \qquad \frac{F_{EB}}{\sqrt{2}} + \frac{3F_{ED}}{5} = 0 \qquad F_{EC} + F_{ED} = 0$$

Solving the equations gives us

$$F_{EB} = -L/\sqrt{2} \qquad F_{EC} = -5L/6 \qquad F_{ED} = 5L/6 \qquad \textit{Ans.}$$

Thus we conclude that F_{EB} and F_{EC} are compressive forces and F_{ED} is tension.

② A negative numerical value for a force indicates compression.

Unless we have computed the external reactions first, we must next analyze joint C with the known value of F_{EC} and the three unknowns F_{CB}, F_{CA}, and F_{CD}. The procedure is identical with that used for joint E. Joints B, D, and A are then analyzed in the same way and in that order, which limits the unknowns to three for each joint. The external reactions computed from these analyses must, of course, agree with the values which can be determined initially from an analysis of the truss as a whole.

In a more general case, if there is no loaded joint which has as few as three unknown forces acting, it is necessary to compute the external reactions first and begin the analysis at one of the external reaction joints upon which no more than three unknown forces act.

PROBLEMS

(In the following problems use plus for tension and minus for compression.)

Problem 4/45

Problem 4/46

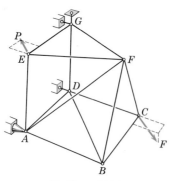

Problem 4/47

4/45 The space truss is built upon the triangular base *ABC*. The locations of joints *D* and *E* are established by the links shown. Show that this configuration is internally stable. Also replace link *AE* by a different link that will preserve the rigidity of the truss.

4/46 The prismatic space truss has a horizontal base *ADE* and a parallel top face *BCF* in the shape of equal equilateral triangles which are connected by three equal vertical legs and braced by three diagonal members as shown. Show that this truss represents a stable configuration.

4/47 The space truss is shown in an intermediate stage of design. The external constraints indicated are sufficient to maintain external equilibrium. How many additional members are needed to prevent internal instability and where can they be placed?

4/48 For the space truss of Sample Problem 4/4 use the result $F_{EC} = -5L/6$ and compute the forces in members *CB*, *CA*, and *CD*.

Ans. $F_{CD} = L/\sqrt{2}$, $F_{CB} = 5L/6$, $F_{CA} = -L$

4/49 The space truss in the form of a tetrahedron is supported by ball-and-socket connections at its base points A and B and is prevented from rotating about AB by the vertical tie bar CD. After noting the vertical components of the reactions under the symmetrical truss at A and B, draw a free-body diagram of the triangular configuration of links BDE and determine the x-component of the force exerted by the foundation on the truss at B.

Ans. $B_x = P$

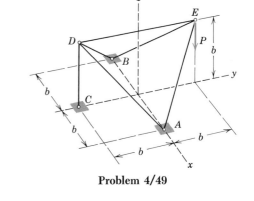

Problem 4/49

4/50 The rectangular space truss 16 m in height is erected on a horizontal square base 12 m on a side. Guy wires are attached to the structure at E and G as shown and are tightened until the tension T in each wire is 9 kN. Calculate the force F in each of the diagonal members.   Ans. $F = -3.72$ kN

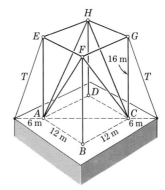

Problem 4/50

4/51 The tetrahedral space truss has a horizontal base ABC in the form of an isosceles triangle and legs AD, BD, and CD which support the mass m from point D. Each vertex of the base is suspended by a vertical wire from overhead supports. Calculate the force induced in members AC and AB.

Ans. $F_{AC} = -\dfrac{5mg}{54}$, $F_{AB} = -\dfrac{4mg}{27}$

Problem 4/51

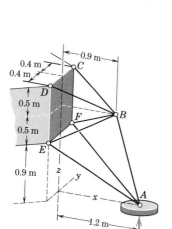

Problem 4/52

▶4/52 A space truss is constructed in the form of a cube with six diagonal members shown. Verify that the truss is internally stable. If the truss is subjected to the compressive forces P applied at F and D along the diagonal FD, determine the forces in members FE and EG.   *Ans.* $F_{FE} = -P/\sqrt{3}$, $F_{EG} = P/\sqrt{6}$

▶4/53 Each of the landing struts for a lunar spacecraft is a space truss symmetrical about the vertical x-z plane as shown. For a landing force of $F = 2.2$ kN, calculate the corresponding force in member BE. The assumption of static equilibrium for the truss is permissible if the mass of the truss is very small. Assume equal loads in the symmetrically placed members.   *Ans.* $F_{BE} = 1.620$ kN

Problem 4/53

4/6 FRAMES AND MACHINES.

Structures composed of joined members any one of which has more than two forces acting on it fall in the categories of frames or machines. Frames are structures which are designed to support applied loads and are usually fixed in position. Machines are structures which contain moving parts and are designed to transmit forces or couples from input values to output values.

Since frames and machines contain multiforce members (three or more forces), these forces in general will *not* be in the directions of the members. Therefore we cannot analyze these structures by the methods developed in Arts. 4/3, 4/4, and 4/5 for simple trusses composed of two-force members where the forces are in the directions of the members.

In the previous chapter the equilibrium of multiforce bodies was discussed and illustrated, but we focused attention on the equilibrium of a *single* rigid body. In this present article attention is focused on the equilibrium of *interconnected* rigid bodies which contain multi-force members. Although most such bodies may be analyzed as two-dimensional systems, there are numerous examples of frames and machines that are three-dimensional.

The forces acting on each member of a connected system are found by isolating the member with a free-body diagram and applying the established equations of equilibrium. The *principle of action and reaction* must be carefully observed when we represent the forces of interaction on the separate free-body diagrams. If the structure contains more members or supports than are necessary to prevent collapse, then, as in the case of trusses, the problem is statically indeterminate, and the principles of equilibrium, although necessary, are not sufficient for solution. Although many frames and machines are statically indeterminate, we will consider in this article only those that are statically determinate.

If the frame or machine constitutes a rigid unit by itself when removed from its supports, as is the A-frame in Fig. 4/13*a*, the analysis is best begun by establishing all the forces external to the

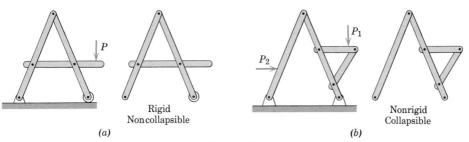

Rigid
Noncollapsible

Nonrigid
Collapsible

(a)   (b)

Figure 4/13

(a)

(b)

Figure 4/14

Vector
Notation

Figure 4/15

structure considered as a single rigid body. We then dismember the structure and consider the equilibrium of each part separately. The equilibrium equations for the several parts will be related through the terms involving the forces of interaction. If the structure is not a rigid unit by itself but depends on its external supports for rigidity, as illustrated in Fig. 4/13b, then the calculation of the external support reactions cannot be completed until the structure is dismembered and the individual parts are analyzed.

In most cases we find that the analysis of frames and machines is facilitated by representing the forces in terms of their rectangular components. This is particularly so when the dimensions of the parts are given in mutually perpendicular directions. The advantage of this representation is that the calculation of moment arms is accordingly simplified. In some three-dimensional problems, particularly when moments are evaluated about axes that are not parallel to the coordinate axes, we find the use of vector notation is an advantage.

It is not always possible to assign every force or its components in the proper sense when drawing the free-body diagrams, and it becomes necessary for us to make an arbitrary assignment. In any event it is *absolutely necessary* that a force be *consistently* represented on the diagrams for interacting bodies which involve the force in question. Thus for two bodies connected by the pin A, Fig. 4/14a, the components when separated must be consistently represented in the *opposite* directions. For a ball-and-socket connection between members of a space frame, we must apply the action-and-reaction principle to all three components as shown in Fig. 4/14b. The assigned directions may prove to be wrong when the algebraic signs of the components are determined upon calculation. If A_x, for instance, should turn out to be negative, it is actually acting in the direction opposite to that originally represented. Accordingly it would be necessary for us to reverse the direction of the force on *both* members and to reverse the sign of its force terms in the equations. Or we may leave the representation as originally made, and the proper sense of the force will be understood from the negative sign. If we choose to use vector notation in labeling the forces, then we must be careful to use a plus sign for an action and a minus sign for the corresponding reaction, as shown in Fig. 4/15.

Finally, situations occasionally arise where it is necessary to solve two or more equations simultaneously in order to separate the unknowns. In most instances, however, we may avoid simultaneous solutions by careful choice of the member or group of members for the free-body diagram and by a careful choice of moment axes which will eliminate undesired terms from the equations. The method of solution described in the foregoing paragraphs is illustrated in the following sample problems.

Sample Problem 4/5

The frame supports the 400-kg load in the manner shown. Neglect the weights of the members compared with the forces induced by the load and compute the horizontal and vertical components of all forces acting on each of the members.

Solution. We observe first that the three supporting members that constitute the frame form a rigid assembly that can be analyzed as a single unit. We also observe that the arrangement of the external supports makes the frame statically determinate.

From the free-body diagram of the entire frame we determine the external reactions. Thus

$$[\Sigma M_A = 0] \qquad 5.5(0.4)(9.81) - 5D = 0 \qquad D = 4.32 \text{ kN}$$

$$[\Sigma F_x = 0] \qquad A_x - 4.32 = 0 \qquad A_x = 4.32 \text{ kN}$$

$$[\Sigma F_y = 0] \qquad A_y - 3.92 = 0 \qquad A_y = 3.92 \text{ kN}$$

① We see that the frame corresponds to the category illustrated in Fig. 4/13a.

Next we dismember the frame and draw a separate free-body diagram of each member. The diagrams are arranged in their approximate relative positions to aid in keeping track of the common forces of interaction. The external reactions just obtained are entered onto the diagram for AD. Other known forces are the 3.92-kN forces exerted by the shaft of the pulley on the member BF, as obtained from the free-body diagram of the pulley. The cable tension of 3.92 kN is also shown acting on AD at its attachment point.

Next, the components of all unknown forces are shown on the diagrams. Here we observe that CE is a two-force member. The force components on CE have equal and opposite reactions, which are shown on BF at E and on AD at C. We may not recognize the actual sense of the components at B at first glance, so they may be arbitrarily but consistently assigned.

② Here we observe that CE is a two-force member. The force components on CE have equal and opposite reactions, which are shown on BF at E and on AD at C.

The solution may proceed by use of a moment equation about B or E for member BF followed by the two force equations. Thus

$$[\Sigma M_B = 0] \qquad 3.92(5) - \tfrac{1}{2}E_x(3) = 0, \qquad E_x = 13.08 \text{ kN} \qquad Ans.$$

$$[\Sigma F_y = 0] \qquad B_y + 3.92 - 13.08/2 = 0, \qquad B_y = 2.62 \text{ kN} \qquad Ans.$$

$$[\Sigma F_x = 0] \qquad B_x + 3.92 - 13.08 = 0, \qquad B_x = 9.15 \text{ kN} \qquad Ans.$$

Positive numerical values of the unknowns mean that we assumed their directions correctly on the free-body diagrams. The value of $C_x = E_x = 13.08$ kN obtained by inspection of the free-body diagram of CE is now entered onto the diagram for AD, along with the values of B_x and B_y just determined. The equations of equilibrium may now be applied to member AD as a check, since all the forces acting on it have already been computed. The equations give

$$[\Sigma M_C = 0] \qquad 4.32(3.5) + 4.32(1.5) - 3.92(2) - 9.15(1.5) = 0$$

$$[\Sigma F_x = 0] \qquad 4.32 - 13.08 + 9.15 + 3.92 - 4.32 = 0$$

$$[\Sigma F_y = 0] \qquad -13.08/2 + 2.62 + 3.92 = 0$$

② Without this critical observation the problem cannot be solved. Note especially that the direction of the line joining the two points of application of force, and not the shape of the member, determines the direction of the force and hence the ratio of the force components acting at C and E.

Sample Problem 4/6

Neglect the weight of the frame and compute the forces acting on all of its members.

Solution. We note first that the frame is not a rigid unit when removed from its supports since $BDEF$ is a movable quadrilateral and not a rigid triangle. Consequently the external reactions cannot be completely determined until the individual members are analyzed. However, we can determine the vertical components of the reactions at A and C from the free-body diagram of the frame as a whole. Thus

$$[\Sigma M_C = 0] \qquad 50(12) + 30(40) - 30A_y = 0 \qquad A_y = 60 \text{ lb} \qquad Ans.$$

$$[\Sigma F_y = 0] \qquad C_y - 50(4/5) - 60 = 0 \qquad C_y = 100 \text{ lb} \qquad Ans.$$

① We see that this frame corresponds to the category illustrated in Fig. 4/13b.

Next we dismember the frame and draw the free-body diagram of each part. Since EF is a two-force member, the direction of the force at E on ED and at F on AB is known. We may assume that the 30-lb force is applied to the pin as a part of member BC. There should be no difficulty in assigning the correct directions for forces E, F, D, and B_x. The direction of B_y, however, may not be assigned by inspection and therefore is arbitrarily shown as downward on AB and upward on BC.

Member ED. The two unknowns are easily obtained by

$$[\Sigma M_D = 0] \qquad 50(12) - 12E = 0 \qquad E = 50 \text{ lb} \qquad Ans.$$

$$[\Sigma F = 0] \qquad D - 50 - 50 = 0 \qquad D = 100 \text{ lb} \qquad Ans.$$

② The directions of A_x and C_x are not obvious initially and can be assigned arbitrarily to be corrected later if necessary.

Member EF. Clearly F is equal and opposite to E with the magnitude of 50 lb.

Member AB. Since F is now known, we solve for B_x, A_x, and B_y from

$$[\Sigma M_A = 0] \qquad 50(3/5)(20) - B_x(40) = 0 \qquad B_x = 15 \text{ lb} \qquad Ans.$$

$$[\Sigma F_x = 0] \qquad A_x + 15 - 50(3/5) = 0 \qquad A_x = 15 \text{ lb} \qquad Ans.$$

$$[\Sigma F_y = 0] \qquad 50(4/5) - 60 - B_y = 0 \qquad B_y = -20 \text{ lb} \qquad Ans.$$

The minus sign shows that we assigned B_y in the wrong direction.

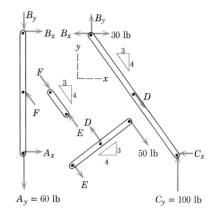

Member BC. The results for B_x, B_y, and D are now transferred to BC, and the remaining unknown C_x is found from

$$[\Sigma F_x = 0] \qquad 30 + 100(3/5) - 15 - C_x = 0 \qquad C_x = 75 \text{ lb} \qquad Ans.$$

We may apply the remaining two equilibrium equations as a check. Thus

$$[\Sigma F_y = 0] \qquad 100 + (-20) - 100(4/5) = 0$$

$$[\Sigma M_C = 0] \qquad (30 - 15)(40) + (-20)(30) = 0$$

③ Alternatively the 30-lb force could be applied to the pin considered a part of BA, with a resulting change in the reaction B_x.

④ Alternatively we could have returned to the free-body diagram of the frame as a whole and found C_x.

Sample Problem 4/7

The machine shown is an overload protection device which releases the load when it exceeds a predetermined value T. A soft metal shear pin S is inserted in a hole in the lower half and is acted upon by the upper half. When the total force on the pin exceeds its strength, it will break. The two halves then rotate about A under the action of the tensions in BD and CD, as shown in the second sketch, and rollers E and F release the eye bolt. Determine the maximum allowable tension T if the pin S will shear when the total force on it is 800 N. Also compute the corresponding force on the hinge pin A.

Dimensions
in Millimeters

Released
Position

Solution. Because of symmetry we analyze only one of the two hinged members. The upper part is chosen, and its free-body diagram along with that for the connection at D is drawn. Because of symmetry the
① forces at S and A have no x-components. The two-force members BD and CD exert forces of equal magnitude $B = C$ on the connection at D. Equilibrium of the connection gives

$[\Sigma F_x = 0] \qquad B \cos \theta + C \cos \theta - T = 0, \qquad 2B \cos \theta = T$

$$B = T/(2 \cos \theta)$$

From the free-body diagram of the upper part we express the equilibrium of moments about point A. Substituting $S = 800$ N and the expression for B gives

② $[\Sigma M_A = 0]$

$$\frac{T}{2 \cos \theta}(\cos \theta)(50) + \frac{T}{2 \cos \theta}(\sin \theta)(36) - 36(800) - \frac{T}{2}(26) = 0$$

Substituting $\sin \theta / \cos \theta = \tan \theta = 5/12$ and solving for T give

$$T\left(25 + \frac{5(36)}{2(12)} - 13\right) = 28\ 800$$

$$T = 1477 \text{ N} \qquad \text{or} \qquad T = 1.477 \text{ kN} \qquad\qquad \textit{Ans.}$$

Finally, equilibrium in the y-direction gives us

$[\Sigma F_y = 0] \qquad S - B \sin \theta - A = 0$

$$800 - \frac{1477}{2(12/13)}\frac{5}{13} - A = 0 \qquad A = 492 \text{ N} \qquad\qquad \textit{Ans.}$$

① Symmetry is always useful to recognize. Here it tells us that the forces acting on the two parts behave as mirror images of each other with respect to the x-axis. Thus we cannot have an action on one member in the plus x-direction and its reaction on the other member in the negative x-direction. Consequently the forces at S and A have no x-components.

② Be careful not to forget the moment of the y-component of B. Note that our units here are newton-millimeters.

PROBLEMS

(Unless otherwise indicated neglect the mass of the various members in the problems that follow.)

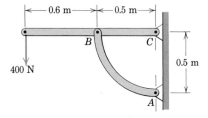

Problem 4/54

4/54 Calculate the magnitude of the force supported by the pin at C for the loaded frame.

Ans. C = 1002 N

Problem 4/55

4/55 Calculate the magnitude of the force supported by the pin at A which secures the loaded bracket to the wall.

Problem 4/56

4/56 Calculate the magnitude of the force supported by the pin B of the frame loaded by the 200-lb-ft couple.

Ans. B = 69.3 lb

4/57 Replace the 200-lb-ft couple in Prob. 4/56 by a 200-lb downward force applied at point D and compute the magnitude of the force acting at B.

Problem 4/57

4/58 Compound-lever snips, shown in the figure, are often used in place of regular tinners' snips when large cutting forces are required. For the gripping force of 150 N, what is the cutting force P at a distance of 30 mm along the blade from the pin at A?

Ans. P = 1467 N

4/59 A small bolt cutter operated by hand for cutting small bolts and rods is shown in the sketch. For a hand grip $P = 150$ N, determine the force Q developed by each jaw on the rod to be cut.

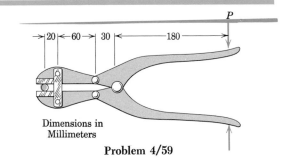

Dimensions in
Millimeters

Problem 4/59

4/60 A limiting torque wrench is designed with a shear pin B which breaks when the force on it exceeds its strength and, hence, limits the torque which can be applied by the wrench. If the limiting strength of the shear pin is 900 N in double shear (i.e., $V/2 = 450$ N), calculate the limiting torque M. What would be the effect on M of increasing b, all other conditions remaining unchanged?

Ans. $M = 75$ N·m

M is decreased as b is increased

Problem 4/60

4/61 Calculate the x- and y-components of the force supported by the pin C for the loaded frame.

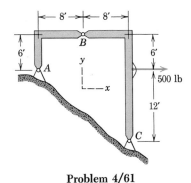

Problem 4/61

4/62 Solve for the x- and y-components of the force supported by the hinge at C for the loaded frame.

Ans. $C_x = 5.25$ kN, $C_y = 1$ kN

Problem 4/62

Problem 4/63

Problem 4/64

Problem 4/65

Problem 4/66

4/63 Calculate the magnitude of the force acting on the pin at D. Pin C is fixed in DE and bears against the smooth slot in the triangular plate.

4/64 Compute the force supported by the pin at A for the channel-lock pliers under a grip of 30 lb.

Ans. $A = 157.6$ lb

4/65 For the paper punch shown find the punching force Q corresponding to a hand grip P.

4/66 The special box wrench with head B swiveled at C to the handle A will accommodate a range of sizes of hexagonal bolt heads. For the nominal size shown where the center O of the bolt and the pin C are in line with the handle, compute the magnitude of the force supported by the pin at C if $P = 160$ N. Assume the surface of the bolt head to be smooth.

Ans. $C = 1367$ N

4/67 The speed reducer consists of the input shaft A and attached pinion B, which drives gear C and its output shaft D with a 2:1 reduction. The center of mass of the 30-kg unit is at G. If a clockwise input torque of 50 N · m is applied to shaft A and if the output shaft D drives a machine at a constant speed, determine the net forces exerted *on* the base flange of the reducer at E and F by the combined action of the bolts and the supporting foundation. (*Caution:* Be careful to apply the reactive torque on the output shaft in the correct direction.)

        Ans. $E = 139$ N down, $F = 433$ N up

Problem 4/67

4/68 The two spur gears A and B drive the bevel gears C and D. For a given torque M_0 express the torque M on the output shaft required to maintain equilibrium.

$$r_1/r_2 = 2/3$$
$$\tan \alpha = 3/4$$

Problem 4/68

4/69 Compute the force acting in the link AB of the lifting tongs which cross without touching.

        Ans. $F_{AB} = 1650$ lb tension

Problem 4/69

4/70 Determine the vertical clamping force at E in terms of the force P applied to the handle of the toggle clamp which holds the workpiece F in place.

Problem 4/70

Problem 4/71

4/71 The toggle pliers are used for a variety of clamping purposes. For the handle position given by $\alpha = 10°$ and for a handle grip of $P = 150$ N calculate the clamping force C produced.   *Ans. C* = 1368 N

Problem 4/72

4/72 Find the x- and y-components of the pin reaction at B for the loaded frame.

Problem 4/73

4/73 Calculate the magnitude of the force supported by the pin at B for the loaded frame.
Ans. B = 233 lb

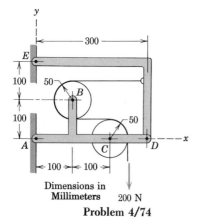

Dimensions in Millimeters   200 N

Problem 4/74

4/74 Calculate the x- and y-components of all forces acting on each member of the loaded frame.

4/75 Calculate the *x*- and *y*-components of the force supported by the pin at *B*. The cables are securely wrapped around the pulleys which are fastened together.   *Ans.* $B_x = 809$ N, $B_y = 785$ N

Problem 4/75

4/76 Compute the *x*- and *y*-components of the force acting on the pin *B* connecting the two members of the machine subjected to the 200-N load.
    Ans. $B_x = 287$ N, $B_y = 215$ N

Problem 4/76

4/77 Replace the 200-N force at *E* of Prob. 4/76 by a 50-N·m clockwise couple and compute the *x*- and *y*-components of all forces on each of the two members of the machine.

4/78 Calculate the *x*- and *y*-components of the force supported by the pin at *E* for the loaded frame.
    Ans. $E_x = 5$ lb, $E_y = 5\sqrt{3}$ lb

Problem 4/78

Problem 4/79

4/79 Find the x- and y-components of the forces at A and C of the loaded frame.

Problem 4/80

4/80 The symmetrical 40-lb stepladder is mounted on wheels so that it may be moved easily. If a 180-lb man stands on the step at C, calculate the tension T in the connecting link AB. The top hinge may be assumed to be on the centerlines of the legs.

Ans. $T = 47.2$ lb

Problem 4/81

4/81 The upper jaw D of the toggle press slides with negligible frictional resistance along the fixed vertical column. Calulate the compressive force R exerted on the cylinder E and the force supported by the pin at A if a force $F = 200$ N is applied to the handle at an angle of $\theta = 75°$.

4/82 A double-axle suspension for use on small trucks is shown in the figure. The mass of the central frame F is 40 kg, and the mass of each wheel and attached link is 35 kg with center of mass 680 mm from the vertical centerline. For a load of $L = 12$ kN transmitted to the frame F, compute the total shear force supported by the pin at A.   *Ans. A = 1.75 kN*

Problem 4/82

4/83 The nose-wheel assembly is raised by the application of a torque M to link BC through the shaft at B. If the arm and wheel AO have a combined weight of 100 lb with center of gravity at G, find the value of M necessary to lift the wheel when D is directly under B at which position angle θ is 30°.

Problem 4/83

4/84 The aircraft landing gear consists of a spring- and hydraulically-loaded piston and cylinder D and the two pivoted links OB and CB. If the gear is moving along the runway at a constant speed with the wheel supporting a stabilized constant load of 24 kN, calculate the total force that the pin at A supports.   *Ans. A = 44.7 kN*

Problem 4/84

Problem 4/85

4/85 Determine the punching force P in terms of the gripping force F for the rivet squeezer shown.

$$\text{Ans. } P = \frac{2Fe}{c\left(1 - \dfrac{a}{b}\right)}$$

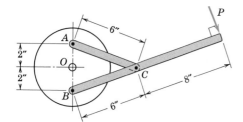

Problem 4/86

4/86 The device shown is an adjustable form of spanner wrench in which the pins at A and B fit into holes in the face of the disk that is to be screwed onto its fixed shaft O. If a torque (moment) of 600 lb-in. about O is required to tighten the disk on the shaft, compute the force supported by each of the pins at A and B when the applied force P has the required value.

Problem 4/87

4/87 Determine the force supported by the pin at C for the loaded frame.   *Ans. C = 2160 N*

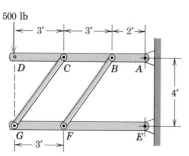

Problem 4/88

4/88 Calculate the magnitude of the force supported by the pin at A for the loaded frame.

4/89 Arm CE and the toothed wheel are mounted independently about the same axis at C. A torque
$M_0 = 30$ N·m on the shaft of arm AB is required to
hold the wheel against a resisting torque M acting
on it for the position shown where link BE is normal
to AB and CE. Calculate the magnitude of the
corresponding force acting on end C of the arm EC.
　　　　　　　　　　　　　　　　　　Ans. $C = 908$ N

Dimensions in
Millimeters

Problem 4/89

4/90 The portable hoist shown is used to raise construction materials to the roof of a building. The hoist has
a weight of 100 lb with mass center at G. If a 180-lb
load is supported by applying the force F on the
handle of the drum, calculate the magnitude of the
force supported by the pin at A.
　　　　　　　　　　　　　　　　　　Ans. $A = 352$ lb

Problem 4/90

4/91 Determine the shearing force Q applied to the bar if
a 400-N force is applied to the handle for
$\theta = 30°$. For a given applied force what value of θ
gives the greatest shear?　　　Ans. $Q = 13.18$ kN

Problem 4/91

Problem 4/92

Dimensions in Millimeters

Problem 4/93

Dimensions in
Meters

Problem 4/94

4/92 The backhoe is controlled by the three hydraulic cylinders, and, in the particular position shown, the hoe can apply a horizontal force $P = 10$ kN. Neglect the masses of the members and compute the magnitude of the forces supported by the pins at A and E.   *Ans.* $A = 22.4$ kN, $E = 36.7$ kN

4/93 The power unit A of the tractor-scraper has a mass of 4 Mg with mass center at G_1. The trailer-scraper unit B fully loaded has a mass of 24 Mg with mass center at G_2. Positioning of the scraper is controlled by two hydraulic cylinders EF, one on each side of the machine. Calculate the compression F in each of the cylinders and the magnitude of the force supported by each of the pins at H, one on each side. Assume that the wheels are free to turn so that there are no horizontal components of force under the wheels.   *Ans.* $F = 131.8$ kN, $H = 113.9$ kN

4/94 The figure shows a special rig for erecting vertical sections of a construction tower. The assembly A has a mass of 1.5 Mg and is elevated by the platform B which itself has a mass of 2 Mg. The platform is guided up the fixed vertical column by rollers and is activated by the hydraulic cylinder CD and links EDF and FH. For the particular position shown calculate the force R exerted by the hydraulic cylinder at D and the magnitude of the force supported by the pin at E. Neglect the mass of the cylinder and links.   *Ans.* $R = 59.7$ kN, $E = 40.0$ kN

▶4/95 The hoisting mechanism for the dump truck is shown in the enlarged view. Determine the compression P in the hydraulic cylinder BE and the magnitude of the force supported by the pin at A for the particular position shown where BA is perpendicular to OAE and link DC is perpendicular to AC. The dump and its load together weigh 20,000 lb with center of mass at G. All dimensions for the indicated geometry are given on the figure.

Ans. $P = 26{,}930$ lb, $A = 14{,}600$ lb

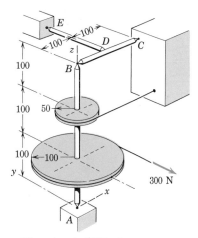

Detail of Hoisting Mechanism

Problem 4/95

4/96 Calculate the x-, y-, and z-components of all forces on each of the three parts of the mechanism. The two disks are rigidly mounted on their shaft, and the larger one is loaded by the 300-N force in the cord wrapped around it and leading in the negative y-direction. The cord around the smaller disk leads in the x-direction and is secured to prevent rotation. Treat all joints as ball-and-socket connections and neglect the weights of all members.

Dimensions in Millimeters

Problem 4/96

Problem 4/97

Problem 4/98

4/97 Calculate the x-, y-, and z-components of all forces acting on each member of the light space frame which is loaded as shown.

▶4/98 The construction A-frame shown in the figure is hinged about the y-axis through points E and F, which cannot offer restraint in the y-direction. The connections at A, B, C, D, and G may be treated as ball-and-socket joints. Link CG is the only two-force member. Compute the x-, y-, and z-components of all forces acting on each member of the frame. The weights of the members may be neglected compared with the loads transmitted.

Ans. $A_x = A_y = B_y = D_y = E_x = F_x = 1.25$ kN
$B_x = B_z = D_x = D_z = E_z = F_z = 2.5$ kN
$C_x = C_z = 5$ kN, $A_z = 0$

4/7 PROBLEM FORMULATION AND REVIEW. In Chapter 4 we have applied the principles of equilibrium to two classes of problems (*a*) simple trusses and (*b*) frames and machines. Basically, no new theory was needed since we merely drew the necessary free-body diagrams and applied our familiar equations of equilibrium. The structures dealt with in Chapter 4, however, have given us the opportunity to develop further our appreciation for the systematic approach to mechanics problems which has been frequently illustrated and emphasized.

The most essential features of the analysis of these two classes of structures are reviewed in the statements which follow.

(*a*) *Simple trusses*

1. Simple trusses are composed of two-force members joined at their ends and capable of supporting tension or compression. Each internal force, therefore, is always in the direction of its member.

2. Simple trusses are built around the basic rigid (noncollapsible) unit of the triangle for plane trusses and the tetrahedron for space trusses. Additional units of a truss are formed by adding new members, two for plane trusses and three for space trusses, attached to existing joints and joined at their ends to form a new joint.

3. The joints for simple trusses are assumed to be pin connections for plane trusses and ball-and-socket connections for space trusses. Thus the joints can transmit force but not moment.

4. External loads on simple trusses are applied only at the joints.

5. Trusses are statically determinate externally when the external constraints are not in excess of those required to maintain an equilibrium position.

6. Trusses are statically determinate internally when constructed in the manner described in item (2) where internal members are not in excess of those required to prevent collapse.

7. The *method of joints* utilizes the force equations of equilibrium for each joint. Analysis must begin at a joint where at least one force is known and not more than two forces are unknown for plane trusses or not more than three forces are unknown for space trusses.

8. The *method of sections* utilizes a free body of an entire section of a truss containing two or more joints and, in general, will involve the equilibrium of a nonconcurrent system of forces. The moment equation of equilibrium is of special value in using the method of sections. In general the forces acting on a section which cuts more than three mem-

bers whose forces are unknown for a plane truss cannot be solved completely since there are only three independent equations of equilibrium.

9. The vector representing a force acting on a joint or a section is drawn on the same side of the joint or section as is the member which transmits the force. With this convention tension is indicated when the force arrow is away from the joint or section, and compression is indicated when the arrow points toward the joint or section.

10. When the two diagonal members which brace a quadrilateral panel are flexible members incapable of supporting compression, only the one in tension is retained, and the panel remains statically determinate.

11. When two joined members under load are collinear and a third member with a different direction is joined with their connection, the force in the third member must be zero, unless an external force is applied at the joint with a component in the direction of the third member.

(b) *Frames and machines*

1. Frames and machines are multiforce structures, that is, structures which contain one or more members subjected to more than two forces.

2. Frames are structures designed to support loads generally under static conditions. Machines are structures which transform input forces and moments to output forces and moments and generally involve one or more moving parts. Some structures may be classified in either category.

3. Only frames and machines which are statically determinate externally and internally are considered in this treatment.

4. The same procedures for the analysis of frames apply to machines.

5. If a frame or machine as a whole is a rigid (noncollapsible) unit when its external supports are removed, then the analysis is begun by computing the external reactions on the entire unit. If a frame or machine as a whole is a nonrigid (collapsible) unit when its external supports are removed, then the analysis of the external reactions cannot be completed until the structure is dismembered.

6. Forces acting in the internal connections of frames and machines are calculated by dismembering the structure and constructing a separate free-body diagram of each part. The principle of action and reaction must be *strictly* observed; otherwise error will result.

7. The force and moment equations of equilibrium are applied to the members as needed to compute the desired unknowns.

REVIEW PROBLEMS

4/99 A frame used to test the compressive strength of concrete blocks is composed of the four pressure pads, the four compression links, and the three outer links arranged in a square as shown. Express the compressive force C acting on each side of the square concrete block in terms of the applied forces P.

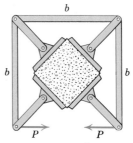

Problem 4/99

4/100 Calculate the forces in members BH, CD, and GD for the truss loaded by the 40- and 60-kN forces.

Ans. BH = 47.1 kN C
$CD = 6.7$ kN C
$GD = 0$

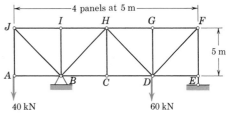

Problem 4/100

4/101 Determine the force in each member of the two trusses that support the 10-kN load at their common pin K.

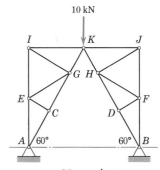

Problem 4/101

4/102 Calculate the force in member BG using a free-body diagram of the rigid member ABC.

Ans. BG = 1800 lb C

Problem 4/102

Problem 4/103

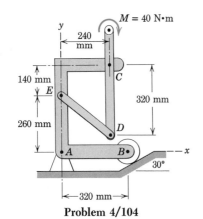

Problem 4/104

4/103 Each of the three loads L on the bridge truss is 20 tons (1 U.S. ton = 2000 lb). Determine the forces in members EG and DF with a minimum of calculation.   *Ans. EG = 60 tons T, DF = 82.5 tons C*

4/104 Calculate the x- and y-components of the forces acting at A and C for the frame loaded by the couple $M = 40$ N·m.

4/105 An adjustable tow bar connecting the tractor unit H with the landing gear J of a large aircraft is shown in the figure. Adjusting the height of the hook F at the end of the tow bar is accomplished by the hydraulic cylinder CD activated by a small hand pump (not shown). For the nominal position shown of the triangular linkage ABC, calculate the force P supplied by the cylinder to the pin C to position the tow bar. The rig has a total weight of 100 lb and is supported by the tractor hitch at E.

Ans. P = 60.8 lb

Problem 4/105

Problem 4/106

4/106 The top of the folding workbench has a mass of 50 kg with mass center at G. Calculate the x- and y-components of the force supported by the pin at E.

4/107 Solve for the force in member *GL* of the loaded tower truss. Is the truss statically determinate?

Ans. GL = 37.5 kN T

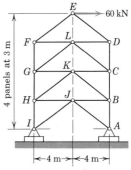

Problem 4/107

4/108 Determine the force in member *AC* in terms of the mass *m* supported by the truss. All interior acute angles are either 30° or 60°. *Ans. AC = mg/3, C*

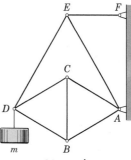

Problem 4/108

4/109 Verify the fact that each of the loaded trusses shown is unstable internally (nonrigid), and indicate at least two alternative ways to insure internal stability (rigidity) for each truss by the addition of one or more members without introducing redundancy.

Problem 4/109

Problem 4/110

4/110 An antitorque wrench is designed for use by a crewman of a spacecraft where he has no stable platform against which to push as he tightens a bolt. The pin A fits into an adjacent hole in the structure which contains the bolt to be turned. Successive oscillations of the gear and handle unit turns the socket in one direction through the action of a ratchet mechanism. The reaction against the pin A provides the "antitorque" characteristic of the tool. For a gripping force of $P = 150$ N determine the torque M transmitted to the bolt and the external reaction R against the pin A normal to the line AB. (One side of the tool is used for tightening and the opposite side for loosening a bolt.)

Ans. $M = 7.88$ N·m, $R = 137.0$ N

Problem 4/111

4/111 In the special position shown for the log hoist, booms AF and EG are at right angles to one another and AF is perpendicular to AB. If the hoist is handling a log weighing 4800 lb, compute the force supported by the pins at A and D in this one position due to the weight of the log.

Ans. $A = 34{,}000$ lb, $D = 17{,}100$ lb

Dimensions in
Millimeters

Problem 4/112

4/112 The loader has a capacity of 4 m³ and is handling shale having a density of 2.6 Mg/m³. For the particular position shown, where the arm EB is horizontal, find the compressive force in the hydraulic piston rod JL and the total shear force supported by the pin at A. The machine is symmetrical about a central vertical plane in the fore-and-aft direction and has two sets of the linkages shown.

Ans. $L = 52.0$ kN, $A = 247$ kN

4/113 The A-frame hoist supports the 100-kg load with a force P on the crank handle. Calculate the x- and y-components of all forces acting on each of the three members of the frame for the vertical position of the crank handle.

Problem 4/113

4/114 Compute the magnitude of the force acting on the member BD at D for the space frame loaded by the two forces shown. Each of the connections may be treated as a ball-and-socket joint.

Ans. $D = 4.90$ kN

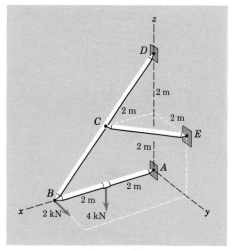

Problem 4/114

DISTRIBUTED FORCES

5

5/1 INTRODUCTION. In the previous chapters we have treated all forces as concentrated along their lines of action and at their points of application. This treatment has provided a reasonable model for the forces with which we have dealt. Actually, however, "concentrated" forces do not exist in the exact sense, since every external force applied mechanically to a body is distributed over a finite contact area however small. The force exerted by the pavement on an automobile tire, for instance, is applied to the tire over its entire area of contact, Fig. $5/1a$, which may be appreciable if the tire is soft. When the dimension b of the contact area is negligible compared with the other pertinent dimensions, such as the distance between wheels, then replacement of the actual distributed forces of contact by their resultant R considered a concentrated force raises no question when we are analyzing the forces acting on the car as a whole. Even the force of contact between a hardened steel ball and its race in a loaded ball bearing, Fig. $5/1b$, will be applied over a finite contact area the dimensions of which, of course, are extremely small. The forces applied to a two-force member of a truss, Fig. $5/1c$, are applied over an actual area of contact of the pin against the hole and internally across the cut section in some such manner as indicated. In these and other similar examples we have no hesitation in treating the forces as concentrated when analyzing their external effects on bodies as a whole.

If, on the other hand, we are interested in finding the distribution of *internal* forces in the material of the body in the neighborhood of the contact location where the internal stresses and strains may be appreciable, then we no longer treat the load as concentrated but would be obliged to take the actual distribution into account. This type of problem requires a knowledge of the properties of the material and belongs in more advanced treatments of the mechanics of materials and the theories of elasticity and plasticity.

When forces are applied over a region whose dimensions are not negligible compared with other pertinent dimensions, then we must account for the actual manner in which the force is distributed by summing up the effects of the distributed force over the entire region. We carry out this process by using the procedures of mathematical

Figure 5/1

(a)

(b)

(c)

Figure 5/2

integration. For this purpose we need to know the intensity of the force at any location. There are three categories into which such problems fall.

Line distribution. When a force is distributed along a line, as in the continuous vertical load supported by a suspended cable, Fig. 5/2*a*, the intensity w of the loading is expressed as force per unit length of line, newtons per meter (N/m), or pounds per foot (lb/ft).

Area distribution. When a force is distributed over an area, as with the hydraulic pressure of water against the inner face of a section of dam, Fig. 5/2*b*, the intensity is expressed as force per unit area. This intensity is known as *pressure* for the action of fluid forces and *stress* for the internal distribution of forces in solids. The basic unit for pressure or stress in SI is the newton per square meter (N/m²), which is also called the *pascal* (Pa). This unit, however, is too small for most applications (6895 Pa = 1 lb/in.²), and the kilopascal (kPa), which equals 10^3 Pa, is more commonly used for fluid pressure and the megapascal, which equals 10^6 Pa, is used for stress. In the U.S. customary system of units both fluid pressure and mechanical stress are commonly expressed in pounds per square inch (lb/in.²).

Volume distribution. A force which is distributed over the volume of a body is known as a *body force.* The most common body force is the force of gravitational attraction which acts on all elements of mass of a body. The determination of the forces on the supports of the heavy cantilevered structure in Fig. 5/2*c*, for example, would require accounting for the distribution of gravitational force throughout the structure. The intensity of gravitational force is the *specific weight* ρg where ρ is the density (mass per unit volume) and g is the acceleration due to gravity. The units for ρg are (kg/m³)(m/s²) = N/m³ in SI and lb/ft³ or lb/in.³ in the U.S. customary system.

The body force due to the earth's gravitational attraction (weight) is by far the most commonly encountered distributed force. Section A of the chapter deals with the determination of the point in a body through which the resultant gravitational force acts. Section B of the chapter treats the important special problems of the distributed forces which act on and in beams and flexible cables and distributed forces which fluids exert on exposed surfaces.

SECTION A. CENTERS OF MASS AND CENTROIDS

5/2 CENTER OF MASS. Consider a three-dimensional body of any size, shape, and mass m. If we suspend the body, as shown in Fig. 5/3, from any point such as A, the body will be in equilibrium under the action of the tension in the cord and the resultant W of the gravitational forces acting on all particles of the body. This resultant is clearly collinear with the cord, and it will be assumed that we mark its position by drilling a hypothetical hole of negligible size along its line of action. We repeat the experiment by suspending the body from other points such as B and C, and in each instance we mark the line of action of the resultant force. For all practical purposes these lines of action will be concurrent at a point G, which is known as the *center of gravity* of the body. An exact analysis, however, would take into account the fact that the directions of the gravity forces for the various particles of the body differ slightly because they converge toward the center of attraction of the earth. Also, since the particles are at different distances from the earth, the intensity of the earth's force field is not exactly constant over the body. These considerations lead to the conclusion that the lines of action of the gravity force resultants in the experiments just described will not quite be concurrent, and therefore no unique center of gravity exists in the exact sense. This condition is of no practical importance as long as we deal with bodies whose dimensions are small compared with those of the earth. We therefore assume a uniform and parallel field of force due to the earth's gravitational attraction, and this condition results in the concept of a unique center of gravity.

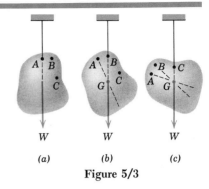

Figure 5/3

To determine mathematically the location of the center of gravity of any body, Fig. 5/4a, we apply the *principle of moments* (Varignon's theorem) to the parallel system of gravitational forces to locate its resultant. The moment of the resultant gravitational force W about any axis equals the sum of the moments about the same axis of the gravitational forces dW acting on all particles considered as infinitesimal elements of the body. The resultant of the gravitational forces acting on all elements is the weight of the body and is given by the sum $W = \int dW$. If we apply the moment principle about the y-axis, for example, the moment about this axis of the elemental weight is $x\,dW$, and the sum of these moments for all elements of the body is $\int x\,dW$. This sum of moments must equal $W\bar{x}$, the moment of the sum. With the substitution of $W = mg$ and $dW = g\,dm$, the moment expressions for all three axes become

$$\bar{x} = \frac{\int x\,dm}{m} \qquad \bar{y} = \frac{\int y\,dm}{m} \qquad \bar{z} = \frac{\int z\,dm}{m} \qquad (5/1)$$

The numerator of each expression represents the *sum of the moments*, and the product of m and the corresponding coordinate of G represents the *moment of the sum*. (The third equation results from reorienting the body and axes together so that $\bar{z}$ is horizontal.)

Equations 5/1 may be expressed in vector form with the aid of Fig. 5/4b, where the elemental mass and the position of G are located by their respective position vectors $\mathbf{r} = \mathbf{i}x + \mathbf{j}y + \mathbf{k}z$ and $\bar{\mathbf{r}} = \mathbf{i}\bar{x} + \mathbf{j}\bar{y} + \mathbf{k}\bar{z}$. Thus Eqs. 5/1 are the components of the single vector equation

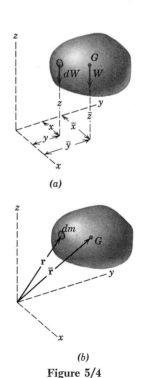

Figure 5/4

$$\bar{\mathbf{r}} = \frac{\int \mathbf{r}\, dm}{m} \qquad (5/2)$$

The density ρ of a body is its mass per unit volume. Thus the mass of a differential element of volume dV becomes $dm = \rho\, dV$. In the event that ρ is not constant throughout the body but can be expressed as a function of the coordinates of the body, it will be necessary to account for this variation in the calculation of both the numerators and denominators of Eqs. 5/1. We may then write these expressions as

$$\bar{x} = \frac{\int x\rho\, dV}{\int \rho\, dV} \qquad \bar{y} = \frac{\int y\rho\, dV}{\int \rho\, dV} \qquad \bar{z} = \frac{\int z\rho\, dV}{\int \rho\, dV} \qquad (5/3)$$

Equations 5/1, 5/2, and 5/3 define the position of the *center of mass*, which is clearly the same point as the center of gravity as long as the gravity field is treated as uniform and parallel. It is meaningless for us to speak of the center of gravity of a body that is removed from the earth's gravitational field, since no gravitational forces would act on the body. It would, however, still possess its unique center of mass. For the most part we will make reference henceforth to the center of mass rather than to the center of gravity. Also, the center of mass has a special significance in calculating the dynamic response of a body to unbalanced forces. This class of problems is discussed at length in Vol. 2 *Dynamics*.

In most problems the calculation of the position of the center of mass may be simplified by an intelligent choice of reference axes. In general the axes should be placed so as to simplify the equations of the boundaries as much as possible. Thus polar coordinates will be useful for bodies having circular boundaries. Another important clue may be taken from considerations of symmetry. Whenever there exists a line or plane of symmetry, a coordinate axis or plane should be chosen to coincide with this line or plane. The center of mass will always lie on such a line or plane, since the moments due to symmetrically located elements will always cancel, and the body can be considered composed of pairs of these elements. Thus the center of mass G of the homogeneous right-circular cone of Fig. 5/5a will lie somewhere on its central axis, which is a line of symmetry. The center of mass of the half right-circular cone lies on its plane of symmetry, Fig. 5/5b. The center of mass of the half ring in Fig. 5/5c lies in both of its planes of symmetry and therefore is situated on line AB. The location of G is always facilitated by the observation of symmetry when it exists.

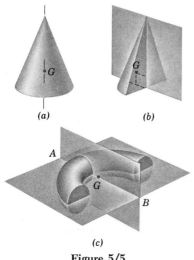

(a)   (b)

(c)

Figure 5/5

5/3 CENTROIDS OF LINES, AREAS, AND VOLUMES. Whenever the density ρ of a body is uniform throughout, it will be a constant factor in both the numerators and denominators of Eqs. 5/3 and will therefore cancel. The expressions that remain define a purely geometrical property of the body, since any reference to its physical properties has disappeared. The term *centroid* is used when the calculation concerns a geometrical shape only. When speaking of an actual physical body, we use the term *center of mass*. If the density is uniform throughout the body, the positions of the centroid and center of mass are identical, whereas if the density varies, these two points will, in general, not coincide.

The calculations of centroids fall within three distinct categories depending on whether the shape of the body involved can be modeled as a line, an area, or a volume.

(*a*) *Lines.* For a slender rod or wire of length L, cross-sectional area A, and density ρ, Fig. 5/6, the body approximates a line segment, and $dm = \rho A \, dL$. If ρ and A are constant over the length of the rod, the coordinates of the center of mass also become the coordinates of the centroid C of the line segment which, from Eqs. 5/1, may be written

$$\bar{x} = \frac{\int x \, dL}{L} \qquad \bar{y} = \frac{\int y \, dL}{L} \qquad \bar{z} = \frac{\int z \, dL}{L} \qquad (5/4)$$

It should be noted that, in general, the centroid C will not lie on the line. If the rod lies in a single plane, such as the *x-y* plane, only two coordinates will require calculation.

(*b*) *Areas.* When a body of density ρ has a small but constant thickness t, it can be modeled as a surface area A, Fig. 5/7. The mass of an element becomes $dm = \rho t \, dA$. Again, if ρ and t are constant over the entire area, the coordinates of the center of mass of the body also become the coordinates of the centroid C of the surface area, and from Eqs. 5/1 may be written

$$\bar{x} = \frac{\int x \, dA}{A} \qquad \bar{y} = \frac{\int y \, dA}{A} \qquad \bar{z} = \frac{\int z \, dA}{A} \qquad (5/5)$$

The numerators in Eqs. 5/5 are known as the *first moments of area.*° If the surface area is curved, as illustrated in Fig. 5/7 with the shell segment, all three coordinates will be involved. Here again the centroid C for the curved surface will in general not lie on the surface. If the area is a flat surface in, say, the *x-y* plane, only the coordinates in that plane will be unknown.

° Second moments of areas (moments of first moments) appear later in our discussion of area moments of inertia in Appendix A.

Figure 5/6

Figure 5/7

(*c*) *Volumes.* For a general body of volume V and density ρ, the element has a mass $dm = \rho\, dV$. The density ρ cancels if it is constant over the entire volume, and the coordinates of the center of mass also become the coordinates of the centroid C of the body. From Eqs. 5/3 or 5/1 they become

$$\bar{x} = \frac{\int x\, dV}{V} \qquad \bar{y} = \frac{\int y\, dV}{V} \qquad \bar{z} = \frac{\int z\, dV}{V} \qquad (5/6)$$

(*d*) *Choice of Element for Integration.* As is often the case the principal difficulty in a theory lies not so much in its concepts but in the procedures for applying it. With mass centers and centroids the concept of the moment principle is simple enough; the difficulties reside primarily with the choice of the differential element and with setting up the integrals. In particular there are five guidelines to be specially observed.

(*1*) *Order of element.* Whenever possible a first-order differential element should be selected in preference to a higher-order element so that only one integration will be required to cover the entire figure. Thus in Fig. 5/8*a* a first-order horizontal strip of area $dA = l\, dy$ will require only one integration with respect to y in order to cover the entire figure. The second-order element $dx\, dy$ will require two integrations, first with respect to x and second with respect to y, to cover the figure. As a further example, for the solid cone in Fig. 5/8*b* we choose a first-order element in the form of a circular slice of volume $dV = \pi r^2\, dy$, which requires only one integration, in preference to choosing a third-order element $dV = dx\, dy\, dz$, which would require three awkward integrations.

(*2*) *Continuity.* Whenever possible we choose an element which can be integrated in one continuous operation to cover the figure.

(a)

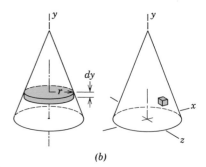

(b)

Figure 5/8

Thus the horizontal strip in Fig. 5/8*a* would be preferable to the vertical strip in Fig. 5/9 which, if used, would require two separate integrals because of the discontinuity in the expression for the height of the strip at $x = x_1$.

(3) *Discarding higher-order terms.* Higher-order terms may always be dropped compared with lower-order terms (see Art. 1/7). Thus the vertical strip of area under the curve in Fig. 5/10 is given by the first-order term $dA = y\,dx$, and the second-order triangular area $\frac{1}{2}dx\,dy$ is discarded. In the limit, of course, there is no error.

(4) *Choice of coordinates.* As a general rule we choose the coordinate system which best matches the boundaries of the figure. Thus the boundaries of the area in Fig. 5/11*a* are most easily described in rectangular coordinates, whereas the boundaries of the circular sector of Fig. 5/11*b* are best suited to polar coordinates.

(5) *Centroidal coordinate of element.* When a first- or second-order differential element is adopted, it is essential to use the *coordinate to the centroid of the element* for the moment arm in setting up the moment of the differential element. Thus, for the horizontal strip of area in Fig. 5/12*a* the moment of dA about the *y*-axis is $x_c\,dA$ where x_c is the *x*-coordinate to the centroid *C* of the element. Note that x_c is *not* the *x* which describes the boundaries of the area. In the *y*-direction for this element the moment arm y_c to the centroid of the element is the same, in the limit, as the *y*-coordinates of the two boundaries.

As a second example, consider the solid half-cone of Fig. 5/12*b* with the semicircular slice of differential thickness as the element of volume. The moment arm for the element in the *x*-direction is the distance x_c to the centroid of the face of the element and not the

Figure 5/9

Figure 5/10

Figure 5/11

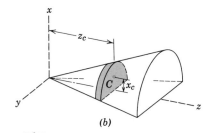

Figure 5/12

x-coordinate to the boundary of the element. On the other hand in the *z*-direction the moment arm z_c to the centroid of the element is the same as the *z*-coordinate of the element.

With these examples in mind we rewrite Eqs. 5/5 and 5/6 in the form

$$\bar{x} = \frac{\int x_c \, dA}{A}$$

$$\bar{y} = \frac{\int y_c \, dA}{A} \qquad (5/5a)$$

$$\bar{z} = \frac{\int z_c \, dA}{A}$$

and

$$\bar{x} = \frac{\int x_c \, dV}{V}$$

$$\bar{y} = \frac{\int y_c \, dV}{V} \qquad (5/6a)$$

$$\bar{z} = \frac{\int z_c \, dV}{V}$$

The subscript c serves as a reminder that the moment arms appearing in the numerators of the integral expressions for moments are *always* the coordinates to the *centroids* of the particular elements chosen.

At this point it would be well for the student to make certain that he understands clearly the principle of moments (Varignon's theorem), which was introduced in Art. 2/4. It is essential that the physical meaning of this principle be recognized as it is applied to the system of parallel weight forces depicted in Fig. 5/4a. With the equivalence clearly in mind between the moment of the resultant weight W and the sum (integral) of the moments of the elemental weights dW, then a mistake in setting up the necessary mathematics is far less likely to occur. Recognition of the principle of moments will provide assurance that the correct expression will be used for the moment arm x_c, y_c, or z_c to the centroid of the particular differential element chosen. Also, with the physical picture of the principle of moments in mind, Eqs. 5/4, 5/5, and 5/6, which are geometric relationships, will be recognized as descriptive also of homogeneous physical bodies where the density ρ has cancelled.

A summary of the centroidal coordinates for some of the commonly used shapes is given in Tables C3 and C4, Appendix C.

Sample Problem 5/1

Centroid of a circular arc. Locate the centroid of a circular arc as shown in the figure.

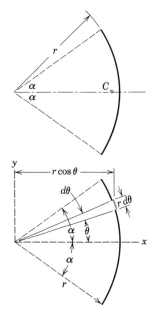

Solution. Choosing the axis of symmetry as the x-axis makes $\bar{y} = 0$. A differential element of arc has the length $dL = r\,d\theta$ expressed in polar coordinates, and the x-coordinate of the element is $r\cos\theta$. Applying the first of Eqs. 5/4 and substituting $L = 2\alpha r$ give

$$[L\bar{x} = \int x\,dL] \qquad (2\alpha r)\bar{x} = \int_{-\alpha}^{\alpha}(r\cos\theta)r\,d\theta$$

$$2\alpha r\bar{x} = 2r^2\sin\alpha$$

$$\bar{x} = \frac{r\sin\alpha}{\alpha} \qquad\qquad Ans.$$

For a semicircular arc $2\alpha = \pi$, which gives $\bar{x} = 2r/\pi$. By symmetry we see immediately that this result also applies to the quarter-circular arc when the measurement is made as shown.

① It should be perfectly evident that polar coordinates are preferable to rectangular coordinates to express the length of a circular arc.

Sample Problem 5/2

Centroid of a triangular area. Determine the distance $\bar{h}$ from the base of a triangle of altitude h to the centroid of its area.

Solution. The x-axis is taken to coincide with the base. A differential strip of area $dA = x\,dy$ is chosen. By similar triangles $x/(h-y) = b/h$. Applying the second of Eqs. 5/5 gives

$$[A\bar{y} = \int y_c\,dA] \qquad \frac{bh}{2}\bar{y} = \int_0^h y\frac{b(h-y)}{h}\,dy = \frac{bh^2}{6}$$

$$\text{and} \qquad \bar{y} = \frac{h}{3} \qquad\qquad Ans.$$

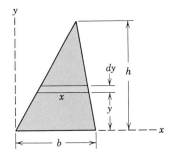

This same result holds with respect to either of the other two sides of the triangle considered a new base with corresponding new altitude. Thus the centroid lies at the intersection of the medians, since the distance of this point from any side is one third the altitude of the triangle with that side considered the base.

① We save one integration here by using the first-order element of area. Recognize that dA must be expressed in terms of the integration variable y; hence, $x = f(y)$ is required.

Sample Problem 5/3

Centroid of the area of a circular sector. Locate the centroid of the area of a circular sector with respect to its vertex.

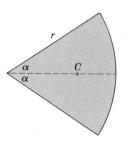

Solution I. The x-axis is chosen as the axis of symmetry, and $\bar{y}$ is therefore automatically zero. We may cover the area by moving an element in the form of a partial circular ring, as shown in the figure, from the center to the outer periphery. The radius of the ring is r_0 and its thickness is dr_0, so that its area is $dA = 2r_0\alpha\,dr_0$.

The x-coordinate to the centroid of the element from Sample Problem 5/1 is $x_c = r_0 \sin\alpha/\alpha$ where r_0 replaces r in the formula. Thus the first of Eqs. 5/5a gives

① $\quad [A\bar{x} = \int x_c\,dA] \qquad \dfrac{2\alpha}{2\pi}(\pi r^2)\bar{x} = \int_0^r \left(\dfrac{r_0 \sin\alpha}{\alpha}\right)(2r_0\alpha\,dr_0)$

② $\qquad\qquad\qquad\qquad r^2\alpha\bar{x} = \tfrac{2}{3}r^3 \sin\alpha$

$$\bar{x} = \frac{2}{3}\frac{r \sin\alpha}{\alpha} \qquad\qquad \textit{Ans.}$$

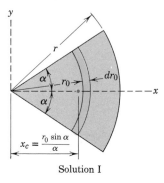

$x_c = \dfrac{r_0 \sin\alpha}{\alpha}$

Solution I

① Note carefully that we must distinguish between the variable r_0 and the constant r.

② Be careful not to use r_0 as the centroidal coordinate for the element.

Solution II. The area may also be covered by swinging a triangle of differential area about the vertex and through the total angle of the sector. This triangle, shown in the illustration, has an area $dA = (r/2)(r\,d\theta)$, where higher-order terms are neglected. From Sample Problem 5/2 the centroid of the triangular element of area is 2/3 of its altitude from its vertex, so that the x-coordinate to the centroid of the element is $x_c = \tfrac{2}{3}r \cos\theta$. Applying the first of Eqs. 5/5a gives

$[A\bar{x} = \int x_c\,dA] \qquad (r^2\alpha)\bar{x} = \int_{-\alpha}^{\alpha} (\tfrac{2}{3}r \cos\theta)(\tfrac{1}{2}r^2\,d\theta)$

$\qquad\qquad\qquad\qquad r^2\alpha\bar{x} = \tfrac{2}{3}r^3 \sin\alpha$

and as before $\qquad \bar{x} = \dfrac{2}{3}\dfrac{r \sin\alpha}{\alpha} \qquad\qquad \textit{Ans.}$

For a semicircular area $2\alpha = \pi$, which gives $\bar{x} = 4r/3\pi$. By symmetry we see immediately that this result also applies to the quarter-circular area where the measurement is made as shown.

It should be noted that, if we had chosen a second-order element $r_0\,dr_0\,d\theta$, one integration with respect to θ would yield the ring with which *Solution I* began. On the other hand integration with respect to r_0 initially would give the triangular element with which *Solution II* began.

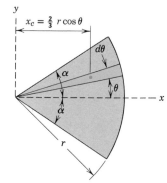

$x_c = \tfrac{2}{3}r \cos\theta$

Solution II

$4r/3\pi$

Sample Problem 5/4

Locate the centroid of the area under the curve $x = ky^3$ from $x = 0$ to $x = a$.

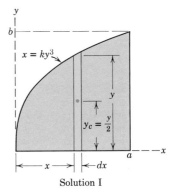

_____

Solution I. A vertical element of area $dA = y\,dx$ is chosen as shown in the figure. The x-coordinate of the centroid is found from the first of Eqs. 5/5a. Thus

$$[A\bar{x} = \int x_c\,dA] \qquad \bar{x}\int_0^a y\,dx = \int_0^a xy\,dx$$

Substituting $y = (x/k)^{1/3}$ and $k = a/b^3$ and integrating give

$$\frac{3ab}{4}\bar{x} = \frac{3a^2b}{7}, \qquad \bar{x} = \tfrac{4}{7}a \qquad\qquad Ans.$$

In solving for $\bar{y}$ from the second of Eqs. 5/5a the coordinate to the centroid of the rectangular element is $y_c = y/2$ where y is the height of the strip governed by the equation of the curve $x = ky^3$. Thus the moment principle becomes

$$[A\bar{y} = \int y_c\,dA] \qquad \frac{3ab}{4}\bar{y} = \int_0^a \left(\frac{y}{2}\right)y\,dx$$

Substituting $y = b(x/a)^{1/3}$ and integrating give

$$\frac{3ab}{4}\bar{y} = \frac{3ab^2}{10}, \qquad \bar{y} = \tfrac{2}{5}b \qquad\qquad Ans.$$

Solution I

① Note that $x_c = x$ for the vertical element.

Solution II. The horizontal element of area shown in the lower figure may be employed in place of the vertical element. The x-coordinate to the centroid of the rectangular element is the average of the coordinates of the ends or $x_c = (a + x)/2$. Hence

$$[A\bar{x} = \int x_c\,dA] \qquad \bar{x}\int_0^b (a - x)\,dy = \int_0^b \left(\frac{a + x}{2}\right)(a - x)\,dy$$

The value of $\bar{y}$ is found from

$$[A\bar{y} = \int y_c\,dA] \qquad \bar{y}\int_0^b (a - x)\,dy = \int_0^b y(a - x)\,dy$$

where $y_c = y$ for the horizontal strip. The evaluation of these integrals will check the previous results for $\bar{x}$ and $\bar{y}$.

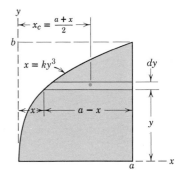

Solution II

Sample Problem 5/5

Hemispherical volume. Locate the centroid of the volume of a hemisphere of radius r with respect to its base.

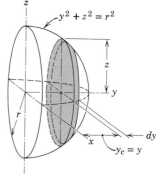

Solution I

Solution I. With the axes chosen as shown in th figure $\bar{x} = \bar{z} = 0$ by symmetry. The most convenient element is a circular slice of thickness dy parallel to the x-z plane. Since the hemisphere intersects the y-z plane in the circle $y^2 + z^2 = r^2$, the radius of the circular slice is $z = +\sqrt{r^2 - y^2}$. The volume of the elemental slice becomes

$$dV = \pi(r^2 - y^2)\,dy$$

The second of Eqs. 5/6a requires

① $[V\bar{y} = \int y_c\,dV]$ $\bar{y}\int_0^r \pi(r^2 - y^2)\,dy = \int_0^r y\pi(r^2 - y^2)\,dy$

where $y_c = y$. Integrating gives

$$\tfrac{2}{3}\pi r^3\bar{y} = \tfrac{1}{4}\pi r^4, \qquad \bar{y} = \tfrac{3}{8}r \qquad\qquad Ans.$$

Solution II

Solution II. Alternatively we may use for our differential element a cylindrical shell of length y, radius z, and thickness dz, as shown in the lower figure. By expanding the radius of the shell from zero to r, we cover the entire volume. By symmetry the centroid of the elemental shell lies at its center, so that $y_c = y/2$. The volume of the element is $dV = (2\pi z\,dz)(y)$. Expressing y in terms of z from the equation of the circle gives $y = +\sqrt{r^2 - z^2}$. Using the value of $\tfrac{2}{3}\pi r^3$ computed in Solution I for the volume of the hemisphere and substituting in the second of Eqs. 5/6a give us

$[V\bar{y} = \int y_c\,dV]$ $(\tfrac{2}{3}\pi r^3)\bar{y} = \int_0^r \dfrac{\sqrt{r^2 - z^2}}{2}(2\pi z\sqrt{r^2 - z^2})\,dz$

$$= \int_0^r \pi(r^2 z - z^3)\,dz = \dfrac{\pi r^4}{4}$$

$$\bar{y} = \tfrac{3}{8}r \qquad\qquad Ans.$$

Solutions I and II are of comparable use since each involves an element of simple shape and requires integration with respect to one variable only.

Solution III. As an alternative we could use the angle θ as our variable with limits of 0 and $\pi/2$. The radius of either element would become $r\sin\theta$, whereas the thickness of the slice would be $dy = (r\,d\theta)\sin\theta$ and that of the shell would be $dz = (r\,d\theta)\cos\theta$. The length of the shell would be $y = r\cos\theta$.

① Can you identify the higher-order element of volume which is omitted from the expression for dV?

Solution III

PROBLEMS

5/1 Determine the coordinates of the centroid of the shaded area.   *Ans.* $\bar{x} = 3b/10, \ \bar{y} = 3b/4$

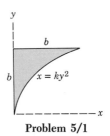

Problem 5/1

5/2 Locate the centroid of the shaded area shown.

Ans. $\bar{x} = 2.09, \ \bar{y} = 1.43$

Problem 5/2

5/3 Determine the coordinates of the centroid of the shaded area.   *Ans.* $\bar{x} = 3a/8, \ \bar{y} = 2b/5$

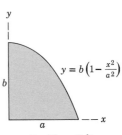

Problem 5/3

5/4 Determine the y-coordinate of the centroid of the area under the sine curve shown.

Problem 5/4

5/5 Determine the y-coordinate of the centroid of the area by direct integration.   *Ans.* $\bar{y} = \dfrac{14R}{9\pi}$

Problem 5/5

Problem 5/6

Problem 5/7

Problem 5/8

Problem 5/9

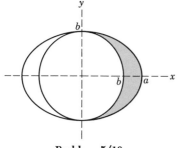

Problem 5/10

5/6 Calculate the x-coordinate of the centroid of the shaded area.

5/7 Calculate the coordinates of the centroid of the area shown.   *Ans.* $\bar{x} = 2a/5, \bar{y} = b/2$

5/8 Calculate the coordinates of the centroid of the segment of the circular area.

5/9 Locate the centroid of the shaded area between the two curves.   *Ans.* $\bar{x} = \frac{12}{25}, \bar{y} = \frac{3}{7}$

5/10 Locate the centroid of the shaded area between the ellipse and the circle.

5/11 Locate the centroid of the area shown in the figure by direct integration. (*Caution:* Observe carefully the proper sign of the radical involved.)

$$Ans.\ \bar{x} = \frac{10 - 3\pi}{4 - \pi}\frac{a}{3}$$

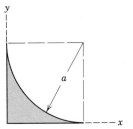

Problem 5/11

5/12 Determine the *x*-coordinate of the centroid of the shaded area shown. (Observe the caution of Prob. 5/11.) $Ans.\ \bar{x} = 0.339a$

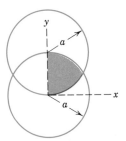

Problem 5/12

5/13 Specify the coordinates of the center of mass of the circular shell section by direct reference to the results of Sample Problem 5/1.

Problem 5/13

5/14 Use the results of Sample Problem 5/3 to compute the coordinates of the mass center of the portion of the solid homogeneous cylinder shown.

$$Ans.\ \bar{x} = \bar{y} = -8/(3\pi)\ in.,\ \bar{z} = 5\ in.$$

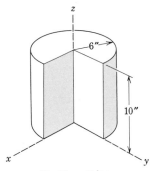

Problem 5/14

5/15 Find the distance $\bar{z}$ from the vertex of the right circular cone to the centroid of its volume.

Problem 5/15

Problem 5/16

Problem 5/17

Problem 5/18

Problem 5/20

5/16 Calculate the distance $\bar{h}$ measured from the base to the centroid of the volume of the frustum of the right-circular cone.   *Ans.* $\bar{h} = \frac{11}{56}h$

5/17 Locate the mass center of the homogeneous solid body whose volume is determined by revolving the shaded area through 360° about the z-axis.

5/18 Determine the z-coordinate of the centroid of the volume obtained by revolving the shaded area under the parabola about the z-axis through 180°.   *Ans.* $\bar{z} = 2a/3$

5/19 Determine the x-coordinate of the centroid of the volume described in Prob. 5/18.

5/20 Determine the y-coordinate of the centroid of the volume obtained by revolving the shaded area about the x-axis through 180°.   *Ans.* $\bar{y} = 15a/(14\pi)$

5/21 Use the results of Sample Problem 5/2 and determine by inspection the distance $\bar{h}$ from the centroid
of the lateral area of any cone or pyramid of altitude
h to the base of the figure.

5/22 Determine the distance $\bar{z}$ from the base of any cone
or pyramid of altitude h to the centroid of its volume.   Ans. $\bar{z} = \dfrac{h}{4}$

Problem 5/22

5/23 Determine the z-coordinate of the mass center of
the homogeneous quarter-spherical shell of radius r.
                                      Ans. $\bar{z} = r/2$

Problem 5/23

5/24 Determine the z-coordinate of the center of mass of
the solid obtained by revolving the quarter-circular
area about the z-axis.   Ans. $\bar{z} = \dfrac{11a}{2(4 + 3\pi)}$

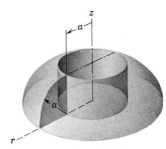

Problem 5/24

5/25 Determine the z-coordinate of the centroid of the
volume generated by revolving the quarter-circular
area about the z-axis through 90°.

Problem 5/25

Problem 5/26

Problem 5/27

Problem 5/28

Problem 5/29

5/26 The homogeneous slender rod has a uniform cross section and is bent into the shape of a parabolic arc with the vertex at the origin. Determine the coordinates of the center of mass of the rod. (*Reminder:* A differential arc length is $dL = \sqrt{(dx)^2 + (dy)^2} = \sqrt{1 + (dy/dx)^2}\, dx$.)

5/27 Locate the centroid of the conical wedge generated by revolving the right triangle of altitude a about its base b through an angle θ.

▶5/28 Determine the position of the center of mass of the homogeneous thin conical shell shown.

$$Ans.\ \bar{x} = 2h/3,\ \bar{z} = 4r/3\pi$$

▶5/29 Locate the center of mass of the homogeneous bell-shaped shell of uniform but negligible thickness.

$$Ans.\ \bar{z} = \frac{a}{\pi - 2}$$

5/30 Determine the position of the centroid of the volume within the bell-shaped shell of Prob. 5/29.

$$\text{Ans. } \bar{z} = \frac{a}{2(10 - 3\pi)}$$

5/31 Determine the x-coordinate of the centroid of the
top half of a bell-shaped shell.

$$\text{Ans. } \bar{x} = \frac{a}{\pi}\left(\frac{3\pi - 8}{\pi - 2}\right)$$

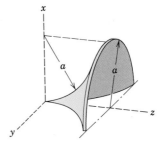

Problem 5/31

▶5/32 Locate the center of mass G of the steel half ring.
(*Hint:* Choose an element of volume in the form of a
cylindrical shell whose intersection with the plane
of the ends is shown.)

$$\text{Ans. } \bar{r} = \frac{a^2 + 4R^2}{2\pi R}$$

Problem 5/32

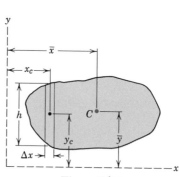

Figure 5/13

5/4 COMPOSITE BODIES AND FIGURES; APPROXIMATIONS.

When a body or figure can be conveniently divided into several parts of simple shape, we may use the principle of Varignon if each part is treated as a finite element of the whole. Thus for such a body, illustrated schematically in Fig. 5/13, whose parts have masses m_1, m_2, m_3 and whose separate coordinates of the centers of mass of these parts in, say, the x-direction are $\bar{x}_1, \bar{x}_2, \bar{x}_3$, the moment principle gives

$$(m_1 + m_2 + m_3)\bar{X} = m_1\bar{x}_1 + m_2\bar{x}_2 + m_3\bar{x}$$

where $\bar{X}$ is the x-coordinate of the center of mass of the whole. Similar relations hold for the other two coordinate directions. We generalize, then, for a body of any number of parts and express the sums in condensed form and obtain the mass-center coordinates

$$\bar{X} = \frac{\Sigma m\bar{x}}{\Sigma m} \qquad \bar{Y} = \frac{\Sigma m\bar{y}}{\Sigma m} \qquad \bar{Z} = \frac{\Sigma m\bar{z}}{\Sigma m} \qquad (5/7)$$

Analogous relations hold for composite lines, areas, and volumes, where the m's are replaced by L's, A's, and V's, respectively. It should be pointed out that if a hole or cavity is considered one of the component parts of a composite body or figure, the corresponding mass represented by the cavity or hole is considered a negative quantity.

Frequently in practice the boundaries of an area or volume are not expressible in terms of simple geometrical shapes or in shapes that can be represented mathematically. For such cases we find it necessary to resort to a method of approximation. As an example consider the problem of locating the centroid C of the irregular area shown in Fig. 5/14. The area may be divided into strips of width Δx and variable height h. The area A of each strip, such as the one shown in color, is $h\,\Delta x$ and is multiplied by the coordinates x_c and y_c to its *centroid* to obtain the moments of the element of area. The sum of the moments for all strips divided by the total area of the strips will give

Figure 5/14

the corresponding centroidal component. A systematic tabulation of the results will permit an orderly evaluation of the total area ΣA, the sums $\Sigma A x_c$ and $\Sigma A y_c$, and the centroidal components

$$\bar{x} = \frac{\Sigma A x_c}{\Sigma A} \qquad \bar{y} = \frac{\Sigma A y_c}{\Sigma A}$$

The accuracy of the approximation will be increased by decreasing the widths of the strips used. In all cases the average height of the strip should be estimated in approximating the areas. Although it is usually of advantage to use elements of constant width, it is not necessary to do so. In fact we may use elements of any size and shape which approximate the given area to satisfactory accuracy.

In locating the centroid of an irregular volume the problem may be reduced to one of determining the centroid of an area. Consider the volume shown in Fig. 5/15 where the magnitudes A of the

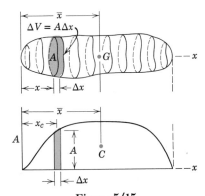

Figure 5/15

cross-sectional areas normal to the x-direction are plotted against x as shown. A vertical strip of area under the curve is $A \, \Delta x$, which equals the corresponding element of volume ΔV. Thus the area under the plotted curve represents the volume of the body, and the x-coordinate to the centroid of the area under the curve is given by

$$\bar{x} = \frac{\Sigma(A \, \Delta x)x_c}{\Sigma A \, \Delta x} \qquad \text{which equals} \qquad \bar{x} = \frac{\Sigma V x_c}{\Sigma V}$$

for the centroid of the actual volume.

Sample Problem 5/6

Locate the center of mass of the bracket-and-shaft combination. The vertical face is made from sheet metal which has a mass of 25 kg/m², the material of the horizontal base has a mass of 40 kg/m², and the steel shaft has a density of 7.83 Mg/m³.

Solution. The composite body may be considered composed of the five elements shown in the lower portion of the illustration. The triangular part will be taken as a negative mass. For the reference axes indicated it is clear by symmetry that the x-coordinate of the center of mass is zero.

The mass m of each part is easily calculated and should need no further explanation. For Part 1 we have from Sample Problem 5/3

$$\bar{z} = \frac{4r}{3\pi} = \frac{4(50)}{3\pi} = 21.2 \text{ mm}$$

For Part 3 from Sample Problem 5/2 we see that the centroid of the triangular mass is $\frac{1}{3}$ of its altitude above its base. Measurement from the coordinate axes becomes

$$\bar{z} = -[150 - 25 - \tfrac{1}{3}(75)] = -100 \text{ mm}$$

The y- and z-coordinates to the mass centers of the remaining parts should be evident by inspection. The terms involved in applying Eqs. 5/7 are best handled in the form of a table as follows:

PART	m kg	$\bar{y}$ mm	$\bar{z}$ mm	$m\bar{y}$ kg·mm	$m\bar{z}$ kg·mm
1	0.098	0	21.2	0	2.08
2	0.562	0	−75.0	0	−42.19
3	−0.094	0	−100.0	0	9.38
4	0.600	50.0	−150.0	30.0	−90.00
5	1.476	75.0	0	110.7	0
Totals	2.642			140.7	−120.73

Equations 5/7 are now applied and the results are

$$\left[\bar{Y} = \frac{\Sigma m\bar{y}}{\Sigma m}\right] \qquad \bar{Y} = \frac{140.7}{2.642} = 53.3 \text{ mm} \qquad\qquad Ans.$$

$$\left[\bar{Z} = \frac{\Sigma m\bar{z}}{\Sigma m}\right] \qquad \bar{Z} = \frac{-120.73}{2.642} = -45.7 \text{ mm} \qquad\qquad Ans.$$

PROBLEMS

5/33 Calculate the *y*-coordinate of the centroid of the triangular area.   *Ans.* $\bar{Y} = 40$ mm

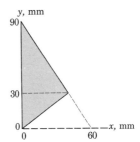

Problem 5/33

5/34 Calculate the *y*-coordinate of the centroid of the figure shown.

Problem 5/34

5/35 Calculate the *x*-coordinate of the centroid of the shaded area shown.   *Ans.* $\bar{X} = 6.54$ mm

Problem 5/35

5/36 Locate the centroid of the shaded area of Prob. 5/8, repeated here, by the method of this article.

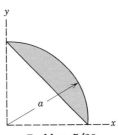

Problem 5/36

5/37 Locate the centroid of the area of Prob. 5/5, repeated here, by the method of this article.

$$Ans.\ \bar{Y} = \frac{14R}{9\pi}$$

Problem 5/37

Problem 5/38

Problem 5/39

Problem 5/40

Problem 5/41

Problem 5/42

5/38 Determine by the method of this article the x-coordinate of the centroid of the shaded area between the ellipse and the circle of Prob. 5/10, shown again here. (Refer to Table C3 in Appendix C for the properties of an elliptical area.)

5/39 Locate the centroid of the shaded area of Prob. 5/11, repeated here, by the method of this article.

$$Ans.\ \bar{X} = \bar{Y} = \frac{10 - 3\pi}{4 - \pi}\frac{a}{3}$$

5/40 Compute the distance $\bar{Y}$ from the x-axis to the centroid of the shaded area.

5/41 Calculate the coordinates of the centroid of the shaded area.   *Ans.* $\bar{X} = \bar{Y} = 4.29$ in.

5/42 Determine the distance $\bar{H}$ from the bottom of the base plate to the centroid of the built-up structural section shown.   *Ans.* $\bar{H} = 39.3$ mm

5/43 A uniform rod is bent into the shape shown and pivoted about O. Find the value of a in terms of the radius r so that the straight section will remain horizontal.

Problem 5/43

5/44 Calculate the coordinates of the center of mass of the slender rod bent into the shape shown.

     Ans. $\bar{X} = 31.1$ mm, $\bar{Y} = 48.9$ mm, $\bar{Z} = 31.1$ mm

Problem 5/44

5/45 By the method of this article determine the height $\bar{H}$ from the base to the mass center of the frustum of the solid cone of Prob. 5/16, repeated here.

Problem 5/45

5/46 The hemispherical shell and its semicircular base are formed from the same piece of sheet metal of small thickness. Use the results of Prob. 5/23 and calculate the coordinates of the mass center of the shell and base combined.   *Ans.* $\bar{X} = 0.475r$, $\bar{Y} = r/3$

Problem 5/46

Dimensions in Millimeters
Problem 5/47

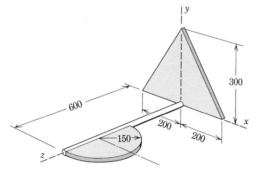

Dimensions in Millimeters
Problem 5/48

Problem 5/49

Problem 5/50

5/47 Calculate the coordinates of the mass center of the metal die casting shown.

5/48 The masses of the three parts of the welded assembly consisting of the triangular plate, uniform rod, and semicircular plate are, respectively, 4.2, 2.2, and 2.5 kg. Calculate the coordinates of the mass center.

Ans. $\bar{X} = 17.9$ mm, $\bar{Y} = 47.2$ mm, $\bar{Z} = 200.6$ mm

5/49 A tubular framework consisting of a semicircular member and a brace is welded to a flat triangular plate as shown. If the tubular material weighs 5 lb per foot of length and the base material weighs 20 lb per square foot of area, calculate the coordinates of the center of gravity of the combined base and frame.

5/50 Determine the position of the center of mass of the cylindrical shell with a closed semicircular end. The shell is made from sheet metal with a mass of 24 kg/m², and the end is made from metal plate with a mass of 36 kg/m².

Ans. $\bar{X} = 348$ mm, $\bar{Y} = -90.5$ mm

5/51 Determine the coordinates of the center of mass of the bracket, which is made from a plate of uniform thickness.   *Ans.* $\bar{X} = -8.3$ mm
$$\bar{Y} = -31.4 \text{ mm}$$
$$\bar{Z} = 10.3 \text{ mm}$$

Problem 5/51

5/52 As an example of the accuracy involved in graphical approximations calculate the percentage error e in determining the x-coordinate of the centroid of the triangular area by using the five approximating rectangles of width $a/5$ in place of the triangle.
    Ans. $e = 1.00$ percent low

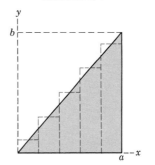

Problem 5/52

5/53 A sheet-metal pattern has the shape shown. Estimate the location of the centroid of the area visually and note its coordinates. Then check your estimate by a calculation using the superimposed grid.

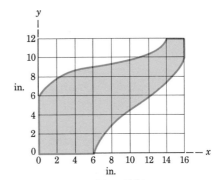

Problem 5/53

5/54 The circular disk rotates about an axis through its center O and has three holes of diameter d positioned as shown. A fourth hole is to be drilled in the disk at the same radius r so that the disk will be in balance (mass center at O). Determine the required diameter D of the new hole and its angular position.   *Ans.* $D = 1.227d$, $\theta = 84.9°$

Problem 5/54

Problem 5/55

Problem 5/56

Problem 5/57

Problem 5/58

5/55 A rectangular piece is removed from the square metal plate of side a. Determine the value of h which will result in the mass center of the remaining plate being as far to the left as possible.

Ans. $h = 0.586a$

5/56 Locate the center of mass of the thin plate formed into the shape shown.

Ans. $\bar{X} = 0$, $\bar{Y} = -14.5$ mm, $\bar{Z} = 73.0$ mm

▶**5/57** A cylindrical container with an extended rectangular back and semicircular ends is all fabricated from the same sheet-metal stock. Calculate the angle α made by the back with the vertical when the container rests in an equilibrium position on a horizontal surface.

Ans. $\alpha = 39.6°$

▶**5/58** A homogeneous charge of explosive is to be formed as a circular cylinder of length L and diameter d_2 with an axial hole of diameter d_1 and depth h as shown in section. Determine the value of h which will place the center of mass of the final charge a maximum distance from the open end.

Ans. $h = \left(\dfrac{d_2}{d_1}\right)^2 L\left[1 - \sqrt{1 - \left(\dfrac{d_1}{d_2}\right)^2}\right]$

5/5 THEOREMS OF PAPPUS. A very simple method exists for calculating the surface area generated by revolving a plane curve about a nonintersecting axis in the plane of the curve. In Fig. 5/16 the line segment of length L in the x-y plane generates a surface when revolved about the x-axis. An element of this surface is the ring generated by dL. The area of this ring is its circumference times its slant height or $dA = 2\pi y\, dL$ and the total area is then

$$A = 2\pi \int y\, dL$$

But since $\bar{y}L = \int y\, dL$, the area becomes

$$\boxed{A = 2\pi\bar{y}L} \qquad (5/8)$$

Figure 5/16

where $\bar{y}$ is the y-coordinate of the centroid C for the line of length L. Thus the generated area is the same as the lateral area of a right circular cylinder of length L and radius $\bar{y}$.

In the case of a volume generated by revolving an area about a nonintersecting line in its plane an equally simple relation exists for finding the volume. An element of the volume generated by revolving the area A about the x-axis, Fig. 5/17, is the elemental ring of cross section dA and radius y. The volume of the element is its circumference times dA or $dV = 2\pi y\, dA$, and the total volume is

$$V = 2\pi \int y\, dA$$

But since $\bar{y}A = \int y\, dA$, the volume becomes

$$\boxed{V = 2\pi\bar{y}A} \qquad (5/9)$$

Figure 5/17

where $\bar{y}$ is the y-coordinate of the centroid C of the revolved area A. Thus the generated volume is obtained by multiplying the generating area by the circumference of the circular path described by its centroid.

The two theorems of Pappus, expressed by Eqs. 5/8 and 5/9, are useful not only in determining areas and volumes of generation, but they are also employed to find the centroids of plane curves and plane areas when the corresponding areas and volumes due to revolution of these figures about a nonintersecting axis are known. Dividing the area or volume by 2π times the corresponding line segment length or plane area will give the distance from the centroid to the axis.

In the event that a line or an area is revolved through an angle θ less than 2π, the generated surface or volume may be found by replacing 2π by θ in Eqs. 5/8 and 5/9. Thus the more general relations are

$$A = \theta\bar{y}L \qquad \text{and} \qquad V = \theta\bar{y}A$$

where θ is expressed in radians.

° Attributed to Pappus of Alexandria, a Greek geometer who lived in the third century A.D. The theorems often bear the name of Guldinus (Paul Guldin, 1577–1643), who claimed original authorship, although the works of Pappus were apparently known to him.

PROBLEMS

5/59 Determine the volume V and lateral area A of a right circular cone of base radius r and altitude h.

5/60 From the known surface area $A = 4\pi r^2$ of a sphere of radius r, determine the radial distance $\bar{r}$ to the centroid of the semicircular arc used to generate the surface.

5/61 From the known volume $V = \frac{4}{3}\pi r^3$ of a sphere of radius r, determine the radial distance $\bar{r}$ to the centroid of the semicircular area used to generate the sphere.

5/62 Use the notation of the half torus of Prob. 5/32, shown again here, and determine the volume V and the surface area A of a complete torus.

Problem 5/62

5/63 A shell structure has the form of a surface obtained by revolving the circular arc through 360° about the z-axis. Determine the surface area of one side of the complete shell.

Problem 5/63

5/64 Compute the volume V of the solid generated by revolving the right triangle about the z-axis through 180°.   *Ans.* $V = 3619$ mm³

Problem 5/64

5/65 The shaded triangle of base b and altitude h is revolved about its base through an angle θ to generate a portion of a complete solid of revolution. Write the expression for the volume V of the solid generated.

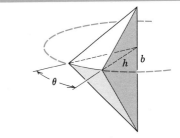

Problem 5/65

5/66 The two circular arcs AB and BC are revolved around the vertical axis to obtain the surface of revolution shown. Compute the area of this surface.

Ans. $A = 157.9$ in.^{2}

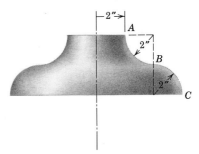

Problem 5/66

5/67 Determine the volume V generated by revolving the quarter-circular area about the z-axis through an angle of 90°.

Problem 5/67

5/68 Determine the volume V generated by revolving the quarter-circular area about the z-axis through 90°.

$$Ans.\ V = \frac{\pi a^3}{12}(3\pi - 2)$$

Problem 5/68

5/69 Compute the volume V and total surface area A of the complete ring whose square cross section is shown.

Problem 5/69

Problem 5/70

5/70 Determine the surface area of one side of the bell-shaped shell of Prob. 5/29, shown again here, using the theorem of Pappus.

5/71 Determine the volume within the bell-shaped shell shown with Prob. 5/70. Use the results cited in Prob. 5/39.

Problem 5/72

5/72 The water storage tank is a shell of revolution and is to be sprayed with two coats of paint which has a coverage of 500 ft² per gallon. The engineer (who remembers mechanics) consults a scale drawing of the tank and determines that the curved line ABC has a length of 34 ft and that its centroid is 8.2 ft from the centerline of the tank. How many gallons of paint will be used for the tank including the vertical cylindrical column?   *Ans.* 8.82 gal

Problem 5/73

5/73 A hand-operated control wheel made of aluminum has the proportions shown in the cross-sectional view. The area of the total section shown is 15 200 mm², and the wheel has a mass of 10.0 kg. Calculate the distance $\bar{r}$ to the centroid of the half-section. The aluminum has a density of 2.69 Mg/m³.

5/74 A surface is generated by revolving the circular arc of 0.8-m radius and subtended angle of 120° completely about the *z*-axis. The diameter of the neck is 0.6 m. Determine the area *A* generated.

Ans. A = 4.62 m²

Problem 5/74

5/75 A styrofoam ring, shown in section, is designed for packing a mass-produced item. Calculate the volume of styrofoam used in each ring.

Ans. V = 27.2(10⁶) mm³

Dimensions in Millimeters

Problem 5/75

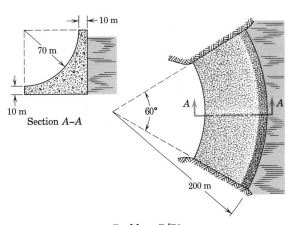

Section *A–A*

5/76 Calculate the mass *m* of concrete required to construct the arched dam shown. Concrete has a density of 2.40 Mg/m³.   *Ans. m* = 1.126 (10⁶) Mg

Problem 5/76

SECTION B. SPECIAL TOPICS

5/6 BEAMS. Structural members that offer resistance to bending caused
by applied loads are known as beams. Most beams are long pris-
matical bars, and the loads are usually applied normal to the axes of
the bars. Beams are undoubtedly the most important of all structural
members, and the basic theory underlying their design must be
thoroughly understood. The analysis of the load-carrying capacities
of beams consists, first, in establishing the equilibrium requirements
of the beam as a whole and any portion of it considered separately.
Second, the relations between the resulting forces and the accompanying internal resistance of the beam to support these forces are
established. The first part of this analysis requires the application of
the principles of statics, while the second part of the problem involves the strength characteristics of the material and is usually

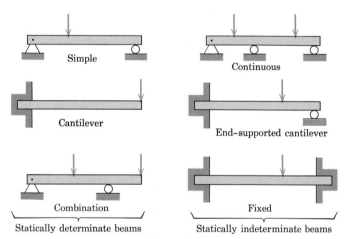

Figure 5/18

treated under the heading of the mechanics of solids or the mechanics
of materials. This article concerns the first aspect of the problem only
and will enable us to calculate the distribution along the beam of the
internal force and moment acting on sections of the beam.

(*a*) *Types of Beams.* Beams supported in such a way that their
external support reactions can be calculated by the methods of statics
alone are called *statically determinate* beams. A beam that has more
supports than are necessary to provide equilibrium is said to be
statically indeterminate, and it is necessary to consider the load-
deformation properties of the beam in addition to the equations of
statical equilibrium to determine the support reactions. In Fig. 5/18

are shown examples of both types of beams. In this article we will analyze only statically determinate beams.

Beams may also be identified by the type of external loading they support. The beams in Fig. 5/18 are supporting concentrated loads, whereas the beam in Fig. 5/19 is supporting a distributed load. The intensity w of a distributed load may be expressed as force per unit length of beam. The intensity may be constant or variable, continuous or discontinuous. The intensity of the loading in Fig. 5/19 is constant from C to D and variable from A to C and from D to B. The intensity is discontinuous at D where it changes magnitude abruptly. Although the intensity itself is not discontinuous at C, the rate of change of intensity dw/dx is discontinuous.

(*b*) *Shear, Bending, and Torsion.* In addition to tension or com-pression a beam has resistance to shear, bending, and torsion. These three effects are illustrated in Fig. 5/20. The force V is called the *shear* force, the couple M is known as the *bending moment,* and the couple T is called a *torsional moment.* These effects represent the vector components of the resultant of the forces acting on a trans-verse section of the beam as shown in the lower part of the figure.

We direct our attention now primarily to the shear force V and bending moment M caused by forces applied to the beam in a single plane. The conventions for positive values of shear V and bending moment M shown in Fig. 5/21 are the ones generally used. By the principle of action and reaction we note that the directions of V and M are reversed on the two sections. It is frequently impossible to tell without calculation whether the shear and moment at a particular section are positive or negative. For this reason it will be found advisable to represent V and M in their positive directions on the free-body diagrams and let the algebraic signs of the calculated values indicate the proper directions.

As an aid to the physical interpretation of the bending couple M consider the beam shown in Fig. 5/22 bent by the two equal and opposite positive moments applied at the ends. The cross section of the beam is taken to be that of an H-section with a very narrow center web and heavy top and bottom flanges. For this beam we may

Figure 5/19

Figure 5/20

Figure 5/22

Figure 5/21

(a)

(b)

Figure 5/23

(c)

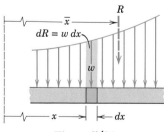

Figure 5/24

neglect the load carried by the small web compared with that carried by the two flanges. It should be perfectly clear that the upper flange of the beam is shortened and is under compression while the lower flange is lengthened and is under tension. The resultant of the two forces, one tensile and the other compressive, acting on any section is a couple and has the value of the bending moment on the section. If a beam of some other cross section were loaded in the same way, the distribution of force over the cross section would be different, but the resultant would be the same couple.

The variation of shear force V and bending moment M over the length of a beam provides information necessary for the design analysis of the beam. In particular, the maximum magnitude of the bending moment is usually the primary consideration in the design or selection of a beam, and its value and position should be determined. The variations in shear and moment are best shown graphically, and the expressions for V and M when plotted against distance along the beam give the *shear-force* and *bending-moment diagrams* for the beam.

The first step in the determination of the shear and moment relations is to establish the values of all external reactions on the beam by applying the equations of equilibrium to a free-body diagram of the beam as a whole. Next, we isolate a portion of the beam, either to the right or to the left of an arbitrary transverse section, with a free-body diagram, and apply the equations of equilibrium to this isolated portion of the beam. These equations will yield expressions for the shear force V and bending moment M acting at the cut section on the part of the beam isolated. The part of the beam which involves the smaller number of forces, either to the right or to the left of the arbitrary section, usually yields the simpler solution. We should avoid using a transverse section that coincides with the location of a concentrated load, as such a position represents a point of discontinuity in the variation of shear or bending moment. Finally, it is important to note that the calculations for V and M on each section chosen should be consistent with the positive convention illustrated in Fig. 5/21.

(c) *Distributed Loads.* Loading intensities which are constant or which vary linearly are easily handled. Figure 5/23 illustrates the three most common cases and the resultants of the distributed loads in each case. In each instance we see that the resultant passes through the centroid of the figure formed by the intensity w and the length L over which the force is distributed.

For a more general loading relationship, Fig. 5/24, we must start with a differential increment of force $dR = w\,dx$. The integral $\int w\,dx$ gives the resultant R and the integral $\int xw\,dx$ gives the moment of the distributed force. The principle of moments gives the location of R by $R\bar{x} = \int xw\,dx$.

(d) *General Loading, Shear, and Moment Relationships.* Certain general relationships may be established for any beam with distributed loads which will aid greatly in the construction of the shear and moment distributions. Figure 5/25 represents a portion of a loaded beam, and an element dx of the beam is isolated. The loading w represents the force per unit length of beam. At the location x the shear V and moment M acting on the element are drawn in their positive directions. On the opposite side of the element where the coordinate is $x + dx$ these quantities are also shown in their positive directions but must be labeled $V + dV$ and $M + dM$, since the changes in V and M with x are required. The applied loading w may be considered constant over the length of the element, since this length is a differential quantity and the effect of any change in w disappears in the limit compared with the effect of w itself.

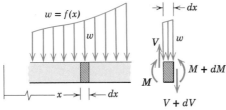

Figure 5/25

Equilibrium of the element requires that the sum of the vertical forces be zero. Thus we have

$$V - w\,dx - (V + dV) = 0$$

or

$$w = -\frac{dV}{dx} \qquad\qquad (5/10)$$

We see from Eq. 5/10 that the slope of the shear diagram must everywhere be equal to the negative of the value of the applied loading. Equation 5/10 holds on either side of a concentrated load but not at the concentrated load by reason of the discontinuity produced by the abrupt change in shear.

Equilibrium of the element in Fig. 5/25 also requires that the moment sum be zero. Taking moments about the left side of the element gives

$$M + w\,dx\frac{dx}{2} + (V + dV)\,dx - (M + dM) = 0$$

The two M's cancel, and the terms $w(dx)^2/2$ and $dV\,dx$ may be dropped, since they are differentials of higher order than those that remain. This leaves merely

$$V = \frac{dM}{dx} \tag{5/11}$$

which expresses the fact that the shear everywhere is equal to the slope of the moment curve. We may now express the moment M in terms of the shear V by integrating Eq. 5/11. Thus

$$\int_{M_0}^{M} dM = \int_{x_0}^{x} V\,dx$$

or

$$M = M_0 + (\text{area under shear diagram from } x_0 \text{ to } x)$$

In this expression M_0 is the bending moment at x_0 and M is the bending moment at x. For beams where there is no externally applied moment M_0 at $x_0 = 0$, the total moment at any section equals the area under the shear diagram up to that section. Summing up the area under the shear diagram is usually the simplest way to construct the moment diagram.

When V passes through zero and is a continuous function of x with $dV/dx \neq 0$, the bending moment M will be a maximum or a minimum, since $dM/dx = 0$ at such a point. Critical values of M also occur when V crosses the zero axis discontinuously, which occurs for beams under concentrated loads.

We observe from Eqs. 5/10 and 5/11 that the degree of V in x is one higher than that of w. Also M is of one higher degree in x than is V. Furthermore M is two degrees in x higher than w. Thus for a beam loaded by $w = kx$ which is of the first degree in x, the shear V is of the second degree in x and the bending moment M is of the third degree in x.

Equations 5/10 and 5/11 may be combined to yield

$$\frac{d^2M}{dx^2} = -w \tag{5/12}$$

Thus, if w is a known function of x, the moment M may be obtained by two integrations, provided that the limits of integration are properly evaluated each time. This method is usable only if w is a continuous function of x.°

When bending in a beam occurs in more than a single plane, a separate analysis in each plane may be carried out. The results may then be combined vectorially.

° When w is a discontinuous function of x, it is possible to introduce a special set of expressions called *singularity functions* which permit writing analytical expressions for shear V and moment M over a range of discontinuities. These functions are not discussed in this book.

Sample Problem 5/7

Determine the shear and moment distributions produced in the simple beam by the 4-kN concentrated load.

Solution. From the free-body diagram of the entire beam we find the support reactions which are

$$R_1 = 1.6 \text{ kN} \qquad R_2 = 2.4 \text{ kN}$$

A section of the beam of length x is next isolated with its free-body diagram on which we show the shear V and the bending moment M in their positive directions. Equilibrium gives

$$[\Sigma F_y = 0] \qquad 1.6 - V = 0 \qquad V = 1.6 \text{ kN}$$

$$[\Sigma M_{R_1} = 0] \qquad M - 1.6x = 0 \qquad M = 1.6x$$

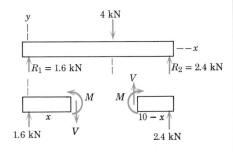

These values of V and M apply to all sections of the beam to the left of the 4-kN load.

A section of the beam to the right of the 4-kN load is next isolated with its free-body diagram upon which V and M are shown in their positive directions. Equilibrium requires

$$[\Sigma F_y = 0] \qquad V + 2.4 = 0 \qquad V = -2.4 \text{ kN}$$

$$[\Sigma M_{R_2} = 0] \qquad -(2.4)(10 - x) + M = 0 \qquad M = 2.4(10 - x)$$

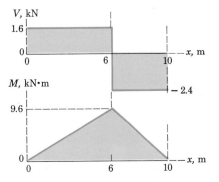

These results apply only to sections of the beam to the right of the 4-kN load.

The values of V and M are plotted as shown. The maximum bending moment occurs where the shear changes direction. As we move in the positive x-direction starting with $x = 0$, we see that the moment M is merely the accumulated area under the shear diagram.

222

Sample Problem 5/8

Determine the shear and moment distributions induced in the beam by the distributed load whose intensity varies linearly with the length as shown.

0.8 kN/m

10 m

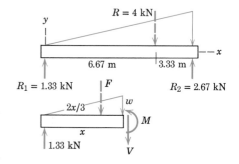

Solution. The resultant R of the linear distribution of load equals the area of the triangular load-distribution diagram and passes through the centroid of this area. Thus $R = \frac{1}{2}(0.8)(10) = 4$ kN. Equilibrium of the beam as a whole, then, gives

$[\Sigma M_{R_1} = 0]$ $0.4(6.67) - R_2(10) = 0$, $R_2 = 2.67$ kN

$[\Sigma F_y = 0]$ $R_1 + 2.67 - 4 = 0$ $R_1 = 1.33$ kN

A representative section of the beam is next isolated with its free-body diagram with V and M shown in their positive directions. The linear load intensity is $w = (x/10)(0.8) = 0.08x$ kN/m. The resultant F of the linear loading on the beam section equals the area of the triangular distribution and passes through the centroid of this area. Thus

$$F = \tfrac{1}{2}wx = \tfrac{1}{2}(0.08x)x = 0.04x^2$$

A moment equation about the cut section and a force equation in the y-direction establish the equilibrium of the section and give us

$[\Sigma M = 0]$ $1.33x - 0.04x^2(x/3) - M = 0$, $M = 1.33x - 0.0133x^3$

$[\Sigma F_y = 0]$ $1.33 - 0.04x^2 - V = 0$, $V = 1.33 - 0.04x^2$

Since the load intensity is a continuous function of x, both expressions for V and M hold throughout the length of the beam.

The shear-force and bending-moment diagrams are now obtained by plotting V and M as shown. We note that M is a maximum when $V = 0$, which occurs when $0 = 1.33 - 0.04x^2$ or $x = \sqrt{1.33/0.04} = 5.77$ m. Substituting this value of x into the expression for M gives

$$M_{max} = 1.33(5.77) - 0.0133(5.77)^3 = 5.13 \text{ kN}\cdot\text{m}$$

Again we observe that the value of M at any section is the area under the shear curve up to that section.

If we differentiate the expressions for V and M with respect to x, we have $-dV/dx = 0.08x$ and $dM/dx = 1.33 - 0.04x^2$, which equal w and V, respectively, as required by Eqs. 5/10 and 5/11.

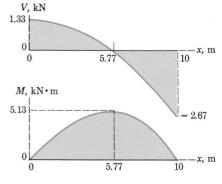

Sample Problem 5/9

Draw the shear-force and bending-moment diagrams for the loaded beam and determine the maximum moment M and its location x from the left end.

Solution. The support reactions are most easily obtained by considering the resultants of the distributed loads as shown on the free-body diagram of the beam as a whole. The first interval of the beam is analyzed from the free-body diagram of the section for $0 < x < 4$ ft. A vertical summation of forces and a moment summation about the cut section yield

$$[\Sigma F_y = 0] \qquad V = 247 - 12.5x^2$$

$$[\Sigma M = 0] \quad M + (12.5x^2)\frac{x}{3} - 247x = 0 \quad M = 247x - 4.167x^3$$

These values of V and M hold for $0 < x < 4$ ft and are plotted for that interval in the shear and moment diagrams shown.

From the free-body diagram of the section for which $4 < x < 8$ ft, equilibrium in the vertical direction and a moment sum about the cut section give

$$[\Sigma F_y = 0] \quad V + 100(x - 4) + 200 - 247 = 0 \quad V = 447 - 100x$$

$$[\Sigma M = 0] \quad M + 100(x - 4)\frac{x - 4}{2} + 200[x - \tfrac{2}{3}(4)] - 247x = 0$$

$$M = -266.7 + 447x - 50x^2$$

These values of V and M are plotted on the shear and moment diagrams for the interval $4 < x < 8$ ft.

The analysis of the remainder of the beam is continued from the free-body diagram of the portion of the beam to the right of a section in the next interval. It should be noted that V and M are represented in their positive directions. A vertical force summation and a moment summation about the section yield

$$V = -353 \text{ lb} \qquad \text{and} \qquad M = 2930 - 353x$$

These values of V and M are plotted on the shear and moment diagrams for the interval $8 < x < 10$ ft.

The last interval may be analyzed by inspection. The shear is constant at $+300$ lb, and the moment follows a straight-line relation beginning with zero at the right end of the beam.

The maximum moment occurs at $x = 4.47$ ft, where the shear curve crosses the zero axis, and the magnitude of M is obtained for this value of x by substitution into the expression for M for the second interval. The maximum moment is

$$M = 731 \text{ lb-ft} \qquad\qquad \textit{Ans.}$$

As before, note that the moment M at any section equals the area under the shear diagram up to that section. For instance, for $x < 4$ ft,

$$\left[\Delta M = \int V\, dx\right] \quad M - 0 = \int_0^x (247 - 12.5x^2)\, dx, \quad M = 247x - 4.167x^3$$

which agrees with the results already computed.

Problem 5/77

3 kN

2 kN

2 m   3 m   1 m

Problem 5/78

P

l

Problem 5/79

y

4 kN   2 kN

4 m   3 m   3 m

Problem 5/80

600 N

4 m

2 m   4 m

A   B

600 N

Problem 5/81

800 lb   1000 lb

12″

A C

600 lb   B

3′   3′   3′   3′

Problem 5/82

PROBLEMS

5/77 Draw the shear and moment diagrams for the diving board which supports the 175-lb man poised for a dive.

5/78 Draw the shear and moment diagrams for the loaded beam.

5/79 Draw the shear and moment diagrams for the cantilever beam.

5/80 Draw the shear and moment diagrams for the horizontal beam shown and determine the maximum magnitude of the bending moment and its location.
Ans. $|M|_{max} = 6.00$ kN·m at $x = 7$ m

5/81 Draw the shear and moment diagrams for the beam shown.

5/82 Draw the shear and moment diagrams for the beam shown and find the bending moment M at section C.
Ans. $M_c = -1667$ lb-ft

5/83 Draw the shear and moment diagrams for the beam loaded at its center by the couple C.

Problem 5/83

5/84 Draw the shear and moment diagrams for the beam subjected to the end couple. What is the moment M at a section 0.5 m to the right of B?

Ans. $M = -120$ N·m

Problem 5/84

5/85 The angle strut is welded to the beam AB and supports the 1-kN load. Draw the shear and moment diagrams for the beam.

Problem 5/85

5/86 Construct the moment diagram for the two beams loaded as shown and connected by the hinge joint B.

Problem 5/86

5/87 Draw the shear and moment diagrams for the beam loaded as shown. Determine the bending moment M with the maximum magnitude.

Ans. $M = -2000$ lb-ft

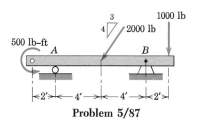

Problem 5/87

5/88 The resistance of a beam of uniform width to bending is found to be proportional to the square of the beam depth y. For the cantilever beam shown the depth is h at the support. Find the required depth y as a function of the length x in order for all sections to be equally effective in their resistance to bending.   Ans. $y = h\sqrt{x/l}$

Problem 5/88

Problem 5/89

Problem 5/90

Problem 5/91

Problem 5/92

Problem 5/93

Problem 5/94

5/89 A curved cantilever beam has the form of a quarter-circular arc. Determine the expressions for the shear V and the bending moment M as functions of θ.

5/90 Construct the bending-moment diagram for the cantilevered shaft AB of the rigid unit shown.

5/91 Write expressions for the torsional moment T and bending moment M in the curved quarter-circular beam under the end load L. Use a notation consistent with the right-handed r-θ-z coordinate system where positive moment vectors are taken in the direction of the positive coordinates.

$$Ans.\ M = -La \cos \theta,\ T = -La(1 - \sin \theta)$$

5/92 Draw the shear and moment diagrams for the cantilever beam with the uniform load w per unit of length.

5/93 Draw the shear and moment diagrams for the uniformly loaded beam and find the maximum bending moment M.

$$Ans.\ M = \frac{wl^2}{8}$$

5/94 Draw the shear and moment diagrams for the beam, which supports the load of 50 lb per foot of beam length distributed over its midsection.

5/95 Draw the shear and moment diagrams for the loaded beam and find the maximum magnitude M of the bending moment.   Ans. $M = \frac{5}{6}Pl$

Problem 5/95

5/96 Draw the shear and moment diagrams for the cantilever beam with the linear loading and find the maximum magnitude M of the bending moment.

Problem 5/96

5/97 Draw the shear and moment diagrams for the loaded cantilever beam where the end couple M_1 is adjusted so as to produce zero moment at the fixed end of the beam. Find the bending moment M at $x = 4$ ft.   Ans. $M = 4800$ lb-ft

Problem 5/97

5/98 Draw the shear and moment diagrams for the linearly loaded simple beam shown. Determine the magnitude M of the maximum bending moment.

$$\text{Ans. } M = \frac{w_0 l^2}{12}$$

Problem 5/98

5/99 A certain beam supports a distributed load of constant intensity $w = 3$ kN/m between A and B as shown. If the shear force and bending moment at A are $+2$ kN and -1.5 kN·m, respectively, calculate the shear force and bending moment at B.

Problem 5/99

5/100 The distributed load increases linearly from 2 to 4 kN/m in a distance of 2 m along a certain beam in equilibrium. If the shear force and bending moment at section A are $+3$ kN and $+2$ kN·m, respectively, calculate the shear force and bending moment at section B.

Ans. $V_B = -3$ kN, $M_B = 2.67$ kN·m

Problem 5/100

Problem 5/101

Problem 5/102

Problem 5/103

Problem 5/104

Problem 5/105

$$w = w_0 \sin \frac{\pi x}{l}$$

Problem 5/106

5/101 Sketch the bending moment diagrams for the upper support beam for the values of overhang $x = 0$, $x = 0.4b$, and $x = l$.

5/102 Draw the shear and moment diagrams for the beam shown. Determine the distance b, measured from the left end, to the point where the bending moment is zero between the supports.

Ans. $b = 1.5$ m

5/103 Determine the maximum bending moment M and the corresponding value of x in the crane beam and indicate the section where this moment acts.

$$\text{Ans. } M_A = \frac{L}{4l}(l - a)^2, \ x = \frac{a + l}{2}$$

5/104 The end-supported shaft is subjected to the linearly varying loads in mutually perpendicular planes. Determine the expression for the resultant bending moment M in the shaft.

$$\text{Ans. } M = \frac{p_0}{6l}x(l - x)\sqrt{5l^2 - 2lx + 2x^2}$$

▶**5/105** The curved cantilever beam in the form of a quarter-circular arc supports a load of w N/m applied along the curve of the beam on its upper surface. Determine the magnitudes of the torsional moment T and bending moment M in the beam as functions of θ.

$$\text{Ans. } T = wr^2\left(\frac{\pi}{2} - \theta - \cos \theta\right)$$

$$M = -wr^2(1 - \sin \theta)$$

▶**5/106** The cantilever beam supports the half-sine-wave loading $w = w_0 \sin (\pi x/l)$ per unit of length. Determine the bending moment M in terms of x by direct use of Eq. 5/12.

$$\text{Ans. } M = \frac{w_0 l}{\pi}\left(\frac{l}{\pi}\sin \frac{\pi x}{l} + x - l\right)$$

5/7 FLEXIBLE CABLES.   One important type of structural member is
the flexible cable which is used in suspension bridges, transmission
lines, messenger cables for supporting heavy trolley or telephone
lines, and many other applications. In the design of these structures it
is necessary to know the relations involving the tension, span, sag, and
length of the cables. We determine these quantities by examining the
cable as a body in equilibrium. In the analysis of flexible cables we
assume that any resistance offered to bending is negligible. This
assumption means that the force in the cable is always in the direction of the cable.

   Flexible cables may support a series of distinct concentrated
loads, as shown in Fig. 5/26a, or they may support loads that are
continuously distributed over the length of the cable, as indicated by
the variable-intensity loading w in Fig. 5/26b. In some instances the
weight of the cable is negligible compared with the loads it supports,
and in other cases the weight of the cable may be an appreciable load
or the sole load, in which case it cannot be neglected. Regardless of
which of these conditions is present, the equilibrium requirements of
the cable may be formulated in the same manner.

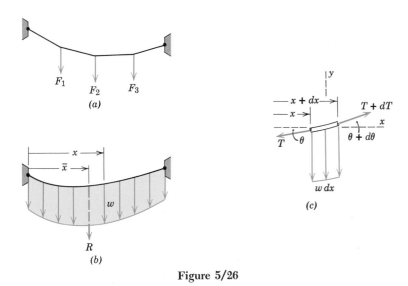

Figure 5/26

   (*a*) *General Relationships.* If the intensity of the variable and
continuous load applied to the cable of Fig. 5/26b is expressed as w
units of force per unit of horizontal length x, then the resultant R of
the vertical loading is

$$R = \int dR = \int w \, dx$$

where the integration is taken over the desired interval. We find the
position of R from the moment principle, so that

$$R\bar{x} = \int x \, dR \qquad \bar{x} = \frac{\int x \, dR}{R}$$

The elemental load $dR = w \, dx$ is represented by an elemental strip of vertical length w and width dx of the shaded area of the loading diagram, and R is represented by the total area. It follows from the foregoing expressions that R passes through the *centroid* of the shaded area.

The equilibrium condition of the cable will be satisfied if each infinitesimal element of the cable is in equilibrium. The free-body diagram of a differential element is shown in Fig. 5/26c. At the general position x the tension in the cable is T, and the cable makes an angle θ with the horizontal x-direction. At the section $x + dx$ the tension is $T + dT$, and the angle is $\theta + d\theta$. Note that the changes in both T and θ are taken positive with positive change in x. The vertical load $w \, dx$ completes the free-body diagram. The equilibrium of vertical and horizontal forces requires, respectively, that

$$(T + dT) \sin (\theta + d\theta) = T \sin \theta + w \, dx$$

$$(T + dT) \cos (\theta + d\theta) = T \cos \theta$$

The trigonometric expansion for the sine and cosine of the sum of two angles and the substitutions $\sin d\theta = d\theta$ and $\cos d\theta = 1$, which hold in the limit as $d\theta$ approaches zero, yield

$$(T + dT)(\sin \theta + \cos \theta \, d\theta) = T \sin \theta + w \, dx$$

$$(T + dT)(\cos \theta - \sin \theta \, d\theta) = T \cos \theta$$

Dropping the second-order terms and simplifying give us

$$T \cos \theta \, d\theta + dT \sin \theta = w \, dx$$

$$-T \sin \theta \, d\theta + dT \cos \theta = 0$$

which we may write as

$$d(T \sin \theta) = w \, dx \qquad \text{and} \qquad d(T \cos \theta) = 0$$

The second relation expresses the fact that the horizontal component of T remains unchanged, which is clear from the free-body diagram. If we introduce the symbol $T_0 = T \cos \theta$ for this constant horizontal force, we may then substitute $T = T_0/\cos \theta$ into the first of the two equations just obtained and get $d(T_0 \tan \theta) = w \, dx$. But $\tan \theta = dy/dx$, so that the equilibrium equation may be written in the form

$$\boxed{\frac{d^2y}{dx^2} = \frac{w}{T_0}} \qquad \qquad (5/13)$$

Equation 5/13 is the *differential equation* for the flexible cable. The solution to the equation is that functional relation $y = f(x)$

which satisfies the equation and also satisfies the conditions at the fixed ends of the cable, called *boundary conditions*. This relationship defines the shape of the cable, and we will use it to solve two important and limiting cases of cable loading.

(*b*) *Parabolic Cable.* When the intensity of vertical loading w is constant, the description closely approximates a suspension bridge where the uniform weight of the roadway may be expressed by the constant w. The mass of the cable is not distributed uniformly with the horizontal but is relatively small and its weight is neglected. For this limiting case we will prove that the cable hangs in a parabolic arc. Figure 5/27 shows such a suspension bridge of span L and sag h with origin of coordinates taken at the midpoint of the span. With

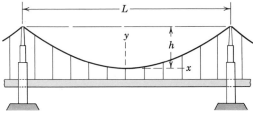

Figure 5/27

both w and T_0 constant, we may integrate Eq. 5/13 once with respect to x to obtain

$$\frac{dy}{dx} = \frac{wx}{T_0} + C$$

where C is a constant of integration. For the coordinate axes chosen, $dy/dx = 0$ when $x = 0$, so that $C = 0$. Hence

$$\frac{dy}{dx} = \frac{wx}{T_0}$$

which defines the slope of the curve as a function of x. One further integration yields

$$\int_0^y dy = \int_0^x \frac{wx}{T_0}\,dx \qquad \text{or} \qquad \boxed{y = \frac{wx^2}{2T_0}} \qquad (5/14)$$

Readers should make certain that they can obtain the identical results with the indefinite integral together with the evaluation of the constant of integration. Equation 5/14 gives the shape of the cable, which we see is a vertical parabola. The constant horizontal component of cable tension becomes the cable tension at the origin.

Inserting the corresponding values $x = L/2$ and $y = h$ in Eq. 5/14 gives

$$T_0 = \frac{wL^2}{8h} \qquad \text{and} \qquad y = \frac{4hx^2}{L^2}$$

The tension T is found from a free-body diagram of a finite portion of the cable, shown in Fig. 5/27, which requires that

$$T = \sqrt{T_0{}^2 + w^2x^2}.$$

Elimination of T_0 gives

$$T = w\sqrt{x^2 + \left(\frac{L^2}{8h}\right)^2} \tag{5/15}$$

The maximum tension occurs when $x = L/2$ and is

$$T_{\text{max}} = \frac{wL}{2}\sqrt{1 + \frac{L^2}{16h^2}} \tag{5/15a}$$

We obtain the length S of the complete cable from the differential relation $ds = \sqrt{(dx)^2 + (dy)^2}$. Thus

$$\int_0^{S/2} ds = \frac{S}{2} = \int_0^{L/2}\sqrt{1 + \left(\frac{dy}{dx}\right)^2}\,dx = \int_0^{L/2}\sqrt{1 + \left(\frac{wx}{T_0}\right)^2}\,dx$$

For convenience in computation we change this expression to a convergent series and then integrate it term by term. From the expansion for a variable x

$$(1 + x)^n = 1 + nx + \frac{n(n-1)}{2!}x^2 + \frac{n(n-1)(n-2)}{3!}x^3 + \cdots$$

the integral may be written as

$$S = 2\int_0^{L/2}\left(1 + \frac{w^2x^2}{2T_0{}^2} - \frac{w^4x^4}{8T_0{}^4} + \cdots\right)dx$$

$$= L\left(1 + \frac{w^2L^2}{24T_0{}^2} - \frac{w^4L^4}{640T_0{}^4} + \cdots\right)$$

Substitution of $w/T_0 = 8h/L^2$ yields

$$S = L\left[1 + \frac{8}{3}\left(\frac{h}{L}\right)^2 - \frac{32}{5}\left(\frac{h}{L}\right)^4 + \cdots\right] \tag{5/16}$$

When we examine the properties of this series, we find that it converges for all values of $h/L \leq 1/4$. In most cases h is much smaller than $L/4$, so that the three terms of Eq. 5/16 give a sufficiently accurate approximation.

(c) *Catenary Cable.*   Consider now a uniform cable, Fig. 5/28, suspended at two points in the same horizontal plane and hanging under the action of its own weight only. The free-body diagram of a finite portion of the cable of length s is shown in the right-hand part of the figure. This free-body diagram differs from that in Fig. 5/27 in that the total vertical force supported is equal to the weight of the section of cable of length s in place of the load distributed uniformly

Figure 5/28

with respect to the horizontal. If the cable has a weight μ per unit of its length, the resultant R of the load is $R = \mu s$, and the incremental vertical load $w\,dx$ of Fig. 5/26c is replaced by $\mu\,ds$. Thus we may substitute $w = \mu\,ds/dx$ into the differential relation, Eq. 5/13, for the cable and obtain

$$\frac{d^2y}{dx^2} = \frac{\mu}{T_0}\frac{ds}{dx} \qquad\qquad (5/17)$$

Since $s = f(x, y)$, it is necessary to change this equation to one containing only the two variables.

We may substitute the identity $(ds)^2 = (dx)^2 + (dy)^2$ to obtain

$$\frac{d^2y}{dx^2} = \frac{\mu}{T_0}\sqrt{1 + \left(\frac{dy}{dx}\right)^2} \qquad\qquad (5/18)$$

Equation 5/18 is the differential equation of the curve (catenary) assumed by the cable. Solution of this equation is facilitated by the substitution $p = dy/dx$, which gives

$$\frac{dp}{\sqrt{1 + p^2}} = \frac{\mu}{T_0}\,dx$$

Integrating this equation gives us

$$\ln\left(p + \sqrt{1 + p^2}\right) = \frac{\mu}{T_0}x + C$$

The constant C is zero since $dy/dx = p = 0$ when $x = 0$. Substituting $p = dy/dx$, changing to exponential form, and clearing the equation of the radical give

$$\frac{dy}{dx} = \frac{e^{\mu x/T_0} - e^{-\mu x/T_0}}{2} = \sinh\frac{\mu x}{T_0}$$

where the hyperbolic function° is introduced for convenience. The slope may be integrated to obtain

$$y = \frac{T_0}{\mu}\cosh\frac{\mu x}{T_0} + K$$

° See Tables B8 and B10, Appendix B, for the definition and integral of hyperbolic functions.

The integration constant K is evaluated from the boundary condition $x = 0$ when $y = 0$. This substitution requires that $K = -T_0/\mu$, and hence

$$y = \frac{T_0}{\mu}\left(\cosh\frac{\mu x}{T_0} - 1\right) \qquad (5/19)$$

Equation 5/19 is the equation of the curve (catenary) assumed by the cable hanging under the action of its weight only.

From the free-body diagram in Fig. 5/28 we see that $dy/dx = \tan\theta = \mu s/T_0$. Thus, from the previous expression for the slope,

$$s = \frac{T_0}{\mu}\sinh\frac{\mu x}{T_0} \qquad (5/20)$$

We obtain the tension T in the cable from the equilibrium triangle of the forces in Fig. 5/28. Thus

$$T^2 = \mu^2 s^2 + T_0{}^2$$

which, upon combination with Eq. 5/20, becomes

$$T^2 = T_0{}^2\left(1 + \sinh^2\frac{\mu x}{T_0}\right) = T_0{}^2\cosh^2\frac{\mu x}{T_0}$$

or

$$T = T_0\cosh\frac{\mu x}{T_0} \qquad (5/21)$$

We may also express the tension in terms of y with the aid of Eq. 5/19, which, when substituted into Eq. 5/21, gives

$$T = T_0 + \mu y \qquad (5/22)$$

Equation 5/22 shows us that the increment in cable tension from that at the lowest position depends only on μy.

Most problems dealing with the catenary involve solutions of Eqs. 5/19 through 5/22, which may be handled by a graphical approximation or solved by computer. The graphical procedure is illustrated in the sample problem following this article.

The solution of catenary problems where the sag-to-span ratio is small may be approximated by the relations developed for the parabolic cable. A small sag-to-span ratio means a tight cable, and the uniform distribution of weight along the cable is not much different from the same load intensity distributed uniformly along the horizontal.

Many problems dealing with both the catenary and parabolic cable involve suspension points that are not on the same level. In such cases we may apply the relations just developed to the part of the cable on each side of the lowest point.

Sample Problem 5/10

The light cable supports a mass of 12 kg per meter of horizontal length and is suspended between the two points on the same level 300 m apart. If the sag is 60 m, find the tension at midlength, the maximum tension, and the total length of the cable.

Solution. With a uniform horizontal distribution of load, the solution of part (b) of Art. 5/7 applies, and we have a parabolic shape for the cable. For $h = 60$ m, $L = 300$ m, and $w = 12(9.81)(10^{-3})$ kN/m the relation following Eq. 5/14 gives for the midlength tension

$$\left[T_0 = \frac{wL^2}{8h} \right] \qquad T_0 = \frac{0.1177(300)^2}{8(60)} = 22.07 \text{ kN} \qquad \textit{Ans.}$$

The maximum tension occurs at the supports and is given by Eq. 5/15a. Thus

$$\left[T_{max} = \frac{wL}{2} \sqrt{1 + \left(\frac{L}{4h}\right)^2} \right]$$

$$T_{max} = \frac{0.1177(300)}{2} \sqrt{1 + \left[\frac{300}{4(60)}\right]^2} = 28.27 \text{ kN} \qquad \textit{Ans.}$$

① *Suggestion:* Check the value of T_{max} directly from the free-body diagram of the right-hand half of the cable from which a force polygon may be drawn.

The sag-to-span ratio is $60/300 = 1/5 < 1/4$. Therefore the series expression developed in Eq. 5/16 is convergent, and we may write for the total length

$$S = 300 \left[1 + \frac{8}{3}\left(\frac{1}{5}\right)^2 - \frac{32}{5}\left(\frac{1}{5}\right)^4 + \cdots \right]$$

$$= 300[1 + 0.1067 - 0.01024 + \cdots] = 329 \text{ m} \qquad \textit{Ans.}$$

Sample Problem 5/11

Replace the cable of Sample Problem 5/10, which is loaded uniformly along the horizontal, by a cable which has a mass of 12 kg per meter of its own length and supports its own weight only. The cable is suspended between two points on the same level 300 m apart and has a sag of 60 m. Find the tension at midlength, the maximum tension, and the total length of the cable.

Solution. With a load distributed uniformly along the length of the cable, the solution of part (*c*) of Art. 5/7 applies, and we have a catenary shape of the cable. Equations 5/20 and 5/21 for the cable length and tension both involve the minimum tension T_0 at midlength which must be found from Eq. 5/19. Thus for $x = 150$ m, $y = 60$ m, and $\mu = 12(9.81)(10^{-3}) = 0.1177$ kN/m, we have

$$60 = \frac{T_0}{0.1177}\left[\cosh\frac{(0.1177)(150)}{T_0} - 1\right]$$

or
$$\frac{7.063}{T_0} = \cosh\frac{17.66}{T_0} - 1$$

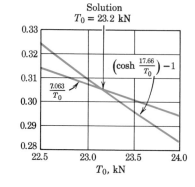

Solution
$T_0 = 23.2$ kN

This equation is most easily solved graphically. We compute the expression on each side of the equals sign and plot it as a function of various values of T_0. The intersection of the two curves establishes the equality and determines the correct value of T_0. This plot is shown in the figure accompanying this problem and yields the solution

$$T_0 = 23.2 \text{ kN}$$

The maximum tension occurs for maximum y and from Eq. 5/22 is

$$T_{\text{max}} = 23.2 + (0.1177)(60) = 30.2 \text{ kN} \qquad Ans.$$

① From Eq. 5/20 the total length of the cable becomes

$$2s = 2\frac{23.2}{0.1177}\sinh\frac{(0.1177)(150)}{23.2} = 330 \text{ m} \qquad Ans.$$

① Note that the solution of Sample Problem 5/10 for the parabolic cable gives a very close approximation to the values for the catenary even though we have a fairly large sag. The approximation is even better for smaller sag-to-span ratios.

PROBLEMS

5/107 The two cables of a suspension bridge with a span of 1000 m and a sag of 150 m support a vertical load of 480 MN distributed uniformly with respect to the horizontal distance between its towers. Compute the tension T_0 in each cable at midspan and the angle θ made by the cables with the horizontal as they approach the support at the top of either tower.   Ans. $T_0 = 200$ MN, $\theta = 31.0°$

5/108 The Golden Gate Bridge in San Francisco has a main span of 4200 ft, a sag of 470 ft, and a total static loading of 21,300 lb per lineal foot of horizontal measurement. The weight of both of the main cables is included in this figure and is assumed uniformly distributed along the horizontal. The angle made by the cable with the horizontal at the top of the tower is the same on each side of each tower. Calculate the midspan tension T_0 in each of the main cables and the compressive force C exerted by each cable on the top of each tower.

5/109 The cable from A to B carries a load of 80 Mg distributed uniformly along the horizontal. The slope of the cable is zero at A, and its weight is small compared with the load it supports. Calculate the maximum tension T_{max} in the cable.
Ans. $T_{max} = 1173$ kN

Problem 5/109

5/110 Expand Eq. 5/19 in a power series for $\cosh(\mu x/T_0)$ and show that the equation for the parabola, Eq. 5/14, is obtained by taking only the first two terms in the series. (See Table B8, Appendix B, for the series expansion of the hyperbolic function.)

5/111 Strain-gage measurements made on the cables of the suspension bridge at position A indicate an increase of 400,000 lb of tension in *each* of the two main cables because the bridge has been repaved. Determine the total weight w' of added paving material used per foot of roadway.   Ans. $w' = 594$ lb/ft

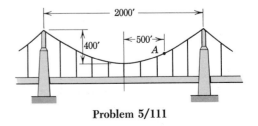

Problem 5/111

5/112 The cable of a suspension bridge with a span of 1000 m is 150 m below the top of the supporting towers at the quarter-span position. Calculate the total length S of each cable between the two identical towers.

5/113 A small suspension bridge for a walkway spans 100 m between supports on the same level. Each of its two cables carries a uniform load of 20 kg per meter of horizontal walkway. If the maximum tension in the cable is found to be 18 kN, compute the tension T_0 at midspan, the sag h, and the angle θ made by the cable with the horizontal at the supports.

$Ans.\ T_0 = 15.09\ \text{kN},\ h = 16.25\ \text{m},\ \theta = 33.0°$

5/114 A cable which carries a load uniformly distributed along the horizontal is 112 ft long and is suspended between two points on the same level 100 ft apart. Determine the sag h of the cable.

5/115 A cable supports a load of 50 kg/m uniformly distributed with respect to the horizontal and is suspended from the two fixed points located as shown. Determine the maximum and minimum tensions T and T_0 in the cable.

$Ans.\ T = 35.61\ \text{kN},\ T_0 = 21.04\ \text{kN}$

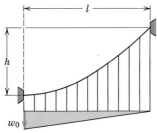

Problem 5/115

5/116 A cable of negligible mass is suspended from the fixed points shown and has a zero slope at its lower end. If the cable supports a unit load w which decreases uniformly with x from w_0 to zero as indicated, determine the equation of the curve assumed by the cable.

$Ans.\ y = \dfrac{3hx^2}{2l^2}\left(1 - \dfrac{x}{3l}\right)$

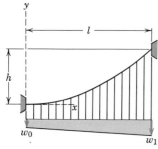

Problem 5/116

5/117 The light cable has zero slope at its lower point of attachment and supports the unit load which varies linearly with x from w_0 to w_1 as shown. Derive the expression for the tension T_0 in the cable at the origin.

Problem 5/117

5/118 The light cable is suspended from two points a distance L apart and on the same horizontal line. If the load per unit of horizontal distance supported by the cable varies from w_0 at the center to w_1 at the ends in accordance with the relation $w = a + bx^2$, derive the equation for the sag h of the cable in terms of the midspan tension T_0.

$$Ans.\ h = \frac{L^2}{48\,T_0}(5w_0 + w_1)$$

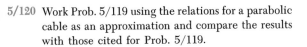

Problem 5/118

5/119 A floating dredge is anchored in position with a single stern cable which has a horizontal direction at the attachment A and extends a horizontal distance of 250 m to an anchorage B on shore. A tension of 300 kN is required in the cable at A. If the cable has a mass of 22 kg per meter of its length, compute the required height H of the anchorage above water level and find the length of cable between A and B.   *Ans. $H = 24.5$ m, $s = 251$ m*

Problem 5/119

5/120 Work Prob. 5/119 using the relations for a parabolic cable as an approximation and compare the results with those cited for Prob. 5/119.

5/121 A flexible cable is secured to point A and passes over a small pulley at B, which is 600 ft higher than A. If it requires a tension $T = 12,000$ lb at B to make $\alpha = 0$ at A, determine the weight μ of the cable per foot of its length.   *Ans. $\mu = 6.35$ lb/ft*

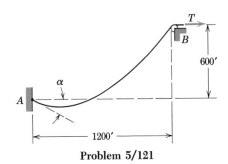

Problem 5/121

5/122 Solve Prob. 121 using the relations for the parabolic cable as an approximation.

5/123 A rope 40 m in length is suspended between two points which are separated by a horizontal distance of 10 m. Compute the distance h to the lowest part of the loop.   *Ans. $h = 18.5$ m*

Problem 5/123

5/124 A cable hanging under the action of its own weight is suspended between two points on the same level and 400 m apart. If the sag is 100 m, determine the total length S of the cable. What error is involved if the length is computed from the expression for the parabolic cable using three terms in the series?

5/125 A power line is suspended from two towers 200 m apart on the same horizontal line. The cable has a mass of 18.2 kg per meter of length and has a sag of 32 m at midspan. If the cable can support a maximum tension of 60 kN, determine the mass ρ of ice per meter which can form on the cable without exceeding the maximum cable tension.

Ans. $\rho = 13.4$ kg of ice per meter

5/126 Solve Prob. 5/125 using the relations for the parabolic cable as an approximation.

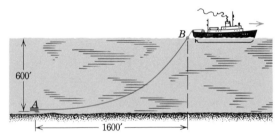

Problem 5/127

▶5/127 A cable ship tows a plow A during a survey of the ocean floor for later burial of a telephone cable. The ship maintains a constant low speed with the plow at a depth of 600 ft and with a sufficient length of cable so that it leads horizontally from the plow, which is 1600 ft astern of the ship. The tow cable has an effective weight of 3.10 lb/ft when the buoyancy of the water is accounted for. Also, the forces on the cable due to movement through the water are neglected at the low speed. Compute the horizontal force T_0 applied to the plow and the maximum tension in the cable. Also find the length of the tow cable from point A to point B.

Ans. $T_0 = 6900$ lb, $T_{\text{max}} = 8760$ lb, $s = 1740$ ft

Problem 5/128

▶5/128 The blimp is moored to the ground winch in a gentle wind with 100 m of 12-mm cable which has a mass of 0.51 kg/m. A torque of 400 N·m on the drum is required to start winding in the cable. At this condition the cable makes an angle of 30° with the vertical as it approaches the winch. Calculate the height H of the blimp. The diameter of the drum is 0.5 m.   *Ans.* $H = 90$ m

5/8 FLUID STATICS. In the work so far, we have directed attention primarily to the action of forces on and between rigid bodies. In this article we shall consider the equilibrium of bodies subjected to forces due to the action of fluid pressures. A fluid is any continuous substance which, when at rest, is unable to support shear force. A shear force is one tangent to the surface upon which it acts and is developed when differential velocities exist between adjacent layers of fluids. Thus a fluid at rest can exert only normal forces on a bounding surface. Fluids may be either gaseous or liquid. The statics of fluids is generally referred to as *hydrostatics* when the fluid is a liquid and as *aerostatics* when the fluid is a gas.

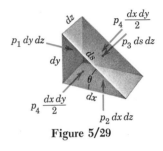

Figure 5/29

(*a*) *Fluid Pressure.* The pressure at any given point in a fluid is the same in all directions (Pascal's law). We may prove this fact by considering the equilibrium of an infinitesimal triangular prism of fluid as shown in Fig. 5/29. The fluid pressures normal to the faces of the element are taken to be p_1, p_2, p_3, and p_4 as shown. With force equal to pressure times area, the equilibrium of forces in the *x*- and *y*-directions gives

$$p_2 \, dx \, dz = p_3 \, ds \, dz \cos \theta \qquad p_1 \, dy \, dz = p_3 \, ds \, dz \sin \theta$$

Since $ds \sin \theta = dy$ and $ds \cos \theta = dx$, these equations require that

$$p_1 = p_2 = p_3 = p$$

By rotating the element through 90° we see that p_4 is also equal to the other pressures. Hence the pressure at any point in a fluid is the same in all directions. In this analysis it is unnecessary to account for the weight of the fluid element since, when the weight per unit volume (density ρ times g) is multiplied by the volume of the element, a differential quantity of third order results which disappears when passing to the limit compared with the second-order pressure force terms.

In all fluids at rest the pressure is a function of the vertical dimension. To determine this function we consider the forces acting on a differential element of a vertical column of fluid of cross-sectional area dA, as shown in Fig. 5/30. The positive direction of vertical measurement h is taken down. The pressure on the upper face is p, and that on the lower face is p plus the change in p, or $p + dp$. The weight of the element equals ρg multiplied by its volume. The normal forces on the lateral surface, which are horizontal and do not affect the balance of forces in the vertical direction, are not shown. Equilibrium of the fluid element in the *h*-direction requires

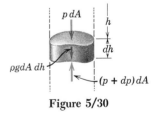

Figure 5/30

$$p \, dA + \rho g \, dA \, dh - (p + dp) \, dA = 0$$

$$dp = \rho g \, dh \qquad\qquad (5/23)$$

This differential relation shows us that the pressure in a fluid increases with depth or decreases with increased elevation. Equation 5/23 holds for both liquids and gases and agrees with common observation of air and water pressures.

Fluids that are essentially incompressible are called liquids, and for most practical purposes we may consider their density ρ constant for every part of the liquid.[†] With ρ a constant Eq. 5/23 may be integrated as it stands, and the result is

$$p = p_0 + \rho gh \qquad\qquad (5/24)$$

The pressure p_0 is the pressure on the surface of the liquid where $h = 0$. If p_0 is due to atmospheric pressure and the measuring instrument records only the increment above atmospheric pressure,[°] the measurement gives what is known as "gage pressure" and is $p = \rho gh$.

The common unit for pressure in SI is the kilopascal (kPa), which is the same as a kilonewton per square meter (10^3 N/m^2). In computing pressure, if we use Mg/m^3 for ρ, m/s^2 for g, and m for h, then the product ρgh gives us pressure in kPa directly. For example, the pressure at a depth of 10 m in fresh water is

$$p = \rho gh = \left(1.0 \frac{\text{Mg}}{\text{m}^3}\right)\left(9.81 \frac{\text{m}}{\text{s}^2}\right)(10 \text{ m}) = 98.1 \left(10^3 \frac{\text{kg} \cdot \text{m}}{\text{s}^2} \frac{1}{\text{m}^2}\right)$$

$$= 98.1 \text{ kN/m}^2 = 98.1 \text{ kPa}$$

In the U.S. customary system fluid pressure is generally expressed in pounds per square inch (lb/in.^{2}) or occasionally in pounds per square foot (lb/ft^2). Thus at a depth of 10 ft in fresh water the pressure is

$$p = \rho gh = \left(62.4 \frac{\text{lb}}{\text{ft}^3}\right)\left(\frac{1}{1728} \frac{\text{ft}^3}{\text{in.}^3}\right)(120 \text{ in.}) = 4.33 \text{ lb/in.}^2$$

(b) *Hydrostatic Pressure on Submerged Rectangular Surfaces.* A surface submerged in a liquid, such as a gate valve in a dam, or the wall of a tank, is subjected to fluid pressure normal to its surface and distributed over its area. In problems where fluid forces are appreciable we must determine the resultant force due to the distribution of pressure on the surface and the position at which this resultant acts. For systems that are open to the earth's atmosphere, the atmospheric pressure p_0 acts over all surfaces and, hence, yields a zero resultant. In such cases, then, we need to consider only the gage pressure $p = \rho gh$ which is the increment above atmospheric pressure.

Consider the special but common case of the action of hydrostatic pressure on the surface of a rectangular plate submerged in a liquid. Figure 5/31a shows such a plate 1-2-3-4 with its top edge horizontal and with the plane of the plate making an arbitrary angle

---

[°] Atmospheric pressure at sea level may be taken to be 101.3 kPa or 14.7 lb/in.^{2}
[†] See Table C1, Appendix C, for table of densities.

θ with the vertical plane. The horizontal surface of the liquid is represented by the x-y' plane. The fluid pressure (gage) acting normal to the plate at point 2 is represented by the arrow 6-2 and equals ρg times the vertical depth from the liquid surface to point 2. This same pressure acts at all points along the edge 2-3. At point 1 on the lower edge, the fluid pressure equals ρg times the vertical depth to point 1, and this pressure is the same at all points along edge 1-4. The variation of pressure p over the area of the plate is governed by the linear depth relationship, and we see, therefore, that it is represented by the altitude of the truncated prism 1-2-3-4-5-6-7-8 with the plate as its base. The resultant force produced by this pressure distribution is represented by R, which acts at some point P known as the *center of pressure*.

We see clearly that the conditions that prevail at the vertical section 1-2-6-5 in Fig. 5/31*a* are identical to those at section 4-3-7-8 and at every other vertical section normal to the plate. Thus we may analyze the problem from the two-dimensional view of a vertical section as shown in Fig. 5/31*b* for section 1-2-6-5. For this section the

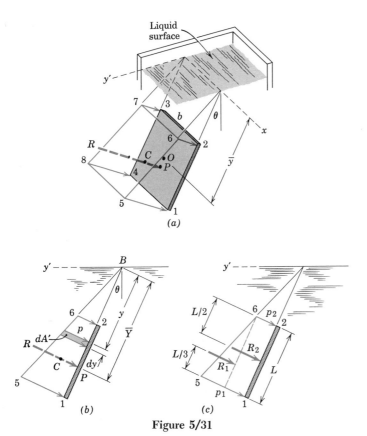

Figure 5/31

pressure distribution is trapezoidal. If b is the horizontal width of the plate, an element of plate area over which the pressure $p = \rho g h$ acts is $dA = b\,dy$, and an increment of the resultant force is $dR = p\,dA = bp\,dy$. But $p\,dy$ is merely the shaded increment of trapezoidal area dA', so that $dR = b\,dA'$. Therefore, we may express the resultant force acting on the entire plate as the trapezoidal area 1-2-6-5 times the width b of the plate,

$$R = b \int dA' = bA'$$

Care must be taken not to confuse the physical area A of the plate with the geometrical area A' defined by the trapezoidal distribution of pressure.

The trapezoidal area representing the pressure distribution is easily expressed by using its average altitude. Therefore the resultant force R may be written in terms of the average pressure $p_{av} = \frac{1}{2}(p_1 + p_2)$ times the plate area A. The average pressure is also the pressure that exists at the average depth to the centroid O of the plate. Therefore an alternative expression for R is

$$R = p_{av}A = \rho g \bar{h} A$$

where $\bar{h} = \bar{y} \cos \theta$.

We obtain the line of action of the resultant force R from the principle of moments. Using the x-axis (point B in Fig. 5/31b) as the moment axis yields $R\bar{Y} = \int y(pb\,dy)$. Substituting $p\,dy = dA'$ and $R = bA'$, and cancelling b give

$$\bar{Y} = \frac{\int y\,dA'}{\int dA'}$$

which is simply the expression for the centroidal coordinate of the trapezoidal area A'. In the two-dimensional view, therefore, the resultant R passes through the centroid C of the trapezoidal area defined by the pressure distribution in the vertical section. Clearly $\bar{Y}$ also locates the centroid C of the truncated prism 1-2-3-4-5-6-7-8 in Fig. 5/31a through which the resultant actually passes.

In treating a trapezoidal distribution of pressure, we may simplify the calculation by considering the resultant composed of two components, Fig. 5/31c. The trapezoid is divided into a rectangle and a triangle, and the force represented by each part is considered separately. The force represented by the rectangular portion acts at the center O of the plate and is $R_2 = p_2 A$ where A is the area 1-2-3-4 of the plate. The force represented by the triangular increment of pressure distribution is $\frac{1}{2}(p_1 - p_2)A$ and acts through the centroid of the triangular portion as shown.

(c) *Hydrostatic Pressure on Cylindrical Surfaces.* For a sub-merged curved surface the resultant R caused by distributed pressure involves more calculation than for a flat surface. As an example consider the submerged cylindrical surface shown in Fig. 5/32a where the elements of the curved surface are parallel to the horizon-tal surface x-y' of the liquid. Vertical sections perpendicular to the surface all disclose the same curve AB and the same pressure distri-bution. Hence the two-dimensional representation in Fig. 5/32b may be used. To find R by a direct integration it is necessary for us to

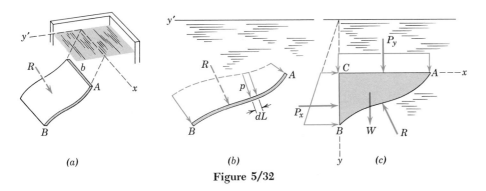

(a)   (b)   (c)

Figure 5/32

integrate the x- and y-components of dR along the curve AB, since the pressure continuously changes direction. Thus

$$R_x = b\int (p\,dL)_x = b\int p\,dy \quad \text{and} \quad R_y = b\int (p\,dL)_y = b\int p\,dx$$

A moment equation would now be required to establish the position of R.

A second method for finding R is usually much simpler. The equilibrium of the block of liquid ABC directly above the surface, shown in Fig. 5/32c, is considered. The resultant R is then disclosed as the equal and opposite reaction of the surface upon the block of liquid. The resultants of the pressures along AC and CB are P_y and P_x, respectively, and are easily obtained. The weight W of the liquid block is calculated from the area ABC of its section multiplied by the constant dimension b and ρg. The weight W passes through the centroid of area ABC. The equilibrant R is then determined com-pletely from the equilibrium equations which we apply to the free-body diagram of the fluid block.

(d) *Hydrostatic Pressure on Flat Surfaces of Any Shape.* Figure 5/33a shows a flat plate of any shape submerged in a liquid. The horizontal surface of the liquid is the plane x-y', and the plane of the plate makes an angle θ with the vertical. The force acting on a differential strip of area dA parallel to the surface of the liquid is $dR = p\,dA = \rho g h\,dA$. The pressure p has the same magnitude throughout the length of the strip, since there is no change of depth along the horizontal strip. We obtain the total force acting on the exposed area A by integration which gives

$$R = \int dR = \int p \, dA = \rho g \int h \, dA$$

Substituting the centroidal relation $\bar{h}A = \int h \, dA$ gives us

$$\boxed{R = \rho g \bar{h} A} \qquad (5/25)$$

The quantity $\rho g \bar{h}$ represents the pressure that exists at the depth of the centroid O of the area and is the average pressure over the area.

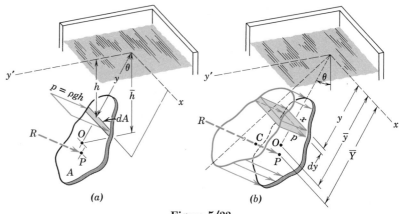

(a)　　　　(b)

Figure 5/33

We may also represent the resultant R geometrically by the volume of the figure shown in Fig. 5/33b. Here the fluid pressure p is represented as an altitude to the plate considered as a base. We see that the resulting volume is a truncated right cylinder. The force dR acting on the differential area $dA = x \, dy$ yields the elemental volume $dV = p \, dA$ shown by the shaded slice, and the total force is, then, represented by the total volume of the cylinder. Hence

$$R = \int dR = \int dV = V$$

We see from Eq. 5/25 that the average altitude of the truncated cylinder is the average pressure $\rho g \bar{h}$ which exists at a depth corresponding to the centroid O of the area exposed to pressure. For problems where the centroid O or the volume V is not readily apparent a direct integration may be performed to obtain R. Thus

$$R = \int dR = \int p \, dA = \int \rho g h x \, dy$$

where the depth h and the length x of the horizontal strip of differential area must be expressed in terms of y to carry out the integration.

The second requirement of the analysis of fluid pressure is the determination of the position of the resultant force in order to account for the moments of the pressure forces. Using the principle of moments with the x-axis of Fig. 5/33b as the moment axis gives

$$R\bar{Y} = \int y\, dR \quad \text{or} \quad \bar{Y} = \frac{\int y\, dV}{V} \qquad (5/26)$$

This second relation satisfies the definition of the coordinate $\bar{Y}$ to the centroid of the volume V, and it is concluded, therefore, that the resultant R passes through the centroid C of the volume described by the plate area as base and the linearly varying pressure as altitude. The point P at which R is applied to the plate is the center of pressure. We note carefully that the center of pressure P and the centroid O of the plate area are *not* the same.

(e) *Buoyancy.* The principle of buoyancy, the discovery of which is credited to Archimedes, is easily explained in the following manner for any fluid, gaseous or liquid, in equilibrium. Consider a portion of the fluid defined by an imaginary closed surface, as illustrated by the irregular dotted boundary in Fig. 5/34*a*. If the body of the fluid could be sucked out from within the closed cavity and replaced simultaneously by the forces that it exerted on the boundary of the cavity, Fig. 5/34*b*, there would be no disturbance of the equilibrium of the surrounding fluid. Furthermore, a free-body diagram of the fluid portion before removal, Fig. 5/34*c*, shows us that

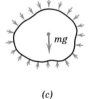

(a)   (b)   (c)

Figure 5/34

the resultant of the pressure forces distributed over its surface must be equal and opposite to its weight mg and must pass through the center of mass of the fluid element. If we replace the fluid element by a body of the same dimensions, the surface forces acting on the body held in this position will be identical with those acting on the fluid element. Thus we see that the resultant force exerted on the surface of an object immersed in a fluid is equal and opposite to the weight of fluid displaced and passes through the center of mass of the displaced fluid. This *resultant force* is called the force of *buoyancy*

$$\boxed{F = \rho g V} \qquad (5/27)$$

where ρ is the density of the fluid, g is the acceleration due to gravity, and V is the volume of the fluid displaced. In the case of a liquid whose density is constant the center of mass of the displaced liquid coincides with the centroid of the displaced volume.

From the foregoing discussion we see that when the density of an object is less than the density of the fluid in which it is fully immersed, there will be an unbalance of force in the vertical direction, and the object will rise. When the immersing fluid is a liquid, the object continues to rise until it comes to the surface of the liquid and then comes to rest in an equilibrium position, assuming that the density of the new fluid above the surface is less than the density of the object. In the case of the surface boundary between a liquid and a gas, such as water and air, the effect of the gas pressure on that portion of the floating object above the liquid is balanced by the added pressure in the liquid due to the action of the gas on its surface.

One of the most important problems involving buoyancy is the determination of the stability of a floating object. This analysis may be illustrated by considering a ship's hull shown in cross section in an upright position in Fig. 5/35a. Point B is the centroid of the displaced volume and is known as the *center of buoyancy*. The resultant of the forces exerted on the hull by the water pressure is the force F.

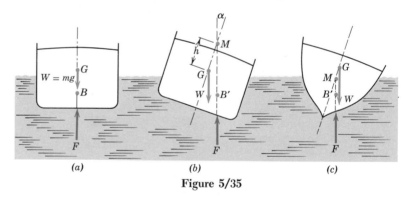

Figure 5/35

Force F passes through B and is equal and opposite to the weight W of the ship. If the ship is caused to list through an angle α, Fig. 5/35b, the shape of the displaced volume changes, and the center of buoyancy will shift to some new position such as B'. The point of intersection of the vertical line through B' with the centerline of the ship is called the *metacenter M*, and the distance h of M above the center of mass G is known as the *metacentric height*. For most hull shapes we find that the metacentric height remains practically constant for angles of list up to about 20°. When M is above G, as in Fig. 5/35b, there is clearly a righting moment which tends to bring the ship back to its original position. The magnitude of this moment for any particular angle of list is a measure of the stability of the ship. If M is below G, as for the hull of Fig. 5/35c, the moment accompanying any list is in the direction to increase the list. This is clearly a condition of instability and must be avoided in the design of any ship.

Sample Problem 5/12

A rectangular plate, shown in vertical section AB, is 4 m high and 6 m wide (normal to the plane of the paper) and blocks the end of a fresh-water channel 3 m deep. The plate is hinged about a horizontal axis along its upper edge through A and is restrained from opening the channel by the fixed ridge at B that bears horizontally against the lower edge of the plate. Calculate the force B exerted against the plate by the ridge.

Solution. The free-body diagram of the plate is shown in section and includes the vertical and horizontal components of the force at A, the unspecified weight $W = mg$ of the plate, the unknown horizontal force B, and the resultant R of the triangular distribution of pressure against the vertical face.

The density of fresh water is $\rho = 1.000$ Mg/m³ so that the average pressure is

① $[p_{av} = \rho g \bar{h}]$ $p_{av} = 1.000(9.81)(\frac{3}{2}) = 14.72$ kPa

The resultant R of the pressure forces against the plate becomes

$[R = p_{av}A]$ $R = (14.72)(3)(6) = 265$ kN

This force acts through the centroid of the triangular distribution of pressure, which is 1 m above the bottom of the plate. A zero moment summation about A establishes the unknown force B. Thus

$[\Sigma M_A = 0]$ $3(265) - 4B = 0,$ $B = 198.7$ kN   *Ans.*

① Note that the units of pressure $\rho g h$ are

$$\left(10^3 \frac{\text{kg}}{\text{m}^3}\right)\left(\frac{\text{m}}{\text{s}^2}\right)(\text{m}) = \left(10^3 \frac{\text{kg} \cdot \text{m}}{\text{s}^2}\right)\left(\frac{1}{\text{m}^2}\right) =$$

$\text{kN/m}^2 = \text{kPa}.$

Sample Problem 5/13

The air space in the closed fresh-water tank is maintained at a pressure of 0.80 lb/in.² (above atmospheric). Determine the resultant force R exerted by the air and water on the end of the tank.

Solution. The pressure distribution on the end surface is shown where $p_0 = 0.80$ lb/in.² The specific weight of fresh water is $\mu = \rho g = 62.4/1728 = 0.0361$ lb/in.³ so that the increment of pressure Δp due to the water is

$$\Delta p = \mu \, \Delta h = 0.0361(30) = 1.083 \text{ lb/in.}^2$$

① The resultant forces R_1 and R_2 due to the rectangular and triangular distributions of pressure, respectively, are

$$R_1 = p_0 A_1 = 0.80(38)(25) = 760 \text{ lb}$$

$$R_2 = \Delta p_{av} A_2 = \frac{1.083}{2}(30)(25) = 406 \text{ lb}$$

The resultant is then $R = R_1 + R_2 = 760 + 406 = 1166$ lb   *Ans.*

We locate R by applying the moment principle about A noting that R_1 acts through the center of the 38-in. depth and that R_2 acts through the centroid of the triangular pressure distribution which is 20 in. below the surface of the water and $20 + 8 = 28$ in. below A. Thus

$[Rh = \Sigma M_A]$ $1166h = 760(19) + 406(28)$ $h = 22.1$ in.   *Ans.*

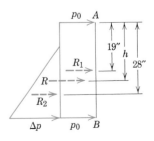

① Dividing the pressure distribution into these two parts is decidedly the simplest way in which to make the calculation.

Sample Problem   5/14

Determine completely the resultant force R exerted on the cylindrical dam surface by the water. The density of fresh water is 1.000 Mg/m³, and the dam has a length b, normal to the paper, of 30 m.

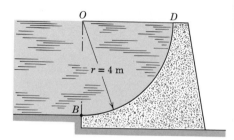

Solution.   The circular block of water BDO is isolated and its free-body diagram is drawn. The force P_x is

① $$P_x = \rho g \bar{h} A = \frac{\rho g r}{2} br = \frac{(1.000)(9.81)(4)}{2}(30)(4) = 2350 \text{ kN}$$

The weight W of the water passes through the mass center G of the quarter-circular section and is

$$mg = \rho g V = (1.000)(9.81)\frac{\pi(4)^2}{4}(30) = 3700 \text{ kN}$$

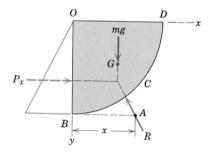

Equilibrium of the section of water requires

$$[\Sigma F_x = 0] \qquad R_x = P_x = 2350 \text{ kN}$$
$$[\Sigma F_y = 0] \qquad R_y = mg = 3700 \text{ kN}$$

The resultant force R exerted by the fluid on the dam is equal and opposite to that shown acting on the fluid and is

$$[R = \sqrt{R_x{}^2 + R_y{}^2}] \qquad R = \sqrt{(2350)^2 + (3700)^2} = 4380 \text{ kN} \quad Ans.$$

The x-coordinate of the point A through which R passes may be found from the principle of moments. Using B as a moment center gives

$$P_x\frac{r}{3} + mg\frac{4r}{3\pi} - R_y x = 0, \qquad x = \frac{2350\left(\frac{4}{3}\right) + 3700\left(\frac{16}{3\pi}\right)}{3700} = 2.55 \text{ m}$$

② **Alternative solution.**   The force acting on the dam surface may be obtained by a direct integration of the components

$$dR_x = p\, dA \cos\theta \qquad \text{and} \qquad dR_y = p\, dA \sin\theta$$

where $p = \rho g h = \rho g r \sin\theta$ and $dA = b(r\, d\theta)$. Thus

$$R_x = \int_0^{\pi/2} \rho g r^2 b \sin\theta \cos\theta\, d\theta = -\rho g r^2 b\left[\frac{\cos 2\theta}{4}\right]_0^{\pi/2} = \tfrac{1}{2}\rho g r^2 b$$

$$R_y = \int_0^{\pi/2} \rho g r^2 b \sin^2\theta\, d\theta = \rho g r^2 b\left[\frac{\theta}{2} - \frac{\sin 2\theta}{4}\right]_0^{\pi/2} = \tfrac{1}{4}\pi\rho g r^2 b$$

Thus $R = \sqrt{R_x{}^2 + R_y{}^2} = \tfrac{1}{2}\rho g r^2 b\sqrt{1 + \pi^2/4}$. Substituting the numerical values gives

$$R = \tfrac{1}{2}(1.00)(9.81)(4^2)(30)\sqrt{1 + \pi^2/4} = 4380 \text{ kN} \qquad Ans.$$

Since dR always passes through point O, we see that R also passes through O and, therefore, the moments of R_x and R_y about O must cancel. So we write $R_x y_1 = R_y x_1$, which gives us

$$x_1/y_1 = R_x/R_y = (\tfrac{1}{2}\rho g r^2 b)/(\tfrac{1}{4}\pi\rho g r^2 b) = 2/\pi$$

By similar triangles we see that

$$x/r = x_1/y_1 = 2/\pi \qquad \text{and} \qquad x = 2r/\pi = 2(4)/\pi = 2.55 \text{ m} \quad Ans.$$

① See note ① in Sample Problem 5/12 if there is any question about the units for $\rho g \bar{h}$.

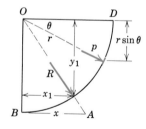

② This approach by integration is feasible here mainly because of the simple geometry of the circular arc.

Sample Problem 5/15

Determine the resultant force R exerted on the semicircular end of the water tank shown in the figure if the tank is filled to capacity. Express the result in terms of the radius r of the circular end and the density ρ of water.

Solution I. We will obtain R first by a direct integration. With a horizontal strip of area $dA = 2x \, dy$ acted upon by the pressure $p = \rho g y$ the increment of the resultant force is $dR = p \, dA$ so that

$$R = \int p \, dA = \int \rho g y (2x \, dy) = 2\rho g \int_0^r y \sqrt{r^2 - y^2} \, dy$$

Integrating gives $\qquad\qquad R = \tfrac{2}{3}\rho g r^3 \qquad\qquad$ *Ans.*

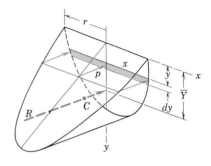

The location of R is determined by using the principle of moments. Taking moments about the x-axis gives

$$[R\bar{Y} = \int y \, dR] \qquad \tfrac{2}{3}\rho g r^3 \bar{Y} = 2\rho g \int_0^r y^2 \sqrt{r^2 - y^2} \, dy$$

Integrating gives $\quad \tfrac{2}{3}\rho g r^3 \bar{Y} = \dfrac{\rho g r^4}{4}\dfrac{\pi}{2} \quad$ and $\quad \bar{Y} = \dfrac{3\pi}{16} r \quad$ *Ans.*

Solution II. We may use Eq. 5/25 directly to find R where the average pressure is $\rho g \bar{h}$ and $\bar{h}$ is the coordinate to the centroid of the area over which the pressure acts. For a semicircular area $\bar{h} = 4r/3\pi$. Thus

$$[R = \rho g \bar{h} A] \qquad R = \rho g \dfrac{4r}{3\pi}\dfrac{\pi r^2}{2} = \tfrac{2}{3}\rho g r^3 \qquad\qquad \textit{Ans.}$$

This calculation amounts to finding the volume of the pressure-area figure.

The resultant R acts through the centroid C of the volume defined by the pressure-area figure. Calculation of the centroidal distance $\bar{Y}$ involves the same integral obtained in *Solution I*.

① Be very careful not to make the mistake of assuming that R passes through the centroid of the area over which the pressure acts.

Sample Problem 5/16

A buoy in the form of a uniform 8-m pole 0.2 m in diameter has a mass of 200 kg and is secured at its lower end to the bottom of a fresh-water lake with 5 m of cable. If the depth of the water is 10 m, calculate the angle θ made by the pole with the horizontal.

Solution. The free-body diagram of the buoy shows its weight acting ① through G, the vertical tension T in the anchor cable, and the buoyancy force B which passes through centroid C of the submerged portion of the buoy. Let x be the distance from G to the waterline. The density of fresh water is $\rho = 10^3$ kg/m^3, so that the buoyancy force is

$[B = \rho g V] \qquad B = 10^3(9.81)\pi(0.1)^2(4 + x) \text{ N}$

Moment equilibrium, $\Sigma M_A = 0$, about A gives

$$200(9.81)(4\cos\theta) - [10^3(9.81)\pi(0.1)^2(4 + x)]\dfrac{4 + x}{2}\cos\theta = 0$$

Thus $\quad x = 3.14 \text{ m} \quad$ and $\quad \theta = \sin^{-1}\left(\dfrac{5}{4 + 3.14}\right) = 44.5° \quad$ *Ans.*

① Since the weight and the buoyancy forces are vertical, the only other force, T, must also be vertical for equilibrium.

PROBLEMS

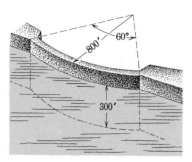

Problem 5/129

5/129 The arched dam has the form of a cylindrical surface of 800-ft radius and subtends an angle of 60°. If
the water is 300 ft deep, determine the total force P
exerted by the water on the face of the dam.

$Ans.\ P = 2246(10^6)\ \text{lb}$

Problem 5/130

5/130 The spring-loaded vertical gate is hinged about a
horizontal axis along its upper edge A and closes the
end of a rectangular fresh-water channel 1.2 m wide
(normal to the plane of the paper). Calculate the
preset spring force F that will limit the depth of the
water to $h = 1.8$ m. $Ans.\ F = 38.1\ \text{kN}$

Problem 5/131

5/131 A concrete retaining wall of rectangular section has
a height h, a thickness b, and a density ρ_c. The wall
holds back mud, which behaves like a liquid of
density ρ_m. Determine the minimum thickness b of
the wall to prevent it from overturning about the
front edge A.

Problem 5/132

5/132 The forms for a small concrete retaining wall are
shown in section. There is a brace BC for every 4 ft
of wall length. Compute the compression in each
brace BC assuming that the joints at A, B, and C act
as hinged connections. Wet concrete may be treated
as a liquid which weighs 150 lb/ft³.

$Ans.\ C = 1622\ \text{lb}$

5/133 The cylindrical caisson of radius r and wall thickness t is lowered into place, and the water is then pumped out. Analyze the forces acting on a semicircular ring of differential height and determine the expression for the compressive stress σ in the cylinder wall for any depth h. The density of water is ρ.

Problem 5/133

5/134 The hinged gate ABC closes an opening of width b (perpendicular to the paper) in a water channel. The water has free access to the under side as well as the right side of the gate. When the water level rises above a certain value of h, the gate will open. Determine the critical value of h. Neglect the mass of the gate.   *Ans.* $h = a\sqrt{3}$

Problem 5/134

5/135 The figure shows the cross section of a rectangular gate 4 m high and 6 m long (perpendicular to the paper) which blocks a fresh-water channel. The gate has a mass of 8.5 Mg and is hinged about a horizontal axis through C. Compute the vertical force P exerted by the foundation on the lower edge A of the gate. Neglect the mass of the frame to which the gate is attached.

Problem 5/135

5/136 The solid concrete cylinder 6 ft long and 4 ft in diameter is supported in a half-submerged position in fresh water by a cable which passes over a fixed pulley at A. Compute the tension T in the cable. The cylinder is waterproofed by a plastic coating. (Consult Table C1, Appendix C, as needed.)   *Ans.* $T = 8957$ lb

Problem 5/136

Problem 5/137

Problem 5/138

5/137 The homogeneous block of density ρ is floating between two liquids of densities $\rho_1 < \rho$ and $\rho_2 > \rho$. Determine an expression for the distance b that the block protrudes into the top liquid.

5/138 The quonset hut is subjected to a horizontal wind, and the pressure p against the circular roof is approximated by $p_0 \cos \theta$. The pressure is positive on the windward side of the hut and is negative on the leeward side. Determine the total horizontal shear Q on the foundation per unit length of roof measured normal to the paper.   *Ans.* $Q = \frac{1}{2}\pi r p_0$

5/139 The hull of a floating oil-drilling platform consists of two rectangular pontoons and six cylindrical columns that support the working platform. When ballasted, the entire structure has a displacement of 26,000 tons (expressed in long tons of 2240 lb). Calculate the total draft h of the structure when moored in the ocean. The weight density of salt water is 64 lb/ft^3. Neglect the vertical components of the mooring forces.   *Ans.* $h = 74.5$ ft

Side View   End View

Problem 5/139

5/140 The sides of a V-shaped fresh-water trough, shown in section, are hinged about their common intersection through O and are held together by a cable and turnbuckle placed every 2 m along the length of the trough. Calculate the tension T supported by each turnbuckle.

Problem 5/140

5/141 A fresh-water channel 9 m wide (normal to the plane of the paper) is blocked by a rectangular barrier, shown by its section ACD. Supporting horizontal struts BC are placed every 0.6 m along the 9-m width. Determine the compression C in each strut BC. The weight of the barrier is assumed to be small compared with the other forces acting.

Ans. $C = 25.4$ kN

Problem 5/141

5/142 A rectangular plate, shown in edge view, is 3 m high and 2.7 m wide (normal to the paper) and separates reservoirs of sea water and oil. The oil has a specific gravity (ratio of density to that of fresh water) of 0.85. Determine the depth h of the water required to make the reaction at B equal to zero.

Problem 5/142

5/143 The rectangular gate shown in section is 10 ft long (perpendicular to the paper) and is hinged about its upper edge B. The gate divides a channel leading to a fresh-water lake on the left and a salt-water tidal basin on the right. Calculate the torque M on the shaft of the gate at B required to prevent the gate from opening when the salt-water level drops to $h = 3$ ft.

Ans. $M = 11.22(10^4)$ lb-ft

Problem 5/143

Problem 5/144

Problem 5/145

Problem 5/146

Problem 5/147

5/144 The form for a concrete footing is hinged to its lower edge A and held in place every 1.5 m by a horizontal brace BC. Neglect the weight of the form and compute the compression C in each brace resulting from the liquid behavior of wet concrete, which has a density of 2.40 Mg/m^3.

5/145 The cover plate for an access opening in a fresh-water tank is bolted in place and the tank is filled to the level shown. Compute the increase in tension in each bolt at A and B due to filling the tank. (Assume no change in the gasket pressure between the cover plate and the tank so that the force due to water pressure is resisted entirely by the increase in the bolt tensions.)   *Ans.* $T_A = 80.0$ N, $T_B = 96.6$ N

5/146 The quarter-circular gate AB, shown in section, has a horizontal width of 6 ft (normal to the plane of the paper) and controls the flow of fresh water over the ledge at B. The gate has a total weight of 6800 lb and is hinged about its upper edge A. Calculate the minimum force P required to keep the gate closed. In locating the center of gravity of the gate neglect its thickness compared with the 9-ft radius.
Ans. $P = 10,830$ lb

5/147 The submersible diving chamber has a total mass of 6.7 Mg including personnel, equipment, and ballast. When the chamber is lowered to a depth of 1.2 km in the ocean, the cable tension is 8 kN. Compute the total volume V displaced by the chamber.

5/148 Steel chain with a nominal shank diameter of 6 mm
has a mass of 93 kg per 100 meters of length and has
a breaking strength of 13.5 kN. What length h of
chain can be lowered into a deep part of the ocean
before the chain breaks under the action of its own
weight?   *Ans. $h = 1704$ m*

5/149 A supertanker moves from a freshwater anchorage
to a berth in salt water and then takes on an additional 2400 m³ of fuel oil (specific gravity 0.88). The
draft marks on the hull of the tanker read the same
in salt water as they did in fresh water before the
transfer. Calculate the final displacement m (total
mass) of the ship in metric tons (Mg).

5/150 The deep submersible research vessel has a passen-
ger compartment in the form of a spherical steel
shell with a mean radius of 1.000 m and a thickness
of 35 mm. Calculate the mass of lead ballast which
the vessel must carry so that the combined weight
of the steel shell and lead ballast exactly cancels
their combined buoyancy. (Consult Table C1, Appendix C, as needed.)
                    Ans. $m = 1.210$ Mg (metric tons)

Problem 5/150

5/151 The solid floating object is composed of a hemisphere and a circular cylinder of equal radii r. If the
object floats with the center of the hemisphere
above the water surface, determine the maximum
height h which the cylinder may have before the
object will no longer float in the upright position
illustrated.   *Ans. $h = r/\sqrt{2}$*

Problem 5/151

5/152 Determine the total force R exerted on the triangular window by the fresh water in the tank. The
water level is even with the top of the window. Also
determine the distance H from R to the water level.

Problem 5/152

Top View

Frontal View
Problem 5/153

5/153 A large butterfly valve in a horizontal pipe consists of a flat circular disk pivoted about a vertical diametral axis O-O. If the difference in fluid pressure between the two sides of the valve varies linearly with distance from the axis from zero to p_0 as shown, determine the torque M on the shaft at O required to hold the valve in place. (*Hint:* Recognize the applicability of the results of Sample Problem 5/15.)

$$\text{Ans. } M = \frac{\pi}{4}p_0 r^3$$

Problem 5/154

5/154 The bottom of a fresh-water channel, shown in section, consists of two uniform rectangular plates AB and BC each having a mass of 1.5 Mg and hinged along their common edge B and hinged also to the base of the fixed channel along their lower edges A and C. The length of the channel is 6 m, measured perpendicular to the plane of the paper. Determine the force P per meter of channel length exerted by each plate on the connecting hinge at B.

Problem 5/155

5/155 A control gate of cylindrical cross section is used to regulate the flow of water over the spillway of a fresh-water dam as shown. The gate has a mass of 7 Mg and has a length normal to the paper of 6 m. A torque applied to the shaft of the gate at O controls the angular position of the gate. For the position shown determine the horizontal and vertical components F_x and F_y of the total force exerted by the shaft on its bearings at O. Account for the effect of fluid pressure by a direct integration over the cylindrical surface.   Ans. $F_x = 127.1$ kN, $F_y = 35.4$ kN

5/156 A flat plate seals a triangular opening in the vertical
wall of a tank of liquid of density ρ. The plate is
hinged about the upper edge O of the triangle.
Determine the force P required to hold the gate in a
closed position against the pressure of the liquid.

$$\text{Ans. } P = \frac{\rho g a b}{6}\left(h + \frac{a}{2}\right)$$

Problem 5/156

5/157 One end of a uniform pole of length l and density ρ_1
is hinged about a point a distance h above the
surface of a liquid of density ρ_2. Find the angle θ
assumed by the pole. Assume that $\rho_2 > \rho_1$.

$$\text{Ans. } \theta = \sin^{-1}\left(\frac{h}{l}\sqrt{\frac{\rho_2}{\rho_2 - \rho_1}}\right)$$

Problem 5/157

5/158 The accurate determination of the vertical position
of the center of mass G of a ship is difficult to
achieve by calculation. It is more easily obtained by
a simple inclining experiment on the loaded ship.
With reference to the figure, a known external mass
m_0 is placed a distance d from the centerline, and
the angle of list θ is measured by means of the
deflection of a plumb bob. The displacement of the
ship and the location of the metacenter M are
known. Calculate the metacentric height $\overline{GM}$ for a
12000-t ship inclined by a 27-t mass placed 7.8 m
from the centerline if a 6-m plumb line is deflected
a distance $a = 0.2$ m. The mass m_0 is at a distance
$b = 1.8$ m above M. [Note that the metric ton (t)
equals 1000 kg and is the same as the megagram
(Mg).] $\text{Ans. } \overline{GM} = 0.530$ m

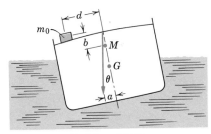

Problem 5/158

5/9 PROBLEM FORMULATION AND REVIEW. In Chapter 5 we have treated various common examples of forces distributed throughout volumes, over areas, and along lines. In all of these problems we are primarily interested in two things—the resultant of the distributed forces and the location of the resultant.

In finding the resultant we begin by multiplying the intensity of the force by the appropriate element of volume, area, or length in terms of which the intensity is expressed. We then sum up (integrate) the incremental forces over the region involved to obtain their resultant.

In finding the location of the line of action of the resultant we depend on the principle of moments (Varignon's theorem). Here we equate the sum of the moments about a convenient axis of all of the increments of force to the moment of the resultant about the same axis, and then solve for the unknown moment arm to the resultant.

When force is distributed throughout a mass, as in the case of gravitational attraction, the intensity is the force of attraction ρg per unit of volume where ρ is the density and g is the gravitational acceleration. For bodies whose density is constant, we saw in Section A that ρg cancels when the moment principle is applied. This leaves us with a strictly geometric problem of finding the centroid of the figure which then coincides with the mass center of the physical body whose boundary defines the figure. For flat plates and shells which are homogeneous and have constant thickness, the problem becomes one of finding the properties of an area. And for slender rods and wires of uniform density and constant cross section, the problem becomes one of finding the properties of a line segment. These mathematical idealizations of the physical bodies are essential elements to recognize in the formulation of actual problems.

For problems that require the integration of differential relationships, we cite four measures which are especially useful. First, select a suitable coordinate system. Generally, the system which provides the simplest description of the boundaries of the region of integration is the best choice. Second, higher-order differential quantities may always be dropped compared with lower-order differential quantities. Third, choose a first-order differential element in preference to a second-order element and a second-order element in preference to a third-order element to reduce the labor of computation. Fourth, wherever possible choose a differential element which avoids discontinuities within the region of integration.

In Section B of Chapter 5 we made use of the foregoing observations along with our principles of equilibrium to solve for the effects of distributed forces in beams, cables, and fluids. In beams and cables we expressed the force intensity as force per unit length. For fluids we expressed the force intensity as force per unit area, or pressure. Although these three problems are physically quite different, in their formulation they embody the common elements cited.

REVIEW PROBLEMS

5/159 Determine the x-coordinate of the centroid of the shaded area between the quadrant of the ellipse and the straight line. (Use Table C3, Appendix C, as needed.)

$$Ans. \ \bar{X} = \frac{2a/3}{\pi - 2}$$

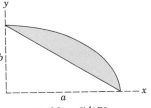

Problem 5/159

5/160 The uniform semicircular bar AB has a mass ρ per unit of its length and rests with its ends on the smooth horizontal surfaces, one higher than the other. Determine the force F_A under end A.

$$Ans. \ F_A = \frac{\rho g \pi r}{2}\left(1 - \frac{2}{\pi}\right)$$

Problem 5/160

5/161 A small balloon for use in recording wind velocity and its direction has a mass of 0.4 kg. If the balloon is inflated with 0.3 kg of helium and exerts an upward force of 5.3 N on its mooring prior to release, determine the diameter d of the spherical balloon. The density of the air is 1.206 kg/m³.

$$Ans. \ d = 1.138 \ m$$

Problem 5/162

5/162 Determine the coordinates of the centroid of the sheet-metal shape.

5/163 Calculate the mass m and total surface area A of the wheel generated by revolving the shaded section about the z-axis. The material of the wheel has a density of 7.21 Mg/m³.

$$Ans. \ m = 291 \ kg, \ A = 1.149 \ m^2$$

Problem 5/163

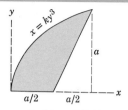

$x = ky^3$

a

$a/2$ $a/2$

Problem 5/164

5/164 Determine the *x*-coordinate of the centroid of the shaded area.

5/165 A cable 300 m long is suspended between two points on the same level 280 m apart. If the cable supports a large load uniformly distributed with respect to the horizontal direction, find the sag *h* in the cable from Eq. 5/16 by successive approximations.   *Ans.* $h = 47.5$ m

0.5 kN/m

←—6 m—→←—6 m—→

Problem 5/166

5/166 Draw the shear and moment diagrams for the cantilever beam loaded as shown.

A

x

$\bar{x}$

•*B*

—50 m—

Problem 5/167

5/167 The center of buoyancy *B* of a ship's hull is the centroid of its submerged volume. The underwater cross-sectional areas *A* of the transverse sections of the tugboat hull shown are tabulated for every five meters of waterline length. Plot these values and determine to the nearest 0.5 m the distance $\bar{x}$ of *B* aft of point *A*.

x, m	A, m^2	x, m	A, m^2
0	0	25	25.1
5	7.1	30	23.8
10	15.8	35	19.5
15	22.1	40	12.5
20	24.7	45	5.1
		50	0

Ans. $\bar{x} = 24$ m

←—4'—→

M

B   Oil   4'

5'   Water

3'

30°

A

Problem 5/168

5/168 The gate *AB* is a 580-lb rectangular plate 5 ft high and 3.5 ft wide and is used to close the discharge channel at the bottom of an oil reservoir. As a result of condensation in the tank, fresh water collects at the bottom of the channel. Calculate the moment *M* applied about the hinge axis *B* required to close the gate against the hydrostatic forces of the water and oil. The specific gravity of the oil is 0.85.

5/169 A structure for observation of sea life beneath the
       ice in polar waters consists of the cylindrical view-
       ing chamber connected to the surface by the cylin-
       drical shaft open at the top for ingress and egress.
       Ballast is carried in the rack below the chamber. To
       ensure a stable condition for the structure, it is
       necessary that its legs bear upon the ice with a force
       that is at least 15 percent of the total buoyancy
       force of the submerged structure. If the structure
       less ballast has a mass of 5.7 Mg, calculate the
       required mass m of lead ballast. The density of lead
       is 11.37 Mg/m^3.   *Ans. m* = 4.24 Mg

Problem 5/169

5/170 Draw the shear and moment diagrams for the
       loaded beam and determine the bending moment M
       that induces the maximum compression in the
       upper fibers of the beam.   *Ans. M* = 52.8 N·m

Problem 5/170

5/171 The beam AD is supported and loaded as shown.
       The cable that holds the 2000-lb load is wrapped
       around the drum at E, and the drum is locked to its
       bracket and cannot rotate. Neglect the dimensions
       of the flanges that attach the brackets to the beam
       at B and D and construct the shear and moment
       diagrams for the loaded beam. Determine the bend-
       ing moment M with the greatest magnitude and the
       distance x from the pivot A at which M occurs.

Problem 5/171

Problem 5/172

5/172 A dam consists of the flat-plate barriers A and B whose masses are small. Supporting struts C and D are placed every 3 m of dam section. A mud sample drawn up to the surface has a density of 1.6 Mg/m³. Determine the compression in C and D. All joints may be assumed to be hinged.

Ans. $C = 474$ kN, $D = 88.3$ kN

Problem 5/173

5/173 The end of a fresh-water channel with a 60° V-section is closed by the slanted triangular plate B. Calculate the resultant force R exerted on B by the water and the height h of the point on B through which R acts.   Ans. $R = 151.0$ kN, $h = 2$ m

Problem 5/174

▶5/174 The surface is generated by revolving the semicircular arc of radius b through 180° around the z-axis. Determine the distance $\bar{r}$ from the z-axis to the centroid of the surface.

$$\text{Ans. } \bar{r} = \frac{\pi(2a^2 + b^2) - 8ab}{\pi(\pi a - 2b)}$$

Problem 5/175

▶5/175 The beam supports a uniformly distributed load as shown. Determine the location x of the two supports so as to minimize the maximum bending moment M in the beam. What is M?

Ans. $M = 0.0215wl^2$, $x = 0.207l$

5/176 While repairs on its cable are being made, the propellers of the cable-laying ship exert a forward thrust of 300 kN to hold a fixed position in calm water. The ocean depth at this location is h, and the cable is observed to make an angle of $\theta = 60°$ with the horizontal where it enters the water at point P. The cable has a mass of 22 kg/m and has a cross-sectional area of 4600 mm². The density of salt water is 1.03 Mg/m³. Calculate the length s of cable from point P to the point where the cable leaves the ocean floor and find the horizontal distance x from this point to a point directly below P. Determine the depth h of the ocean. Also compute the torque M on the 2-m-diameter cable-laying drum to prevent it from turning. (*Note:* The vertical loading is the difference between the cable weight and the weight of the water displaced.)

Ans. $s = 3070$ m, $x = 2330$ m

$h = 1770$ m, $M = 600$ kN·m

Problem 5/176

FRICTION

6

6/1 INTRODUCTION. In the preceding chapters the forces of action and reaction between contacting surfaces have for the most part been assumed to act normal to the surfaces. This assumption characterizes the interaction between smooth surfaces and was illustrated in example 2 of Fig. 3/1. Although in many instances this ideal assumption involves only a relatively small error, there are a great many problems wherein we must consider the ability of contacting surfaces to support tangential as well as normal forces. Tangential forces generated between contacting surfaces are known as friction forces and are present to some degree with the interaction between all real surfaces. Whenever a tendency exists for one contacting surface to slide along another surface, we find that the friction forces developed are always in a direction to oppose this tendency.

In some types of machines and processes we desire to minimize the retarding effect of friction forces. Examples are bearings of all types, power screws, gears, the flow of fluids in pipes, and the propulsion of aircraft and missiles through the atmosphere. In other situations we wish to maximize the use of friction, as in brakes, clutches, belt drives, and wedges. Wheeled vehicles depend on friction for both starting and stopping, and ordinary walking depends on friction between the shoe and the ground. Friction forces are present throughout nature and exist to a considerable extent in all machines no matter how accurately constructed or carefully lubricated. A machine or process in which friction is small enough to be neglected is often referred to as *ideal*. When friction must be taken into account, the machine or process is termed *real*. In all real cases where sliding motion between parts occurs, the friction forces result in a loss of energy which is dissipated in the form of heat. In addition to the generation of heat and the accompanying loss of energy, friction between mating parts will cause wear.

SECTION A. FRICTIONAL PHENOMENA

6/2 TYPES OF FRICTION. There are a number of separate types of frictional resistance encountered in mechanics, and we will mention each of these types briefly in this article prior to a more detailed account of the most common type of friction in the next article.

(*a*) *Dry Friction.* Dry friction is encountered when the unlubricated surfaces of two solids are in contact under a condition of sliding or tendency to slide. A friction force tangent to the surfaces of contact is developed both during the interval leading up to impending slippage and while slippage takes place. The direction of the force always opposes the motion or impending motion. This type of friction is also called *Coulomb* friction. The principles of dry or Coulomb friction were developed largely from the experiments of Coulomb in 1781 and from the work of Morin from 1831 to 1834. Although a comprehensive theory of dry friction is not available, an analytical model sufficient to handle the vast majority of problems in dry friction is available and is described in Art. 6/3 which follows. This model forms the basis for most of this chapter.

(*b*) *Fluid Friction.* Fluid friction is developed when adjacent layers in a fluid (liquid or gas) are moving at different velocities. This motion gives rise to frictional forces between fluid elements, and these forces depend on the relative velocity between layers. When there is no such relative velocity, there is no fluid friction. Fluid friction depends not only on the velocity gradients within the fluid, but also on the viscosity of the fluid, which is a measure of its resistance to shearing action between fluid layers. Fluid friction is treated in the study of fluid mechanics and will not be developed further in this book.

(*c*) *Internal Friction.* Internal friction is found in all solid materials that are subjected to cyclical loading. For highly elastic materials the recovery from deformation occurs with very little loss of energy caused by internal friction. For materials which have low limits of elasticity and which undergo appreciable plastic deformations during loading, the amount of internal friction that accompanies this deformation may be considerable. The mechanism of internal friction is associated with the action of shear deformation, and the student should consult a reference on materials science for a detailed description of this shear mechanism. Since this book deals primarily with the external effects of forces, we shall not be concerned with internal friction in the work which follows.

6/3 DRY FRICTION.   The remainder of this chapter will be devoted to a description of the effects of dry friction acting on the exterior surfaces of rigid bodies. We shall explain the mechanism of dry friction with the aid of a very simple experiment.

(*a*) *Mechanism of Friction.* Consider a solid block of mass *m* resting on a horizontal surface, as shown in Fig. 6/1*a*. The contacting surfaces possess a certain amount of roughness. The experiment will involve the application of a horizontal force *P* which will vary continuously from zero to a value sufficient to move the block and give it an appreciable velocity. The free-body diagram of the block for any value of P is shown in Fig. 6/1*b*, and the tangential friction force exerted by the plane on the block is labeled F. This friction force will *always* be in a direction to oppose motion or the tendency toward motion of the body on which it acts. There is also a normal force N which in this case equals mg, and the total force R exerted by the supporting surface on the block is the resultant of N and F. A magnified view of the irregularities of the mating surfaces, Fig. 6/1*c*, will aid in visualizing the mechanical action of friction. Support is necessarily intermittent and exists at the mating humps. The direction

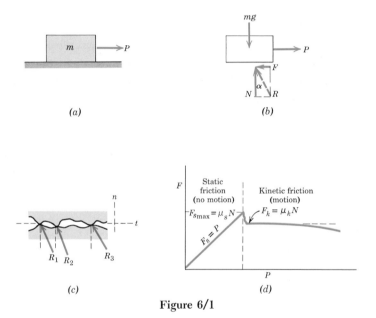

Figure 6/1

of each of the reactions on the block R_1, R_2, R_3, etc., will depend not only on the geometric profile of the irregularities but also on the extent of local deformation at each contact point. The total normal force N is merely the sum of the n-components of the R's, and the total frictional force F is the sum of the t-components of the R's. When the surfaces are in relative motion, we can see that the contacts are more nearly along the tops of the humps, and the t-components of the R's will be smaller than when the surfaces are at rest relative to one another. This consideration helps to explain the well-known fact that the force P necessary to maintain motion is generally less than that required to start the block when the irregularities are more nearly in mesh.

Assume now that we perform the experiment indicated and

record the friction force F as a function of P. The resulting experimental relation is indicated in Fig. 6/1d. When P is zero, equilibrium requires that there be no friction force. As P is increased, the friction force must be equal and opposite to P as long as the block does not slip. During this period the block is in equilibrium, and all forces acting on the block must satisfy the equilibrium equations. Finally we reach a value of P which causes the block to slip and to move in the direction of the applied force. At this same time the friction force drops slightly and rather abruptly to a somewhat lower value. Here it remains essentially constant for an interval but then drops off still more with higher velocities.

The region up to the point of slippage or impending motion is known as the range of *static friction*, and the value of the friction force is determined by the *equations of equilibrium*. This force may have any value from zero up to and including, in the limit, the maximum value. For a given pair of mating surfaces we find that this maximum value of static friction $F_{s_{\max}}$ is proportional to the normal force N. Hence we may write

$$F_{s_{\max}} = \mu_s N$$

where μ_s is the proportionality constant, known as the *coefficient of static friction*. We must observe carefully that this equation describes only the *limiting* or *maximum* value of the static friction force and not any lesser value. Thus the equation applies *only* to cases where it is known that motion is impending with the friction force at its peak value.

After slippage occurs a condition of *kinetic friction* accompanies the ensuing motion. Kinetic friction force is usually somewhat less than the maximum static friction force. We find also that the kinetic friction force F_k is proportional to the normal force. Hence

$$F_k = \mu_k N$$

where μ_k is the *coefficient of kinetic friction*. It follows that μ_k is generally less than μ_s. As the velocity of the block increases, the kinetic friction coefficient decreases somewhat, and when high velocities are reached the effect of lubrication by an intervening fluid film may become appreciable. Coefficients of friction depend greatly on the exact condition of the surfaces as well as on the velocity and are subject to a considerable measure of uncertainty.

It is customary to write the two friction force equations merely as

$$\boxed{F = \mu N} \tag{6/1}$$

We will understand from the problem at hand whether limiting static friction with its corresponding coefficient of static friction or whether kinetic friction with its corresponding kinetic coefficient is involved. We emphasize again that many problems involve a static friction force which is *less* than the maximum value at impending motion, and therefore under these conditions the friction equation *cannot* be used.

From Fig. 6/1c we observe that for rough surfaces there is a greater possibility for large angles between the reactions and the *n*-direction than for smoother surfaces. Thus a friction coefficient reflects the roughness of a pair of mating surfaces and incorporates a geometric property of both mating contours. It is meaningless to speak of a coefficient of friction for a single surface.

The direction of the resultant R in Fig. 6/1b measured from the direction of N is specified by $\tan \alpha = F/N$. When the friction force reaches its limiting static value, the angle α reaches a maximum value ϕ_s. Thus

$$\tan \phi_s = \mu_s$$

When slippage occurs, the angle α will have a value ϕ_k corresponding to the kinetic friction force. In like manner

$$\tan \phi_k = \mu_k$$

We usually write merely

$$\boxed{\tan \phi = \mu} \tag{6/2}$$

where application to the limiting static case or to the kinetic case is inferred from the problem at hand. The angle ϕ_s is known as the *angle of static friction*, and the angle ϕ_k is called the *angle of kinetic friction*. This friction angle ϕ for each case clearly defines the limiting position of the total reaction R between two contacting surfaces. If motion is impending, R must be one element of a right circular cone of vertex angle $2\phi_s$, as shown in Fig. 6/2. If motion is not impending, R will be within the cone. This cone of vertex angle $2\phi_s$ is known as the *cone of static friction* and represents the locus of possible positions for the reaction R at impending motion. If motion occurs, the angle of kinetic friction applies, and the reaction must lie on the surface of a slightly different cone of vertex angle $2\phi_k$. This cone is the *cone of kinetic friction*.

Further experiment shows that the friction force is essentially independent of the apparent or projected area of contact. The true contact area is much smaller than the projected value, since only the peaks of the contacting surface irregularities support the load. Relatively small normal loads result in high stresses at these contact points. As the normal force increases, the true contact area also increases as the material undergoes yielding, crushing, or tearing at the points of contact. A comprehensive theory of dry friction must go

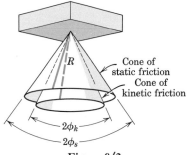

Figure 6/2

beyond the mechanical explanation presented here. For example, there is some evidence to support the theory that molecular attraction may be an important cause of friction under conditions where the mating surfaces are in very intimate contact. Other factors that influence dry friction are the generation of high local temperatures and adhesion at contact points, relative hardness of mating surfaces, and the presence of thin surface films of oxide, oil, dirt, or other substances.

(*b*) *Types of Friction Problems.* From what has been said we may now recognize three distinct types of problems which are encountered in dry friction. The very first step in the solution of a friction problem is to identify which of these categories applies.

(1) In the *first* type the condition of impending motion exists. Here a body which is in equilibrium is on the verge of slipping, and the friction force equals the limiting static friction $F_{s_{\max}} = \mu_s N$. The equations of equilibrium will, of course, also hold.

(2) In the *second* type of problem impending motion need not exist, and therefore the friction force may be smaller than that given by Eq. 6/1 with the static coefficient. In this event the friction force will be determined only by the equations of equilibrium. In such a problem we may be asked whether or not the existing friction is sufficient to maintain the body at rest. To answer the question, we may assume equilibrium and calculate the corresponding friction force F necessary to maintain this condition from the equations of equilibrium. We then compare this friction force with the maximum static friction that the surfaces can support as calculated from Eq. 6/1 with $\mu = \mu_s$. If F is less than that given by Eq. 6/1, we may conclude that the assumed friction force can be supported and therefore the body is at rest. If the calculated value of F is greater than the limiting value, then we conclude that the given surfaces cannot support that much friction force, and therefore motion exists and the friction becomes kinetic.

(3) In the *third* type of problem relative motion exists between the contacting surfaces, and here the kinetic coefficient of friction clearly applies. For this case Eq. 6/1 will always give the kinetic friction force directly.

The foregoing discussion applies to all dry contacting surfaces and, to a limited extent, to moving surfaces which are partially lubricated. Some typical values of the coefficients of friction are given in Table C1, Appendix C. These values are only approximate and are subject to considerable variation, depending on the exact conditions prevailing. They may be used, however, as typical examples of the magnitudes of frictional effects. When a reliable calculation involving friction is required, it is often desirable to determine the appropriate friction coefficient by experiment wherein the surface conditions of the problem are duplicated as closely as possible.

Sample Problem 6/1

Determine the maximum angle θ which the adjustable incline may have with the horizontal before the block of mass m begins to slip. The coefficient of static friction between the block and the inclined surface is μ_s.

Solution. The free-body diagram of the block shows its weight $W = mg$, the normal force N, and the friction force F exerted by the incline on the block. The friction force acts in the direction to oppose the slipping which would occur if no friction were present.

① Equilibrium in the x- and y-directions requires

$[\Sigma F_x = 0] \qquad mg \sin\theta - F = 0 \qquad F = mg \sin\theta$

$[\Sigma F_y = 0] \qquad -mg \cos\theta + N = 0 \qquad N = mg \cos\theta$

Dividing the first equation by the second gives $F/N = \tan\theta$. Since the maximum angle occurs when $F = F_{s_{\max}} = \mu_s N$, for impending motion we have

$$\mu_s = \tan\theta_{\max} \qquad \text{or} \qquad \theta_{\max} = \tan^{-1}\mu_s \qquad \textit{Ans.}$$

① We choose reference axes along and normal to the direction of F to avoid resolving both F and N into components.

② This problem describes a very simple way to determine a static coefficient of friction. The maximum value of θ is known as the *angle of repose*.

Sample Problem 6/2

Determine the range of values which the mass m_0 may have so that the 100-kg block shown in the figure will neither start moving up the plane nor slip down the plane. The coefficient of static friction for the contact surfaces is 0.30.

Solution. The maximum value of m_0 will be given by the requirement for motion impending up the plane. The friction force on the block therefore acts down the plane as shown in the free-body diagram of the block for Case I in the figure. With the weight $mg = 100(9.81) = 981$ N, the equations of equilibrium give

$[\Sigma F_y = 0] \qquad N - 981 \cos 20° = 0, \qquad N = 922$ N

$[F = \mu N] \qquad F = 0.30(922) = 277$ N

$[\Sigma F_x = 0] \quad m_0(9.81) - 277 - 981 \sin 20° = 0, \qquad m_0 = 62.4$ kg $\quad$ *Ans.*

① The minimum value of m_0 is determined when motion is impending down the plane. The friction force on the block will act up the plane to oppose the tendency to move as shown in the free-body diagram for Case II. Equilibrium in the x-direction requires

Case I

Case II

$[\Sigma F_x \doteq 0] \quad m_0(9.81) + 277 - 981 \sin 20° = 0, \qquad m_0 = 6.0$ kg $\quad$ *Ans.*

Thus m_0 may have any value from 6.0 to 62.4 kg, and the block will remain at rest.

In both cases equilibrium requires that the resultant of F and N will be concurrent with the 981-N weight and the applied tension T.

① We see from the results of Sample Problem 6/1 that the block would slide down the incline without the restraint of attachment to m_0 since $\tan 20° > 0.30$. Thus a value of m_0 will be required to maintain equilibrium.

274

Sample Problem 6/3

Determine the magnitude and direction of the friction force acting on the 100-kg block shown if, first, $P = 500$ N, and, second, $P = 100$ N. The coefficient of static friction is 0.20, and the coefficient of kinetic friction is 0.17. The forces are applied with the block initially at rest.

Solution. There is no way of telling from the statement of the problem whether the block will remain in equilibrium or whether it will begin to slip following the application of P. It is therefore necessary that we make an assumption, so we will take the friction force to be up the plane, as shown by the solid arrow. From the free-body diagram a balance of forces in both x- and y-directions gives

$$[\Sigma F_x = 0] \qquad P \cos 20° + F - 981 \sin 20° = 0$$

$$[\Sigma F_y = 0] \qquad N - P \sin 20° - 981 \cos 20° = 0$$

Case I. $P = 500$ N
Substitution into the first of the two equations gives

$$F = -134 \text{ N}$$

The negative sign tells us that *if* the block is in equilibrium, the friction force acting on it is in the direction opposite to that assumed and therefore is down the plane as represented by the dotted arrow. We cannot reach a conclusion on the magnitude of F, however, until we verify that the surfaces are capable of supporting 134 N of friction force. This may be done by substituting $P = 500$ N into the second equation, which gives

$$N = 1093 \text{ N}$$

The maximum static friction force that the surfaces can support is then

$$[F = \mu N] \qquad F_{max} = 0.20(1093) = 219 \text{ N}$$

Since this force is greater than that required for equilibrium, we conclude that the assumption of equilibrium was correct. The answer is, then,

$$F = 134 \text{ N down the plane} \qquad\qquad Ans.$$

Case II. $P = 100$ N
Substitution into the two equilibrium equations gives

$$F = 231 \text{ N}, \qquad N = 956 \text{ N}$$

But the maximum possible static friction force is

$$[F = \mu N] \qquad F_{max} = 0.20(956) = 191 \text{ N}$$

It follows that 231 N of friction cannot be supported. Therefore equilibrium cannot exist, and we obtain the correct value of the friction force by using the kinetic coefficient of friction accompanying the motion down the plane. Hence the answer is

$$[F = \mu N] \qquad F = 0.17(956) = 163 \text{ N up the plane} \qquad Ans.$$

① We should note that even though ΣF_x is no longer equal to zero, equilibrium does exist in the y-direction, so that $\Sigma F_y = 0$. Therefore the normal force N is 956 N whether or not the block is in equilibrium.

Sample Problem 6/4

The homogeneous rectangular block of mass m, width b, and height H is placed on the horizontal surface and subjected to a horizontal force P which moves the block along the surface with a constant velocity. The coefficient of kinetic friction between the block and the surface is μ. Determine (a) the greatest value that h may have so that the block will slide without tipping over and (b) the location of a point C on the bottom face of the block through which the resultant of the friction and normal forces acts if $h = H/2$.

Solution. (a) With the block on the verge of tipping, we see that the entire reaction between the plane and the block will be at A. The free-body diagram of the block shows this condition. Since slipping occurs, the friction force is the limiting value μN, and the angle θ becomes $\theta = \tan^{-1}\mu$. The resultant of F and N passes through a point B through which P must also pass, since three coplanar forces in equilibrium are concurrent. Hence from the geometry of the block

$$\tan\theta = \mu = \frac{b/2}{h}, \quad h = \frac{b}{2\mu} \qquad Ans.$$

If h were greater than this value, moment equilibrium about A would not be satisfied, and the block would tip over.

(b) With $h = H/2$ we see from the free-body diagram for case (b) that the resultant of F and N passes through a point C which is a distance x to the left of the vertical centerline through G. The angle θ is still $\theta = \phi = \tan^{-1}\mu$ as long as the block is slipping. Thus, from the geometry of the figure we have

$$\frac{x}{H/2} = \tan\theta = \mu \quad \text{so} \quad x = \mu H/2 \qquad Ans.$$

If μ represents the coefficient of static friction, then our solutions describe the condition under which the block is (a) on the verge of tipping and (b) on the verge of slipping, both from a rest position.

① Recall that the equilibrium equations apply to a body moving with a constant velocity (zero acceleration) just as well as to a body at rest.

② Alternatively, we could equate the moments about G to zero which would give us $F(H/2) - Nx = 0$. Thus with $F = \mu N$ we get $x = \mu H/2$.

Sample Problem 6/5

The three flat blocks are positioned on the 30° incline as shown, and a force P parallel to the incline is applied to the middle block. The upper block is prevented from moving by a wire which attaches it to the fixed support. The coefficient of static friction for each of the three pairs of mating surfaces is shown. Determine the maximum value which P may have before any slipping takes place.

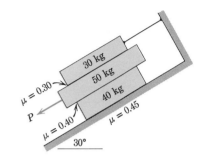

Solution. The free-body diagram of each block is drawn. The friction forces are assigned in the directions to oppose the relative motion which ① would occur if no friction were present. There are two possible conditions for impending motion. Either the 50-kg block slips and the 40-kg block remains in place, or the 50- and 40-kg blocks move together with slipping occurring between the 40-kg block and the incline.

The normal forces, which are in the y-direction, may be determined without reference to the friction forces, which are all in the x-direction. Thus

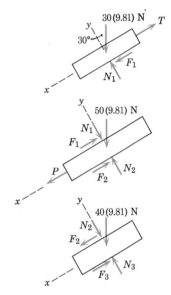

$[\Sigma F_y = 0]$ (30-kg) $N_1 - 30(9.81)\cos 30° = 0$ $N_1 = 255$ N

(50-kg) $N_2 - 50(9.81)\cos 30° - 255 = 0$ $N_2 = 680$ N

(40-kg) $N_3 - 40(9.81)\cos 30° - 680 = 0$ $N_3 = 1019$ N

We shall assume arbitrarily that only the 50-kg block slips, so that the 40-kg block remains in place. Thus for impending slippage at both surfaces of the 50-kg block we have

$[F = \mu N]$ $F_1 = 0.30(255) = 76.5$ N $F_2 = 0.40(680) = 272$ N

Equilibrium of forces at impending motion for the 50-kg block gives

$[\Sigma F_x = 0]$ $P - 76.5 - 272 + 50(9.81)\sin 30° = 0$, $P = 103.1$ N

We now check on the validity of our initial assumption. For the 40-kg block with $F_2 = 272$ N the friction force F_3 would be given by

$[\Sigma F_x = 0]$ $272 + 40(9.81)\sin 30° - F_3 = 0$ $F_3 = 468$ N

But the maximum possible value of F_3 is $F_3 = \mu N_3 = 0.45(1019) = 459$ N. Thus 468 N cannot be supported and our initial assumption was wrong. We conclude, therefore, that slipping occurs between the 40-kg block and the incline. With the corrected value $F_3 = 459$ N equilibrium of the 40-kg block requires

② $[\Sigma F_x = 0]$ $F_2 + 40(9.81)\sin 30° - 459 = 0$ $F_2 = 263$ N

Equilibrium of the 50-kg block gives, finally,

$[\Sigma F_x = 0]$ $P + 50(9.81)\sin 30° - 263 - 76.5 = 0$, $P = 93.8$ N   *Ans.*

① In the absence of friction the middle block, under the influence of P, would have a greater movement than the 40-kg block, and the friction force F_2 will be in the direction to oppose this motion as shown.

② We see now that F_2 is less than $\mu N_2 = 272$ N.

PROBLEMS

6/1 The 100-lb block rests on the horizontal surface, and a force P whose direction can be varied is applied to the block. If $P = 50$ lb and if the block begins to move when α is increased to $60°$, calculate the static coefficient of friction μ between the block and the surface.   *Ans. $\mu = 0.577$*

Problem 6/1

6/2 Calculate the force P required to move the 50-kg block up the $30°$ incline.

Problem 6/2

6/3 A horizontal force P of 200 N is applied to the 100-kg crate in an effort to slide it down the incline. If the coefficient of friction between the crate and the incline is 0.70, compute the friction force F acting on the crate.   *Ans. $F = 523$ N*

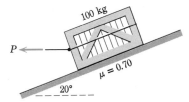

Problem 6/3

6/4 The coefficient of friction between the block and the incline is μ. Determine the value of x for which the force P required to initiate motion up the plane is the least of all possible values. Neglect the radius of the pulley.

Problem 6/4

6/5 If $P = 50$ lb and $\mu = 0.30$ in Prob. 6/1, determine the value of α, as it is gradually increased, for which the block begins to slide.   *Ans. $\alpha = 18.4°$*

6/6 Determine the maximum distance d in terms of l at which the lower end of the prop of negligible mass may be set without slipping as it supports the heavy hinged panel. The coefficient of friction is 0.30.

Problem 6/6

Problem 6/7

6/7 The tongs are used to handle hot steel tubes that are being heat treated in an oil bath. For a 20° jaw opening, what is the minimum coefficient of friction between the jaws and the tube that will enable the tongs to grip the tube without slipping?

Ans. $\mu_{\min} = 0.176$

Problem 6/8

6/8 The figure shows a device, called a jam cleat, that secures a rope under tension by reason of large friction forces developed. For the position shown determie the minimum coefficient of friction between the rope and the cam surfaces for which the cleat will be self-locking. Also find the magnitude of the total reaction R on each of the cam bearings.

Problem 6/9

6/9 The 100-lb wheel rolls on its hub up the circular incline under the action of the 25-lb weight attached to a cord around the rim. Determine the angle θ at which the wheel comes to rest, assuming that friction is sufficient to prevent slippage. What is the minimum coefficient of friction that will permit this position to be reached with no slipping?

Ans. $\theta = 32.2°$, $\mu_{\min} = 0.630$

Problem 6/10

6/10 Determine the horizontal force P required to cause slippage to occur. The friction coefficients for the three pairs of mating surfaces are indicated. The top block is free to move vertically.

6/11 The 900-mm-diameter wheel has a mass of 30 kg. Determine the couple M required to roll the wheel over the 90-mm obstruction on the horizontal surface. Specify the minimum value of the coefficient of friction between the wheel and the obstruction so that the wheel does not slip.

Ans. $M = 79.5$ N·m, $\mu_{min} = 0.75$

Problem 6/11

6/12 The homogeneous rectangular block of mass m rests on the inclined plane that is hinged about a horizontal axis through O. If the coefficient of static friction between the block and the plane is μ, specify the conditions that determine whether the block tips before it slips or slips before it tips as the angle θ is gradually increased.

Problem 6/12

6/13 Determine the maximum angle θ at which the V-block may be tilted with the horizontal before the solid cylinder begins to slide. The coefficient of friction is μ.

Ans. $\theta = \tan^{-1}(\mu/\cos\alpha)$

Problem 6/13

6/14 A uniform block of mass m is at rest on an incline θ. Determine the maximum force P that can be applied to the block in the direction shown before slipping begins. The coefficient of friction between the block and the incline is μ. Also determine the angle β between the horizontal direction of P and the direction of initial movement of the block.

Problem 6/14

6/15 The 300-lb crate with mass center at G is supported on the horizontal surfaces by a skid at A and a roller at B. If a force P of 60 lb is required to initiate motion of the crate, determine the coefficient of friction at A.

Ans. $\mu = 0.33$

Problem 6/15

500 mm ⟵2 m⟶⟵2 m⟶ 500 mm

A   B

4 m

G

P

Problem 6/16

6/16 The center of mass of the vertical 800-kg panel is at G. The panel is mounted on wheels which permit ease of horizontal movement along the fixed rail. If the bearing of the wheel at A becomes "frozen" so that the wheel cannot turn, compute the force P required to move the panel. The coefficient of friction between the wheel and the rail is 0.30.

r β A

P

Problem 6/17

6/17 A metal hoop of negligible thickness has a mass m and mean radius r and hangs from a support at A. If the coefficient of friction between the hoop and the support is μ, determine the vertical force P applied to the periphery of the hoop required to slip the hoop on the support. Also determine the angle β at which slipping occurs.

$$Ans. \ P = mg\,\frac{\mu}{\sqrt{1 + \mu^2} - \mu}, \ \beta = \tan^{-1}\mu$$

r A β

P

Problem 6/18

6/18 The hoop of Prob. 6/17 is subjected to a horizontal force P as shown here. If slipping on the support at A occurs when β reaches 60° as P is increased, determine the coefficient of friction μ between the hoop and the support. Also express the value of P in terms of the mass m of the hoop as slipping is initiated.

M 2'

30°

Problem 6/19

6/19 The uniform cylinder weighs 400 lb and is supported by the roller, which turns with negligible friction. If the coefficient of friction between the cylinder and the vertical surface is 0.6, calculate the torque M required to turn the cylinder. Also find the reaction R on the bearing of the roller as M is applied. *Ans.* $M = 206$ lb-ft, $R = 343$ lb

α

P

G

r_2 r_1

Problem 6/20

6/20 The wheel shown will roll to the left when the angle α of the cord is small. When α is large the wheel rolls to the right. Determine by inspection from the geometry of the free-body diagram the angle α for which the wheel will not roll in either direction. If the coefficient of friction is μ and the mass of the wheel is m, determine the value of P for which the wheel will slip for the critical value of α.

6/21 Calculate the force T required to rotate the 200-kg reel of telephone cable that rests on its hubs and bears against a vertical wall. The coefficient of friction for each pair of contacting surfaces is 0.60.
Ans. $T = 727$ N

Problem 6/21

6/22 The roll of paper is to be rolled slowly up the incline by a tension T applied to the paper as it is pulled horizontally off the roll. The coefficient of friction between the roll and the incline is 0.20. Prove whether the paper rolls without slipping or whether it slips.

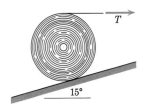

Problem 6/22

6/23 The left-hand jaw of the C-clamp can be slid along the frame to increase the capacity of the clamp. To prevent slipping of the jaw on the frame when the clamp is under load, the dimension x must exceed a certain minimum value. Find this value corresponding to given dimensions a and b and a coefficient of friction μ between the frame and the loose-fitting jaw.
Ans. $x = \dfrac{a - b\mu}{2\mu}$

Problem 6/23

6/24 Determine the torque M required to spin the uniform wheel of mass m in its position against the vertical wall. The coefficient of friction for each pair of contacting surfaces is μ.

Problem 6/24

6/25 The uniform 100-lb wheel with its hubs resting on inclined rails bears against a fixed support at A. A cord which is wrapped securely around the outer periphery leads horizontally to a hook at B. If the support at A is removed, (a) calculate the friction force acting on the wheel. The coefficients of static and kinetic friction between the hub and rail are 0.50 and 0.45, respectively. (b) If these friction coefficients were 0.30 and 0.25, what would happen?

Problem 6/25

Problem 6/26

6/26   The industrial truck is used to move the solid 1200-kg roll of paper up the 30° incline. If the coefficients of friction between the roll and the vertical barrier of the truck and between the roll and the incline are both 0.40, compute the required tractive force P between the tires of the truck and the horizontal surface.   *Ans. P = 22.1 kN*

6/27   Calculate the horizontal force between the tires and the horizontal surface for the industrial truck in Prob. 6/26 if the roll of paper is on the verge of rolling down the incline.

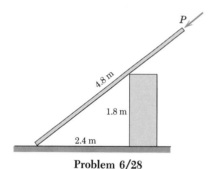

Problem 6/28

6/28   Determine the force P required to move the uniform 50-kg plank from its rest position shown if the coefficient of friction at both contact locations is 0.50.   *Ans. P = 19.62 kN*

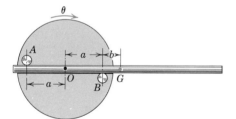

Problem 6/29

6/29   The uniform rod with center of mass at G is supported by the pegs A and B, which are fixed in the wheel. If the coefficient of friction between the rod and the pegs is μ, determine the angle θ through which the wheel may be turned about its horizontal axis through O before the rod begins to slip. Neglect the diameter of the rod compared with the other dimensions.

Problem 6/30

6/30   What force P must the two men exert on the rope to slide the uniform 20-ft plank on the overhead rack? The plank weighs 200 lb, and the coefficient of friction between the plank and each support is 0.50.   *Ans. P = 162.3 lb*

6/31 A small roller on end B of the uniform 60-kg bar is constrained to move in the smooth vertical guides. The coefficient of friction between end A and the horizontal supporting surface is 0.80. Determine the horizontal force P required to initiate slipping at A.

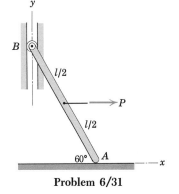

Problem 6/31

6/32 The uniform slender pole of length l and mass m is leaned against the vertical wall at the angle shown. If the coefficient of friction μ is the same at both ends of the pole, calculate the minimum value which μ may have if the pole is not to slip.

Ans. $\mu_{min} = 1/3$

Problem 6/32

6/33 Assuming the worker is strong enough, can he maintain equilibrium of the uniform 7-m plank in the position shown by exerting a horizontal pull on the rope? The coefficient of static friction between the 60-kg plank and the top of the building is 0.30. If the answer is yes, then compute the friction force existing in the equilibrium state.

Problem 6/33

6/34 Find the greatest height h of the step which the 80-kg man can reach without causing the hinged stepladder to collapse. The coefficient of friction at A is 0.40 and at B is 0.50, and each of the two uniform sections of the hinged ladder has a mass of 10 kg.

Ans. $h = 2.37$ m

Problem 6/34

Problem 6/35

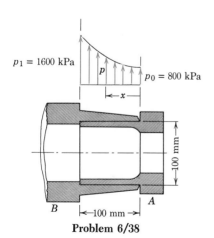

Problem 6/38

6/35 The semicylindrical shell of mass m and radius r is rolled through an angle θ by the horizontal force P applied to its rim. If the coefficient of friction is μ, determine the angle θ at which the shell slips on the horizontal surface as P is gradually increased. What value of μ will permit θ to reach $90°$?

6/36 If the coefficient of friction between the solid semi-circular cylinder and the incline is 0.30, determine the maximum angle θ with the horizontal at which the incline can be tilted before the cylinder slips. For this condition what is the corresponding angle α through which the cylinder has rolled with respect to the incline?   *Ans.* $\theta = 16.70°$, $\alpha = 25.9°$

Problem 6/36

6/37 If the worker in Prob. 6/33 relaxes his pull on the rope but maintains its horizontal direction, what angle θ will the plank make with the vertical when it begins to slip? (*Hint:* Approximate the answer with a graphical solution of the equation which governs θ.)   *Ans.* $\theta = 40.6°$

6/38 The circular collar A shown in section is mated with part B with a shrink fit which sets up a pressure or compressive stress p between the parts. The pressure has the values shown at the extremities of the overlap and is closely approximated by the relation $p = p_0 + kx^2$ in between. If a torque of 1.3 kN·m is required to turn collar A inside of part B, calculate the effective coefficient of friction μ between the two parts.   *Ans.* $\mu = 0.776$

6/39 A uniform bar of mass m and length l rests on a horizontal surface with its mass evenly distributed along its length. If the coefficient of friction between the bar and the supporting surface is μ, write an expression for the horizontal force P, applied at the end of the bar, required to move the bar and find the distance a to axis O about which the bar is observed to rotate. (*Hint:* The normal force under the bar is uniformly distributed over the length of the bar. The friction force acting on the bar is, therefore, uniformly distributed over the length of each of the two sections on either side of O.)

Ans. $P = 0.414\,\mu mg$, $a = 0.293l$

Problem 6/39

6/40 Determine the minimum force P and its corresponding angle α required to tip the uniform 200-lb crate about its front edge. Specify the minimum coefficient of friction which will permit the tipping to take place. *Ans.* $P_{min} = 70.7$ lb, $\mu_{min} = 1/3$

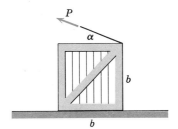

Problem 6/40

6/41 An I-beam of mass m is supported by two fixed horizontal rails as shown. Compute the applied load P that is just sufficient to cause the beam to slip and determine the corresponding friction force at A as slippage begins. The coefficient of friction between the beam and the rails is μ.

Ans. $P = \dfrac{\mu mgb}{2(a + b)}$, $F_A = \dfrac{\mu mga}{2(a + b)}$

Problem 6/41

▶**6/42** The device shown prevents clockwise rotation in the horizontal plane of the central wheel by means of frictional locking of the two small rollers. For given values of R and r and for a common coefficient of friction μ at all contact surfaces, determine the range of values of d for which the device will operate as described.

Ans. $\dfrac{2r + (1 - \mu^2)R}{1 + \mu^2} < d < (R + 2r)$

Problem 6/42

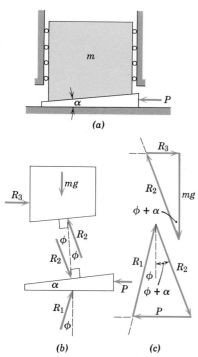

(a)

(b)   (c)

Forces to Raise Load

Figure 6/3

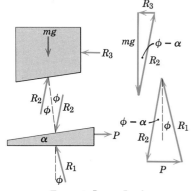

Forces to Lower Load

Figure 6/4

SECTION B. APPLICATIONS OF FRICTION IN MACHINES

6/4 WEDGES. A wedge is one of the simplest and most useful of machines and is used as a means of producing small adjustments in the
position of a body or as a means of applying large forces. Wedges are
largely dependent on friction. When sliding of a wedge is impending,
the resultant force on each sliding surface of the wedge will be
inclined from the normal to the surface by an amount equal to the
friction angle. The component of the resultant along the surface is the
friction force, which is always in the direction to oppose the motion
of the wedge relative to the mating surfaces.

Figure 6/3a shows a wedge that is used to position or lift a large
mass m, where the vertical loading is mg. The coefficient of friction
for each pair of surfaces is $\mu = \tan \phi$. The force P required to start the
wedge is found from the equilibrium triangles of the forces on the
load and on the wedge. The free-body diagrams are shown in Fig.
6/3b, where the reactions are inclined at an angle ϕ from their
respective normals and are in the direction to oppose the motion. The
mass of the wedge is neglected. From these diagrams we may write
the force equilibrium equations

$$\mathbf{W} + \mathbf{R}_2 + \mathbf{R}_3 = \mathbf{0} \quad \text{and} \quad \mathbf{R}_1 + \mathbf{R}_2 + \mathbf{P} = \mathbf{0}$$

where $W = mg$. The solutions of these equations are shown in the
c-part of the figure, where R_2 is found first in the upper diagram using
the known value of mg. The force P is then found from the lower
triangle once the value of R_2 has been established.

If P is removed, the wedge will remain in place as long as α is
less than ϕ, in which case the wedge is said to be self-locking. If the
wedge is self-locking and is to be withdrawn, a pull P on the wedge
would be required. In this event the reactions R_1 and R_2 would act on
the opposite sides of their normals to oppose the new impending
motion, and the solution would proceed along lines similar to those
described for the case of raising the load. The free-body diagrams and
vector polygons for this condition are shown in Fig. 6/4.

Wedge problems lend themselves to graphical solutions as indicated in the two figures. The accuracy of a graphical solution is easily
held to within tolerances consistent with the uncertainty of friction
coefficients. Algebraic solutions may also be obtained from the trigonometry of the equilibrium polygons.

6/5 SCREWS. Screws are used for fastenings and for transmitting power
or motion. In each case the friction developed in the threads largely
determines the action of the screw. For transmitting power or motion
the square thread is more efficient than the V-thread, and the analysis
illustrated here is confined to the square thread.

Consider the square-threaded jack, Fig. 6/5, under the action of the axial load W and a moment M applied about the axis of the screw. The force R exerted by the thread of the jack frame on a small representative portion of the thread of the screw is shown on the free-body diagram of the screw. Similar reactions exist on all segments of the screw thread where contact occurs with the thread of the base. If M is just sufficient to turn the screw, the thread of the screw will slide around and up on the fixed thread of the frame. The angle ϕ made by R with the normal to the thread will be the angle of friction, so that $\tan \phi = \mu$. The moment of R about the vertical axis of the screw is $Rr \sin(\alpha + \phi)$, and the total moment due to all reactions on the threads is $\Sigma Rr \sin(\alpha + \phi)$. Since $r \sin(\alpha + \phi)$ appears in each term, we may factor it out. The moment equilibrium equation for the screw becomes

Figure 6/5

$$M = [r \sin(\alpha + \phi)] \Sigma R$$

Equilibrium of forces in the axial direction further requires

$$W = \Sigma R \cos(\alpha + \phi) = [\cos(\alpha + \phi)] \Sigma R$$

Dividing M by W gives us

$$M = Wr \tan(\alpha + \phi) \qquad (6/3)$$

We determine the helix angle α by unwrapping the thread of the screw for one turn where we see immediately that $\alpha = \tan^{-1}(L/2\pi r)$.

We may use the unwrapped thread of the screw as an alternative model to simulate the action of the entire screw, as shown in Fig. 6/6a. The equivalent force required to push the movable thread up the fixed incline is $P = M/r$, and the triangle of force vectors gives Eq. 6/3 immediately.

If the moment M is removed, the friction force changes direction so that ϕ is measured to the other side of the normal to the thread. The screw will remain in place and be self-locking provided that $\alpha < \phi$ and will be on the verge of unwinding if $\alpha = \phi$.

In order to lower the load by unwinding the screw, we must reverse the direction of M as long as $\alpha < \phi$. This condition is illustrated in Fig. 6/6b for our simulated thread on the fixed incline, and we see that an equivalent force $P = M/r$ must be applied to the thread to pull it down the incline. Therefore, from the triangle of vectors we get the moment required to lower the screw, which is

(a) To Raise Load

(b) To Lower Load ($\alpha < \phi$)

$$M = Wr \tan(\phi - \alpha) \qquad (6/3a)$$

If $\alpha > \phi$ the screw will unwind by itself, and we see from Fig. 6/6c that the moment required to prevent unwinding would be

$$M = Wr \tan(\alpha - \phi) \qquad (6/3b)$$

(c) To Lower Load ($\alpha > \phi$)

Figure 6/6

Sample Problem 6/6

The horizontal position of the 500-kg rectangular block of concrete is adjusted by the 5° wedge under the action of the force **P**. If the coefficient of static friction for both pairs of wedge surfaces is 0.30 and if the coefficient of static friction between the block and the horizontal surface is 0.60, determine the least force *P* required to move the block.

Solution. The free-body diagrams of the wedge and the block are drawn with the reactions R_1, R_2, and R_3 inclined with respect to their normals by the amount of the friction angles for impending motion. The friction angle for limiting static friction is given by $\phi = \tan^{-1}\mu$. Each of the two friction angles is computed and shown on the diagram.

We start our vector diagram expressing the equilibrium of the block at a convenient point *A* and draw the only known vector, the weight **W** of the block. Next we add R_3 whose 31.0° inclination from the vertical is now known. The vector R_2, whose 16.7° inclination from the horizontal is also known, must close the polygon for equilibrium. Thus point *B* on the lower polygon is determined by the intersection of the known directions of R_3 and R_2 and their magnitudes become known.

For the wedge we draw R_2, which is now known, and add R_1 whose direction is known. The directions of R_1 and **P** intersect at *C*, thus giving us the solution for the magnitude of **P**.

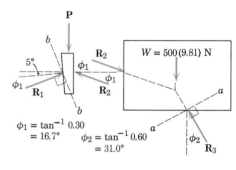

Algebraic solution. The simplest choice of reference axes for calculation purposes is, for the block, in the direction *a-a* normal to R_3 and, for the wedge, in the direction *b-b* normal to R_1. The angle between R_2 and the *a*-direction is 16.7° + 31.0° = 47.7°. Thus for the block

$$[\Sigma F_a = 0] \qquad 500(9.81)\sin 31.0° - R_2\cos 47.7° = 0$$

$$R_2 = 3747 \text{ N}$$

For the wedge the angle between R_2 and the *b*-direction is 90° − (2ϕ_1 + 5°) = 51.6°, and the angle between **P** and the *b*-direction is ϕ_1 + 5° = 21.7°. Thus

$$[\Sigma F_b = 0] \qquad 3747\cos 51.6° - P\cos 21.7° = 0$$

$$P = 2505 \text{ N} \qquad\qquad Ans.$$

Graphical solution. The accuracy of a graphical solution is well within the uncertainty of the friction coefficients and provides a simple and direct result. By laying off the vectors to a reasonable scale following the sequence described, the magnitudes of **P** and the **R**'s are easily scaled directly from the diagrams.

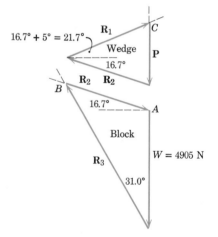

① Be certain to note that the reactions are inclined from their normals in the direction to oppose the motion. Also, we note the equal and opposite reactions R_2.

② It should be evident that we avoid simultaneous equations by eliminating reference to R_3 for the block and R_1 for the wedge.

Sample Problem 6/7

The single-threaded screw of the vise has a mean diameter of 1 in. and has 5 square threads per inch. The coefficient of static friction in the threads is 0.20. A 60-lb pull applied normal to the handle at A produces a clamping force of 1000 lb between the jaws of the vise. (a) Determine the frictional moment M_B developed at B due to the thrust of the screw against the body of the jaw. (b) Determine the force Q applied normal to the handle at A required to loosen the vise.

Solution. From the free-body diagram of the jaw we first obtain the tension T in the screw.

$$[\Sigma M_C = 0] \qquad 1000(16) - 10T = 0 \qquad T = 1600 \text{ lb}$$

The helix angle α and the friction angle ϕ for the thread are given by

①

$$\alpha = \tan^{-1} \frac{L}{2\pi r} = \tan^{-1} \frac{1/5}{2\pi(0.5)} = 3.64°$$

$$\phi = \tan^{-1} \mu = \tan^{-1} 0.20 = 11.31°$$

where the mean radius of the thread is $r = 0.5$ in.

(a) *To tighten.* The isolated screw is simulated by the free-body diagram shown where all of the forces acting on the threads of the screw are represented by a single force R inclined at the friction angle ϕ from the normal to the thread. The moment applied about the screw axis is $60(8) = 480$ lb-in. in the clockwise direction as seen from the front of the vise. The frictional moment M_B due to the friction forces acting on the collar at B is in the counterclockwise direction to oppose the impending motion. From Eq. 6/3 with T substituted for W the net moment acting on the screw is

$$M = Tr \tan (\alpha + \phi)$$

$$480 - M_B = 1600(0.5) \tan (3.64° + 11.31°)$$

$$M_B = 266 \text{ lb-in.} \qquad \textit{Ans.}$$

(b) *To loosen.* The free-body diagram of the screw on the verge of being loosened is shown with R acting at the friction angle from the normal in the direction to counteract the impending motion. Also shown is the frictional moment $M_B = 266$ lb-in. acting in the clockwise direction to oppose the motion. The angle between R and the screw axis is now $\phi - \alpha$, and we use Eq. 6/3a with the net moment equal to the applied moment M' minus M_B. Thus

$$M = Tr \tan (\phi - \alpha)$$

$$M' - 266 = 1600(0.5) \tan (11.31° - 3.64°)$$

$$M' = 374 \text{ lb-in.}$$

Thus the force on the handle required to loosen the vise is

$$Q = M'/d = 374/8 = 46.8 \text{ lb} \qquad \textit{Ans.}$$

① Be careful to calculate the helix angle correctly. Its tangent is the lead L (advancement per revolution) divided by the mean circumference $2\pi r$ and not by the diameter $2r$.

(a) To Tighten

(b) To Loosen

② Note that R swings to the opposite side of the normal as the impending motion reverses direction.

Problem 6/43

Problem 6/44

Problem 6/46

Problem 6/47

PROBLEMS

6/43 A wedge will be self-locking provided that its angle α is less than a critical value. If the coefficient of friction between the wedge and the material being split is μ, what is the critical value of α?

6/44 The precise alignment of a heavy-duty diesel engine on its foundation is accomplished by screw-adjusted wedges under each of its four mounting flanges. Calculate the horizontal thrust P in the adjusting screw necessary to raise the mounting flange if it supports one-fourth of the total engine mass of 4.8 Mg. The coefficient of friction for both sides of the wedge is 0.25.   *Ans.* $P = 6.33$ kN

6/45 A screw jack with square threads having a mean radius of 25 mm supports a load of 5 kN. If the coefficient of friction is 0.25, what is the greatest lead L (advancement per turn) of the screw for which the screw will not unwind by itself? For this condition what torque M applied to the screw would be required to raise the load?

6/46 The turnbuckle supports a tension T of 15,000 lb. Each of the screws has a mean diameter of 1.354 in. and has a single thread with a lead (advancement per turn) of $\frac{1}{3}$ inch, one being right-handed and the other left-handed. If a torque of 270 lb-ft is required to loosen the turnbuckle by turning the body with both screws prevented from rotating, calculate the effective coefficient of friction μ in the threads.
Ans. $\mu = 0.241$

6/47 The two 5° wedges shown are used to adjust the position of the column under a vertical load of 5 kN. Determine the magnitude of the forces P required to raise the column if the coefficient of friction for all surfaces is 0.40.

6/48 If the loaded column of Prob. 6/47 is to be lowered, calculate the horizontal forces P' required to withdraw the wedges. Ans. $P' = 3.51$ kN

6/49 A 5° steel wedge is forced under the end of the 4200 lb machine with a force $P = 1100$ lb. If the coefficient of friction between the wedge and both the machine and the horizontal floor is 0.30, determine the position x of the center of gravity G of the machine. The machine is prevented from sliding horizontally by a rigid ledge at A.

Problem 6/49

6/50 The movable head of a universal testing machine has a mass of 2.2 Mg and is elevated into testing position by two 78-mm-diameter lead screws each with a single thread and a lead of 13 mm. If the coefficient of friction in the threads is 0.25, how much torque M must be supplied to each screw (a) to raise the head and (b) to lower the head? The inner loading columns are not attached to the head during positioning. Ans. (a) $M = 129.3$ N·m
(b) $M = 81.8$ N·m

Problem 6/50

6/51 The device shown is used as a jack. The screw has a double square thread with a mean diameter of $\frac{7}{8}$ in. and a lead of $\frac{1}{3}$ in. Section A of the screw has right-hand threads, and section B has left-hand threads. For $\theta = 30°$ determine (a) the torque M which must be applied to the screw to raise a load $P = 1500$ lb and (b) the torque M' needed to lower the load. The coefficient of friction in the threads is 0.20.

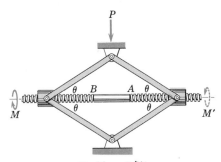

Problem 6/51

6/52 The detent mechanism consists of the spring-loaded plunger with a spherical end which positions the horizontal bar by engaging the spaced notches. If the spring exerts a force of 40 N on the plunger in the position shown and a force $P = 60$ N is required to move the detent bar against the plunger, calculate the coefficient of friction between the plunger and the detent. It is known from earlier tests that the coefficient of friction between the light bar and the horizontal surface is 0.30. Assume that the plunger is well lubricated and accurately fitted so that the friction between it and its guide is negligible. Ans. $\mu = 0.368$

Problem 6/52

Problem 6/53

6/53 The vertical position of the 100-kg block is adjusted by the screw-activated wedge. Calculate the moment M which must be applied to the handle of the screw to raise the block. The single-threaded screw has square threads with a mean diameter of 30 mm and advances 10 mm for each complete turn. The coefficient of friction for the screw threads is 0.25, and the coefficient of friction for all mating surfaces of the block and wedge is 0.40. Neglect friction at the ball joint A.

6/54 Calculate the moment M' which must be applied to the handle of the screw of Prob. 6/53 to withdraw the wedge and lower the 100-kg load.

Ans. $M' = 3.02 \text{ N·m}$

Problem 6/55

6/55 Compute the force P required to move the 20-kg wheel. The coefficient of friction at A is 0.25 and that for both pairs of wedge surfaces is 0.30. Also, the spring S is under a compression of 100 N, and the rod offers negligible support to the wheel.

6/56 Work Prob. 6/55 if the compression in the spring is 200 N. All other conditions remain unchanged.

Ans. $P = 198.8 \text{ N}$

Problem 6/57

▶6/57 The tapered pin is forced into a mating tapered hole in the fixed block with a force $P = 400$ N. If the force required to remove the pin is $P' = 300$ N, calculate the coefficient of friction between the pin and the surface of the hole. (*Hint:* The pressure (stress) normal to the tapered pin surface remains unchanged until the pin actually moves.)

Ans. $\mu = 0.122$

▶6/58 Replace the square thread of the screw jack in Fig. 6/5 by a V-thread as indicated in the figure accompanying this problem and determine the moment M on the screw required to raise the load W. The force R acting on a representative small section of the thread is shown with its relevant projections. The vector R_1 is the projection of R in the plane of the figure containing the axis of the screw. The analysis is begun with an axial force and a moment summation and includes substitutions for the angles γ and β in terms of θ, α, and the friction angle $\phi = \tan^{-1}\mu$. The helix angle of the single thread is exaggerated for clarity.

$$\text{Ans. } M = Wr \, \frac{\tan\alpha + \mu\sqrt{1 + \tan^2\dfrac{\theta}{2}\cos^2\alpha}}{1 - \mu\tan\alpha\sqrt{1 + \tan^2\dfrac{\theta}{2}\cos^2\alpha}}$$

$$\text{where } \alpha = \tan^{-1}\frac{L}{2\pi r}$$

Problem 6/58

Figure 6/7

6/6 JOURNAL BEARINGS. A journal bearing is a bearing that gives lateral support to a shaft in contrast to axial or thrust support. For dry bearings and for many partially lubricated bearings we may apply the principles of dry friction which give us a satisfactory approximation for design purposes. A dry or partially lubricated journal bearing with contact or near contact between the shaft and the bearing is shown in Fig. 6/7, where the clearance between the shaft and bearing is greatly exaggerated. As the shaft begins to turn in the direction shown, it rolls up the inner surface of the bearing until slippage occurs. Here it remains in a more or less fixed position during rotation. The torque M required to maintain rotation and the radial load L on the shaft will cause a reaction R at the contact point A. For equilibrium in the vertical direction R must equal L but will not be collinear with it. We see that the force R will be tangent to a small circle of radius r_f called the *friction circle*. The angle between R and its normal component N is the friction angle ϕ. Equating the sum of the moments about O to zero gives

$$M = Rr_f = Rr \sin \phi \qquad (6/4)$$

For a small coefficient of friction the angle ϕ is small, and the sine and tangent may be interchanged with only small error. Since $\mu = \tan \phi$, a good approximation to the torque is

$$M = \mu Rr \qquad (6/4a)$$

This relation gives the amount of torque or moment applied to the shaft which is necessary to overcome friction for a dry or partially lubricated journal bearing.

6/7 THRUST BEARINGS; DISK FRICTION. Friction between circular surfaces under normal pressure is encountered in pivot bearings, clutch plates, and disk brakes. To examine this application we consider the two flat circular disks of Fig. 6/8 whose shafts are mounted in bearings (not shown) so that they can be brought into contact under the axial force P. The maximum torque that this clutch can transmit will be equal to the torque M required to slip one disk against the other. If p is the normal pressure at any location between the plates, the frictional force acting on an elemental area is $\mu p \, dA$,

Figure 6/8

where μ is the friction coefficient and dA is the area $r\, dr\, d\theta$ of the element. The moment of this elemental friction force about the shaft axis is $\mu pr\, dA$, and the total moment becomes

$$M = \int \mu pr\, dA$$

where we evaluate the integral over the area of the disk. To carry out this integration we must know the variation of μ and p with r.

In the following examples we shall consider that μ is constant. Furthermore, if the surfaces are new, flat, and well supported, it is reasonable to assume that the pressure p is constant and uniformly distributed so that $\pi R^2 p = P$. Substituting the constant value of p in the expression for M gives us

$$M = \frac{\mu P}{\pi R^2} \int_0^{2\pi} \int_0^R r^2\, dr\, d\theta = \tfrac{2}{3}\mu PR \qquad (6/5)$$

We may interpret this result as being equivalent to the moment due to a friction force μP acting at a distance $\tfrac{2}{3}R$ from the shaft center.

If the friction disks are rings, as in the collar bearing of Fig. 6/9, the limits of integration are the inside and outside radii R_i and R_o, respectively, and the frictional torque becomes

$$M = \tfrac{2}{3}\mu P \frac{R_o^{\,3} - R_i^{\,3}}{R_o^{\,2} - R_i^{\,2}} \qquad (6/5a)$$

Figure 6/9

After some wear of the surfaces has taken place, we find that the frictional moment decreases somewhat. When the wearing-in period is over, the surfaces retain their new relative shape and further wear is therefore constant over the surface. This wear depends both on the circumferential distance traveled and the pressure p. Since the distance traveled is proportional to r, the expression $rp = K$ may be written, where K is a constant. The value of K is determined from equilibrium of the axial forces or

$$P = \int p\, dA = K \int_0^{2\pi} \int_0^R dr\, d\theta = 2\pi KR$$

With $pr = K = P/(2\pi R)$, we may write the expression for M as

$$M = \int \mu pr\, dA = \frac{\mu P}{2\pi R} \int_0^{2\pi} \int_0^R r\, dr\, d\theta$$

which becomes $\qquad\qquad M = \tfrac{1}{2}\mu PR \qquad\qquad (6/6)$

The frictional moment for worn-in plates is, therefore, only $(\tfrac{1}{2})/(\tfrac{2}{3})$, or $\tfrac{3}{4}$ as much as for new surfaces.

If the friction disks are rings of inside radius R_i and outside radius R_o, substitution of these limits in the integrations shows that the frictional torque for worn-in surfaces is

$$M = \tfrac{1}{2}\mu P(R_o + R_i) \qquad (6/6a)$$

Sample Problem   6/8

The bell crank fits over a 100-mm-diameter shaft which is fixed and cannot rotate. The horizontal force T is applied to maintain equilibrium of the crank under the action of the vertical force $P = 100$ N. Determine the maximum and minimum values that T may have without causing the crank to rotate in either direction. The coefficient of static friction μ between the shaft and the bearing surface of the crank is 0.20.

Solution.   Impending rotation occurs when the reaction R of the fixed shaft on the bell crank makes an angle $\phi = \tan^{-1}\mu$ with the normal to the bearing surface and is, therefore, tangent to the friction circle. Also, equilibrium requires that the three forces acting on the crank be concurrent at point C. These facts are shown in the free-body diagrams for the two cases of impending motion.

The following calculations are needed:

Friction angle $\phi = \tan^{-1}\mu = \tan^{-1}0.20 = 11.31°$

Radius of friction circle $r_f = r\sin\phi = 50\sin 11.31° = 9.81$ mm

Angle $\theta = \tan^{-1}\dfrac{120}{180} = 33.7°$

Angle $\beta = \sin^{-1}\dfrac{r_f}{OC} = \sin^{-1}\dfrac{9.81}{\sqrt{(120)^2 + (180)^2}} = 2.60°$

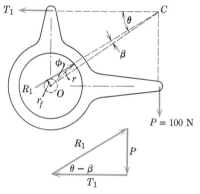

(a) Counterclockwise Motion Impends

(a) *Impending counterclockwise motion.*   The equilibrium triangle of forces is drawn and gives

$$T_1 = P\,\text{ctn}\,(\theta - \beta) = 100\,\text{ctn}\,(33.7° - 2.60°)$$

$$T_1 = T_{\max} = 165.8\ \text{N} \qquad\qquad Ans.$$

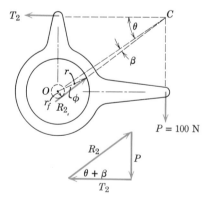

(b) Clockwise Motion Impends

(b) *Impending clockwise motion.*   The equilibrium triangle of forces for this case gives

$$T_2 = P\,\text{ctn}\,(\theta + \beta) = 100\,\text{ctn}\,(33.7° + 2.60°)$$

$$T_2 = T_{\min} = 136.2\ \text{N} \qquad\qquad Ans.$$

PROBLEMS

6/59 The two flywheels are mounted on a common shaft which is supported by a journal bearing between them. Each flywheel weighs 80 lb, and the diameter of the shaft is 1.600 in. If a 25-lb-in. couple M on the shaft is required to maintain rotation of the flywheels and shaft at a constant low speed, compute (*a*) the coefficient of friction μ in the bearing and (*b*) the radius r_f of the friction circle.

Problem 6/59

6/60 A torque M of 1510 N·m must be applied to the 50-mm-diameter shaft of the hoisting drum to raise the 500-kg load at constant speed. The drum and shaft together have a mass of 100 kg. Calculate the coefficient of friction μ for the bearing.

Ans. $\mu = 0.271$

Problem 6/60

6/61 Calculate the torque M on the shaft of the hoisting drum of Prob. 6/60 that is required to lower the 500-kg load at constant speed. Use the value $\mu = 0.271$ calculated in Prob. 6/60 for the coefficient of friction.

6/62 The front wheels of an experimental rear-drive vehicle have a radius of 300 mm and are equipped with disk-type brakes consisting of a ring A with outside and inside radii of 150 mm and 75 mm, respectively. The ring, which does not turn with the wheel, is forced against the wheel disk with a force P. If the pressure between the ring and the disk wheel is uniform over the mating surfaces, compute the friction force F between each front tire and the horizontal road for an axial force $P = 1$ kN when the vehicle is powered at constant speed with the wheels turning. The coefficient of friction between the disk and ring is 0.35.   Ans. $F = 136.1$ N

Problem 6/62

Problem 6/63

Problem 6/64

Problem 6/65

6/63 In the figure is shown a multiple-disk clutch for marine use. The driving disks A are splined to the driving shaft B so that they are free to slip along the shaft but must rotate with it. The disks C drive the housing D by means of the bolts E along which they are free to slide. In the clutch shown there are five pairs of friction surfaces. Assume the pressure to be uniformly distributed over the area of the disks and determine the maximum torque M which can be transmitted if the coefficient of friction is 0.15 and $P = 500$ N.   *Ans.* $M = 335$ N·m

6/64 The shaft A fits loosely in the wrist-pin bearing of the connecting rod with center of mass at G as shown. With the rod initially in the vertical position the shaft is rotated slowly until the rod slips at the angle α. Write the exact expression for the coefficient of friction.

6/65 If the coefficient of kinetic friction between the 30-mm-diameter pin and the pulley is 0.25, calculate the tension T required to raise the 500-kg load. Also find the tension T_0 in the stationary part of the cable. Neglect the mass of the pulley.
                    Ans. $T = 2.52$ kN, $T_0 = 2.38$ kN

6/66 Calculate the tension T required to lower the load in Prob. 6/65. Also find T_0.

6/67 Each of the four wheels of the vehicle weighs 40 lb
and is mounted on a 4-in.-diameter journal (shaft).
The total weight of the vehicle is 960 lb, including
wheels, and is distributed equally on all four wheels.
If a force $P = 16$ lb is required to keep the vehicle
rolling at a constant low speed on a horizontal
surface, calculate the coefficient of friction which
exists in the wheel bearings. (*Hint:* Draw a complete free-body diagram of one wheel.)

Ans. $\mu = 0.204$

Problem 6/67

6/68 Determine the angle θ with the horizontal made by
the steepest slope on which the vehicle of Prob.
6/67 can stand without rolling by itself in the absence of a force P. Take the coefficient of friction in
the wheel bearings to be 0.20.

6/69 The linkage is initially at rest under the action of
the torques M_1 and M_2. If M_1 is increased gradually
until the linkage moves, write the exact expression
for the angle α between the resultant compressive
force R in link AB and its centerline as motion
impends. The coefficient of friction for each bearing
is μ.

Ans. $\alpha = \sin^{-1}\dfrac{\mu d}{l\sqrt{1+\mu^2}}$

Problem 6/69

6/70 The 20-Mg missile is lowered into its protective silo
on a two-screw elevator as shown. Each screw has a
mass of 1.07 Mg, is 130 mm in diameter, and has a
single square thread with a lead of 13 mm. The
screws are turned in synchronism by a motor unit in
the base of the silo. The entire mass of the missile,
screws, and 4.2-Mg elevator platform is supported
equally by flat collar bearings at A, each of which
has an outside diameter of 250 mm and an inside
diameter of 125 mm. The pressure on the bearings is
assumed to be uniformly distributed over the bearing surface. If the coefficient of friction for the
collar bearing and the screws at B is 0.15, calculate
the torque M which must be applied to each screw
(*a*) to raise the elevator and (*b*) to lower the elevator.

Ans. (*a*) $M = 3290$ N·m
(*b*) $M = 2790$ N·m

250 mm

125 mm

Detail of Collar
Bearing at A

Problem 6/70

Problem 6/71

6/71 For the flat sanding disk of radius a the pressure p developed between the disk and the sanded surface decreases linearly with r from a value p_0 at the center to $p_0/2$ at $r = a$. If the coefficient of friction is μ, derive the expression for the torque M required to turn the shaft under an axial force L.

Problem 6/72

6/72 The spherical thrust bearing on the end of the shaft supports an axial load P. Determine the expression for the moment M required to turn the shaft against friction in the bearing. Assume that the pressure p is directly proportional to $\sin \alpha$ and that the coefficient of friction is μ.   *Ans.* $M = \mu Pr$

Problem 6/73

6/73 Determine the expression for the torque M required to turn the shaft whose thrust L is supported by a conical pivot bearing. The coefficient of friction is μ, and the bearing pressure is constant.

$$Ans.\ M = \frac{\mu L}{3 \sin \dfrac{\alpha}{2}} \frac{d_2^{\,3} - d_1^{\,3}}{d_2^{\,2} - d_1^{\,2}}$$

6/8 FLEXIBLE BELTS.

The impending slippage of flexible members such as belts and ropes over sheaves and drums is of importance in the design of belt drives of all types, band brakes, and hoisting rigs. In Fig. 6/10*a* is shown a drum subjected to the two belt tensions T_1 and T_2, the torque M necessary to prevent rotation, and a bearing reaction R. With M in the direction shown, T_2 is greater than T_1. The free-body diagram of an element of the belt of length $r\, d\theta$ is shown in the *b*-part of the figure. We proceed with the force analysis of this element in a manner similar to that which we have illustrated for other variable-force problems where the equilibrium of a differential element is established. The tension increases from T at the angle θ to $T + dT$ at the angle $\theta + d\theta$. The normal force is a differential dN, since it acts on a differential element of area. Likewise the friction force, which must act on the belt in a direction to oppose slipping, is a differential and is $\mu\, dN$ for impending motion. Equilibrium in the *t*-direction gives

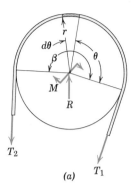

$$T \cos \frac{d\theta}{2} + \mu\, dN = (T + dT) \cos \frac{d\theta}{2}$$

or

$$\mu\, dN = dT$$

since the cosine of a differential quantity is unity. Equilibrium in the *n*-direction requires that

$$dN = (T + dT) \sin \frac{d\theta}{2} + T \sin \frac{d\theta}{2}$$

or

$$dN = T\, d\theta$$

(b)

Figure 6/10

In this reduction we recall that the sine of a differential angle equals the angle and that the product of two differentials must be neglected in the limit compared with the first-order differentials remaining. Combining the two equilibrium relations gives

$$\frac{dT}{T} = \mu\, d\theta$$

Integrating between corresponding limits yields

$$\int_{T_1}^{T_2} \frac{dT}{T} = \int_0^\beta \mu\, d\theta$$

or

$$\ln \frac{T_2}{T_1} = \mu\beta$$

where the $\ln (T_2/T_1)$ is a natural logarithm (base e). Solving for T_2 gives

$$T_2 = T_1 e^{\mu\beta} \tag{6/7}$$

We note that β is the total angle of belt contact and must be expressed in radians. If a rope were wrapped around a drum n times, the angle β would be $2\pi n$ radians. Equation 6/7 holds equally well for a noncircular section where the total angle of contact is β. This conclusion is evident from the fact that the radius r of the circular drum of Fig. 6/10 does not enter into the equations for the equilibrium of the differential element of the belt.

The relation expressed by Eq. 6/7 also applies to belt drives where both the belt and the pulley are rotating at constant speed. In this case the equation describes the ratio of belt tensions for slipping or impending slipping. When the speed of rotation becomes large, there is a tendency for the belt to leave the rim, so that Eq. 6/7 will involve some error.

6/9 ROLLING RESISTANCE. Deformation at the point of contact between a rolling wheel and its supporting surface introduces a resistance to rolling which we will mention only briefly. This resistance is not due to tangential friction forces and therefore is an entirely different phenomenon from that of dry friction.

To describe rolling resistance we consider the wheel of Fig. 6/11 under the action of a load L on the axle and a force P applied at its center to produce rolling. The deformation of the wheel and supporting surfaces as shown is greatly exaggerated. The distribution of pressure p over the area of contact is similar to that indicated, and the resultant R of this distribution will act at some point A and will pass through the center of the wheel for equilibrium. We find the force P necessary to initiate and maintain rolling at constant velocity by equating the moments of all forces about A to zero. This gives us

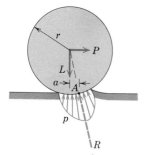

Figure 6/11

$$P = \frac{a}{r}L = \mu_r L$$

where the moment arm of P is taken to be r. The ratio $\mu_r = a/r$ is referred to as the coefficient of rolling resistance. The coefficient as defined is the ratio of resisting force to normal force, and in this respect is analogous to the coefficient of static or kinetic friction. On the other hand there is no slipping or impending slipping in the interpretation of μ_r.

The dimension a depends on many factors which are difficult to quantify, so that a comprehensive theory of rolling resistance is not available. The distance a is a function of the elastic and plastic properties of the mating materials, the radius of the wheel, the speed of travel, and the roughness of the surfaces. Some tests indicate only a small variation with wheel radius, and a is often taken to be independent of the rolling radius. Unfortunately, the quantity a has also been referred to as the coefficient of rolling friction in some references. However, a has the dimension of length and therefore is not a dimensionless coefficient in the usual sense.

Sample Problem 6/9

A flexible cable which supports the 100-kg load is passed over a fixed circular drum and subjected to a force P to maintain equilibrium. The coefficient of static friction μ between the cable and the fixed drum is 0.30. (a) For $\alpha = 0$, determine the maximum and minimum values which P may have in order not to raise or lower the load. (b) For $P = 500$ N, determine the minimum value which the angle α may have before the load begins to slip.

Solution. Impending slipping of the cable over the fixed drum is given by Eq. 6/7 which is $T_2/T_1 = e^{\mu\beta}$.

(a) $\alpha = 0$

(a) With $\alpha = 0$ the angle of contact is $\beta = \pi/2$ rad. For impending upward motion of the load, $T_2 = P_{max}$, $T_1 = 981$ N, and we have

$$P_{max}/981 = e^{0.30(\pi/2)}, \qquad P_{max} = 981(1.602) = 1572 \text{ N} \qquad Ans.$$

For impending downward motion of the load, $T_2 = 981$ N, $T_1 = P_{min}$. Thus

$$981/P_{min} = e^{0.30(\pi/2)}, \qquad P_{min} = 981/1.602 = 612 \text{ N} \qquad Ans.$$

(b) $P = 500$ N

(b) With $T_2 = 981$ N and $T_1 = P = 500$ N, Eq. 6/7 gives us

$$981/500 = e^{0.30\beta}, \qquad 0.30\beta = \ln(981/500) = 0.674$$

$$\beta = 2.247 \text{ rad} \quad \text{or} \quad \beta = 2.247\left(\frac{360}{2\pi}\right) = 128.7°$$

$$\alpha = 128.7° - 90° = 38.7° \qquad Ans.$$

① We are careful to note that β must be expressed in radians.

② In our derivation of Eq. 6/7 be certain to note that $T_2 > T_1$.

③ As was noted in the derivation of Eq. 6/7 the radius of the drum does not enter into the calculations. It is only the angle of contact and the coefficient of friction that determine the limiting conditions for impending motion of the flexible cable over the curved surface.

Problem 6/74

Problem 6/76

Problem 6/77

Problem 6/79

PROBLEMS

6/74 A force $P = 5mg$ is required to raise the load of mass m with the cord making $1\frac{1}{4}$ turns around the fixed shaft. Calculate the coefficient of friction μ between the cord and the shaft.   *Ans. $\mu = 0.205$*

6/75 For a given coefficient of friction and a given number of turns around the shaft for the configuration of Prob. 6/74, a force P of 3 kN is required to raise m and a force P of 0.48 kN is required to lower m. Find m.

6/76 The 20-lb block and the 50-lb block are connected by a rope hung over the fixed curved surface. If the system is on the verge of slipping, calculate the coefficient of friction between the rope and the surface.   *Ans. $\mu = 0.292$*

6/77 A dockworker adjusts a spring line (rope) which keeps a small ship from drifting alongside a wharf. What force T can be supported in the line if he exerts a pull of 240 N on the free end with (a) one complete turn around the mooring bit as shown and (b) two complete turns? The coefficient of friction between the hemp rope and the cast-steel mooring bit is 0.30.

6/78 If the dockworker of Prob. 6/77 is to support a tension of 16 kN in the rope leading to the ship, how many turns around the mooring bit are necessary? The coefficient of friction between the rope and the bit is 0.30.   *Ans. $n = 2.23$ turns*

6/79 The 180-lb tree surgeon lowers himself with the rope over a horizontal limb of the tree. If the coefficient of friction between the rope and the limb is 0.50, compute the force which the man must exert on the rope to let himself down slowly.

6/80 The tape slides around the two fixed pegs as shown and is under the action of the horizontal tensions $T_1 = 40$ N and $T_2 = 160$ N. Determine the coefficient of friction μ between the tape and the pegs.

Ans. $\mu = 0.313$

Problem 6/80

6/81 The worker hauls a 100-kg log up the 30° ramp by means of the power winch which turns in the direction shown. If the worker exerts a horizontal pull of 160 N on the free end of the rope, compute the coefficient of friction μ_1 between the log and the ramp. The coefficient of friction between the rope and the drum is 0.25.

Problem 6/81

6/82 A counterclockwise moment $M = 1500$ lb-in. is applied to the flywheel. If the coefficient of friction between the band and the wheel is 0.20, compute the minimum force P necessary to prevent the wheel from rotating.

Ans. $P = 142.0$ lb

Problem 6/82

Problem 6/83

6/83 A device for lowering a person in a sling down a rope at a constant controlled rate is shown in the figure. The rope passes around a central shaft fixed to the frame and leads freely out of the lower collar. The number of turns is adjusted by turning the lower collar, which winds or unwinds the rope around the shaft. Entrance of the rope into the upper collar at A is equivalent to $\frac{1}{4}$ of a turn, and passage around the corner at B is also equivalent to $\frac{1}{4}$ of a turn. Friction of the rope through the straight portions of the collars averages 10 N for each collar. If three complete turns around the shaft, in addition to the corner turns, are required for a 75-kg man to lower himself at a constant rate without exerting a pull on the free end of the rope, calculate the coefficient of friction μ between the rope and the contact surfaces of the device. Neglect the small helix angle of the rope around the shaft.

Problem 6/84

6/84 The assembly shown has a mass of 100 kg with center of mass at G and is supported by the two wires, which pass around the fixed pegs and are kept under equal tensions by the equalizer plate A and the adjustable spring S. Calculate the minimum tension T in the spring which will ensure that the assembly remains suspended as shown. The coefficient of friction between the wires and the pegs is 0.30.   *Ans.* $T = 555$ N

Problem 6/85

6/85 The uniform drum A with center of mass at midlength is suspended by a rope which passes over the fixed cylindrical surface B. The coefficient of static friction between the rope and the surface over which it passes is μ. Determine the maximum value that the dimension a may have before the drum tips out of its horizontal position.

6/86 Find the couple M required to turn the pipe in the V-block against the action of the flexible band. A force $P = 25$ lb is applied to the lever which is pivoted about O. The coefficient of friction between the band and the pipe is 0.30, and that between the pipe and the block is 0.40. The weights of the parts are negligible.   *Ans. $M = 1834$ lb-in.*

Problem 6/86

6/87 Replace the flat belt and pulley of Fig. 6/10 by a V-belt and matching grooved pulley as indicated by the cross-sectional view accompanying this problem. Derive the relation among the belt tensions, the angle of contact, and the coefficient of friction for the V-belt when slipping impends. Use of a V-belt with $\alpha = 35°$ would be equivalent to increasing the coefficient of friction for a flat belt of the same material by what factor n?

$$\text{Ans. } T_2 = T_1 e^{\mu\beta'} \text{ where } \beta' = \frac{\beta}{\sin(\alpha/2)}$$

$$n = 3.33$$

V-belt cross section

Problem 6/87

▶6/88 The roller chain is used as a pipe grip. If the coefficient of friction between the chain and the fixed pipe is 0.25, determine the minimum value of h which will ensure that the grip will not slip on the pipe regardless of P. Neglect the masses of the chain and handle, and neglect any friction between the left end of the handle and the pipe.
Ans. $h = 96.9$ mm

Problem 6/88

▶6/89 The chain has a mass ρ per unit length. Determine the overhang h below the fixed cylindrical guide for which the chain will be on the verge of slipping. The coefficient of friction is μ. (*Hint:* The resulting differential equation involving the variable chain tension T at the corresponding angle θ is of the form $dT/d\theta + KT = f(\theta)$, a first-order, linear, nonhomogeneous equation with constant coefficient. The solution is

$$T = Ce^{-K\theta} + e^{-K\theta}\int e^{K\theta}f(\theta)\,d\theta$$

where C and K are constants.)

$$\text{Ans. } h = \frac{2\mu r}{1 + \mu^2}(1 + e^{\mu\pi})$$

Problem 6/89

6/10 PROBLEM FORMULATION AND REVIEW.

In our study of friction we have concentrated our attention on dry or Coulomb friction where a simple mechanical model of surface irregularities between the contacting bodies, Fig. 6/1, suffices to explain the phenomenon adequately for most engineering purposes. By having this model clearly in mind we can easily visualize the three types of dry-friction problems which are encountered. These categories are:

1. Static friction less than the maximum possible value and determined by the equations of equilibrium. (Usually requires a check to see that $F < \mu_s N$)
2. Limiting static friction with impending motion ($F_{max} = \mu_s N$)
3. Kinetic friction where sliding motion occurs between contacting surfaces ($F = \mu_k N$)

A coefficient of friction applies to a given pair of mating surfaces. It is meaningless to speak of a coefficient of friction for a single surface. The static coefficient of friction μ_s for a given pair of surfaces is usually greater than the kinetic coefficient μ_k. We generally drop the subscript and understand from the problem at hand which of the two coefficients applies. The friction force which acts on a body is always in the direction to oppose the slipping of the body which takes place or the slipping which would take place in the absence of friction.

When we encounter friction forces distributed in some prescribed manner over a surface or along a line, we select a representative element of the surface or line and evaluate the force and moment effects of the elemental friction force acting on the element. We then integrate these effects over the entire surface or line.

Friction coefficients are subject to considerable variation depending on the exact condition of the mating surfaces. Computation of coefficients of friction to three significant figures represents an accuracy which cannot easily be duplicated by experiment and when cited are included for purposes of computational check only. For design computations in engineering practice the use of a handbook value for a coefficient of static or kinetic friction must be viewed only as an approximation.

In reviewing the foregoing introduction to frictional problems the reader should bear in mind the existence of the other forms of friction mentioned in the introductory article of the chapter. Problems which involve fluid friction, for example, are among the most important of the friction problems encountered in engineering, and a study of this phenomenon is included in the subject of fluid mechanics.

REVIEW PROBLEMS

6/90 A frictional locking device allows bar *A* to move to the left but prevents movement to the right. If the coefficient of friction between the shoe *B* and the bar *A* is 0.40, specify the maximum length *b* of the link which will permit the device to work as described.   *Ans. b* = 118.5 mm

Problem 6/90

6/91 A sliding element of a machine tool consists of a bar with a V-section that is supported by the grooved block. The bar supports a vertical load *w* per unit of its length. If the coefficient of friction is *μ*, determine the horizontal force *P* required to move the bar.

Problem 6/91

6/92 The coefficients of static and kinetic friction between the 100-kg block and the inclined plane are 0.30 and 0.20, respectively. Determine (*a*) the friction force *F* acting on the block when *P* is applied with a magnitude of 200 N to the block at rest, (*b*) the force *P* required to initiate motion up the incline from rest, and (*c*) the friction force *F* acting on the block if *P* = 600 N.   *Ans.* (*a*) *F* = 66.0 N
(*b*) *P* = 516 N
(*c*) *F* = 148 N

Problem 6/92

6/93 A 3200-lb car with 120-in. wheelbase has a center of gravity 2 ft from the road and midway between the front and rear axles. If the coefficient of friction between the tires and the road is 0.80, find the angle *θ* with the horizontal made by the steepest grade that the car can climb at constant speed before the rear driving wheels slip. What torque *M* is applied to each of the 26-in.-diameter rear wheels by the engine under these conditions? Neglect any friction under the front wheels.

Problem 6/94

6/94 The uniform 3 m beam is suspended by the cable that passes over the large pulley. A locking pin at *A* prevents rotation of the pulley. If the coefficient of friction between the cable and the pulley is 0.25, determine the minimum value of *x* for which the cable will not slip on the pulley.

Ans. x = 0.343 m

Problem 6/95

6/95 The screw of the small press has a mean diameter of 25 mm and has a double square thread with a lead of 8 mm. The flat thrust bearing at *A* is shown in the enlarged view and has surfaces which are well worn. If the coefficient of friction for both the threads and the bearing at *A* is 0.25, calculate the torque *M* on the handwheel required (*a*) to produce a compressive force of 4 kN and (*b*) to loosen the press from the 4-kN compression.

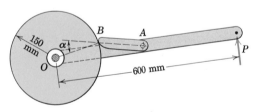

Problem 6/96

6/96 The figure shows a friction silent ratchet for turning the wheel counterclockwise under the action of a force *P* applied to the handle. The floating link *AB* engages the wheel so that $\alpha = 20°$. If a force $P = 150$ N is required to turn the wheel about its bearing at *O*, determine (*a*) the minimum coefficient of friction between the link and the wheel that will ensure no slipping at *B* and (*b*) the magnitude *R* of the force on the pin at *A*.

Ans. (a) $\mu_{min} = 0.364$, (b) R = 1.754 kN

Problem 6/97

6/97 The lower end *A* of the uniform 100-lb plank rests on rollers that are free to move on the horizontal surface. If the coefficients of static and kinetic friction between the plank and the corner *B* are 0.80 and 0.70, respectively, compute the friction force *F* acting at *B* if the plank is released from rest in the position shown.

6/98 Calculate the torque M which the engine must supply to the rear axle of the car to roll it over the curbing from a rest position if the rear wheels do not slip. Determine the minimum coefficient of friction at the rear wheels to prevent slipping. The car has a mass of 1200 kg.

Ans. $M = 2.02$ kN·m, $\mu_{min} = 1.00$

Problem 6/98

6/99 The coefficient of friction between the collar of the drill-press table and the vertical column is 0.30. Will the collar and table slide down the column under the action of the drill thrust if the operator forgets to secure the clamp, or will friction be sufficient to hold it in place? Neglect the weight of the table and collar compared with the drill thrust and assume that contact occurs at the points A and B.

Problem 6/99

6/100 The two 10° wedges are positioned so that a downward force P on the one wedge will result in an elevation of the 1200-lb load. The coefficient of friction for all sliding surfaces is 0.20, and the weights of the wedges are negligible. Determine P.

Ans. $P = 492$ lb

Problem 6/100

6/101 In the figure are shown the elements of a rolling mill. Determine the maximum thickness b which the slab may have and still enter the rolls by means of the friction between the slab and the rolls. The coefficient of friction is μ. (*Hint:* The critical condition for entering occurs when the resultant of the horizontal forces on the slab is zero.)

Ans. $b = a + d\dfrac{\sqrt{1 + \mu^2} - 1}{\sqrt{1 + \mu^2}}$

Problem 6/101

Problem 6/102

Problem 6/103

Problem 6/104

Problem 6/105

6/102 Determine the force P that will begin to rotate the cylinder of mass m against the action of friction. The coefficient of friction for both pairs of contacting surfaces is μ.

6/103 A bulldozer moves the 750-kg log up the 20° incline by pushing with the blade, which is normal to the incline. If the coefficient of friction between the blade and the log is 0.50 and that between the log and the ground is 0.80, calculate the force component P, normal to the blade, that must be exerted against the log.   *Ans. $P = 5.03$ kN*

6/104 The thin-walled tubular shaft of mean diameter D rotates about the vertical axis under an axial load L and bears against a fixed spherical surface of radius r. If the coefficient of friction between the tube and the spherical surface is μ, write the expression for the moment M on the shaft necessary to overcome friction at the support.

6/105 The truck unloads its cargo box by sliding it slowly off the elevated rack. The box has a total weight of 10,000 lb with center of mass at G in the center of the box. The coefficient of friction between the box and the rack is 0.30. Calculate the braking force F between the tires and the level road as the box is on the verge of slipping down the rack from the position shown and the truck is on the verge of rolling forward. No slipping occurs at the lower corner of the box.   *Ans. $F = 2034$ lb*

6/106 A force of 1 kN is developed in the hydraulic cylinder *C* to activate the block brake. If the coefficient of static friction between the blocks and the rim of the wheel is 0.60, compute the maximum torque *M* which can be applied to the wheel without causing rotation. The wheel is mounted in a fixed bearing at its center. Assume that the forces between the blocks and the wheel act at the centers of the contact faces of the blocks.   *Ans. M* = 558 N·m

Problem 6/106

6/107 The two brake shoes and their lining pivot about the points *O* and are expanded against the brake drum through the action of the hydraulic cylinder *C*. The pressure *p* between the drum and the lining may be shown to vary directly as the sine of the angle θ measured from the pin *O* for each shoe and has a value p_0 at $\theta = \beta$. The width of the lining in contact with the drum is *b*. Write the expression for the braking torque M_f on the wheel if the coefficient of friction between the drum and the lining is μ.

Problem 6/107

6/108 The friction tongs are designed to lift 1000-lb crates with a nominal width of 48 in. From the configuration of the links determine whether slippage is more likely for crates that are slightly wider (with contact at *A*) or slightly narrower (with contact at *B*) than the nominal size. Determine the minimum coefficient of friction μ between the tongs and the crate that will prevent slippage for the case where slippage is more likely, and calculate the corresponding tension *T* in the horizontal chain that connects the jaws of the tongs.

Problem 6/108

Problem 6/109

Problem 6/110

6/109 The reel of telephone cable has a mass of 3 Mg and is supported on its shaft in the V-notched blocks on both sides of the reel. The reel is raised off the ground by jacking up the supports so that cable may be pulled off in the horizontal direction as shown. The shaft is fastened to the reel and turns with it. If the coefficient of friction between the shaft and the V-surfaces is 0.30, calculate the pull P in the cable required to turn the reel.   *Ans.* $P = 313$ N

▶6/110 A light flexible cord is passed around the circular disk of mass m and ends in a small pulley that is free to find its equilibrium position on the cord. If the coefficient of friction between the cord and the disk is 0.50, determine the angle α between the normals to the cord at the tangency points for the position where the disk is on the verge of turning under the action of a couple M applied to the disk. (*Suggestion:* Solve the resulting equation for α graphically.)
          Ans. $\alpha = 87.3°$

VIRTUAL WORK

7

7/1 INTRODUCTION.

In the previous chapters we have analyzed the equilibrium of a body by isolating it with a free-body diagram and writing the zero-force and zero-moment summation equations. For the most part this approach has been employed for a body whose equilibrium position was known or specified and where one or more of the external forces was an unknown to be determined.

There is a separate class of problems in which bodies are composed of interconnected members that allow relative motion between the parts, thus permitting various possible equilibrium configurations to be examined. For problems of this type, the force- and moment-equilibrium equations, although valid and adequate, are often not the most direct and convenient approach. Here we find that a method based on the concept of the work done by a force is more useful and direct. Also, the method provides a deeper insight into the behavior of mechanical systems and allows us to examine the question of the stability of systems in equilibrium. We will now develop this approach, called the *method of virtual work*.

7/2 WORK.

We must first define the term work, which is used in a quantitative sense as contrasted to its common nontechnical usage.

(a) Work of a Force. Consider the constant force **F** acting on the body, Fig. 7/1a, whose movement along the plane from A to A' is represented by the vector $\Delta \mathbf{s}$, called the *displacement* of the body. By definition the work U done by the force **F** on the body during this displacement is the component of the force in the direction of the displacement times the displacement or

$$U = (F \cos \alpha)\, \Delta s$$

From Fig. 7/1b we see that the same result is obtained if we multiply the magnitude of the force by the component of the displacement in the direction of the force which is

$$U = F(\Delta s \cos \alpha)$$

Since we obtain the same result regardless of which direction we resolve the vectors, we observe immediately that work U is a scalar quantity.

(a)

(b)

Figure 7/1

Figure 7/2

Path of O

(a)

(b)

Figure 7/3

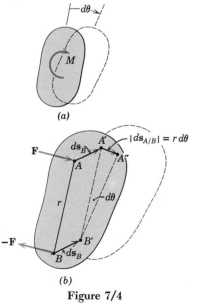

(a)

(b)

Figure 7/4

Work is a positive quantity when the working component of the force is in the same direction as the displacement. When the working component is in the direction opposite to the displacement, Fig. 7/2, the work done is negative. Thus

$$U = (F \cos \alpha) \, \Delta s = -(F \cos \theta) \, \Delta s$$

We now generalize the definition of work to account for conditions under which the direction of the displacement and the magnitude and direction of the force may be variable. Figure 7/3a shows a force **F** acting on a body at a point O which moves along the path shown. During an infinitesimal displacement ds along the path the work done by **F** is

$$\boxed{dU = F \cos \alpha \, ds \qquad \text{or} \qquad dU = \mathbf{F} \cdot d\mathbf{s}} \qquad (7/1)$$

where the dot product of **F** and $d\mathbf{s}$ fits the definition of work. Again, we may interpret this expression as the force component $F \cos \alpha$ in the direction of the displacement times the displacement, or as the displacement component $ds \cos \alpha$ in the direction of the force times the force, as represented in Fig. 7/3b. If we express **F** and $d\mathbf{s}$ in terms of their rectangular components, we have

$$dU = (\mathbf{i}F_x + \mathbf{j}F_y + \mathbf{k}F_z) \cdot (\mathbf{i} \, dx + \mathbf{j} \, dy + \mathbf{k} \, dz)$$
$$= F_x \, dx + F_y \, dy + F_z \, dz$$

To obtain the total work U done by **F** during a finite movement of point O from A to B, Fig. 7/3a, we must integrate dU between these positions. Thus

$$U = \int \mathbf{F} \cdot d\mathbf{s} = \int (F_x \, dx + F_y \, dy + F_z \, dz)$$

or

$$U = \int F \cos \alpha \, ds$$

In order to carry out this integration we must know the relation between the force components and their respective coordinates or the relations between F and s and between $\cos \alpha$ and s.

In the case of concurrent forces acting on a body the work done by their resultant equals the total work done by the several forces. This we conclude from the fact that the component of the resultant in the direction of the displacement equals the sum of the components of the several forces in the same direction.

(b) Work of a Couple. In addition to the work done by forces, couples also may do work. In Fig. 7/4a we have a couple M acting on a body that changes its angular position by an amount $d\theta$. The work done by the couple is easily determined from the combined work of each of the two forces which constitute the couple. In the *b*-part of the figure we represent the couple by two equal and opposite forces **F** and $-\mathbf{F}$ acting at points A and B such that $M = Fr$. During the

infinitesimal movement in the plane of the figure, line AB moves to $A''B'$. We now take the displacement of A in two steps, first, a displacement ds_B equal to that of B and, second, a displacement $ds_{A/B}$ (read the displacement of A with respect to B) due to the rotation about B. We see that the work done by $\mathbf{F}$ during the displacement from A to A' is equal and opposite to that due to $-\mathbf{F}$ acting through the equal displacement from B to B'. Thus we conclude that no work is done by a couple during a translation (movement without rotation). During the rotation, however, $\mathbf{F}$ does work equal to $\mathbf{F} \cdot ds_{A/B} = Fr\,d\theta$ where $d\theta$ is the infinitesimal angle of rotation in radians. Since $M = Fr$ we have

$$\boxed{dU = M\,d\theta} \tag{7/2}$$

The work of the couple is positive if M has the same sense as $d\theta$ (clockwise in this illustration) and negative if M has a sense opposite to that of the rotation. The total work of a couple during a finite rotation in its plane becomes

$$U = \int M\,d\theta$$

Work has the dimensions of (force) $\times$ (distance). In SI units the unit of work is the joule (J), which is the work done by a force of one newton moving through a distance of one meter in the direction of the force ($J = N \cdot m$). In the U.S. customary system the unit of work is the foot-pound (ft-lb), which is the work done by a one-pound force moving through a distance of one foot in the direction of the force. Dimensionally, the work of a force and the moment of a force are the same although they are entirely different physical quantities. We observe carefully that work is a scalar given by the dot product and involves the product of a force and a distance both measured along the same line. Moment, on the other hand, is a vector given by the cross product and involves the product of force and distance measured at right angles to the force. In order to distinguish between these two quantities when we write their units, in SI units we shall use the joule (J) for work and reserve the combined units newton-meter ($N \cdot m$) for moment. In the U.S. customary system we normally use the sequence foot-pound (ft-lb) for work and pound-foot (lb-ft) for moment.

(c) *Virtual Work.* We consider now a particle whose static equilibrium position is determined by the forces which act upon it. Any assumed and arbitrary small displacement δs away from this natural position is called a *virtual displacement*. The term *virtual* is used to indicate that the displacement does not exist in reality but is only assumed so that we may compare various possible equilibrium positions in the process of selecting the correct one. The work done by any force $\mathbf{F}$ acting on the particle during the virtual displacement $\delta\mathbf{s}$ is called *virtual work* and is

$$\delta U = \mathbf{F}\cdot\delta\mathbf{s} \qquad \text{or} \qquad \delta U = F\,\delta s\,\cos\alpha$$

where α is the angle between $\mathbf{F}$ and $\delta\mathbf{s}$. The difference between ds and δs is that ds refers to an infinitesimal change in an actual movement and can be integrated, whereas δs refers to an infinitesimal virtual or assumed movement and cannot be integrated. Mathematically both quantities are first-order differentials.

A virtual displacement may also be a rotation $\delta\theta$ of a body. The virtual work done by a couple M during a virtual angular displacement $\delta\theta$ is, then, $\delta U = M\,\delta\theta$.

We may regard the force $\mathbf{F}$ or couple M as remaining constant during any infinitesimal virtual displacement. If we account for any change in $\mathbf{F}$ or M during the infinitesimal motion, higher-order terms will result which disappear in the limit. This consideration is the same mathematically as that which permits us to neglect the product $dx\,dy$ when writing $dA = y\,dx$ for the element of area under the curve $y = f(x)$.

7/3 **EQUILIBRIUM.**   We now express the equilibrium conditions in terms of virtual work, first for a particle, second for a single rigid body, and third for a system of connected rigid bodies.

(*a*) *Particle.* Consider the particle or small body in Fig. 7/5 which finds its equilibrium position as a result of the forces in the springs to which it is attached. If the mass of the particle is significant, then the weight mg would also be included as one of the forces. For an assumed virtual displacement $\delta\mathbf{s}$ of the particle away from its equilibrium position the total virtual work done on the particle will be

$$\delta U = \mathbf{F}_1\cdot\delta\mathbf{s} + \mathbf{F}_2\cdot\delta\mathbf{s} + \mathbf{F}_3\cdot\delta\mathbf{s} + \cdots = \Sigma\mathbf{F}\cdot\delta\mathbf{s}$$

Figure 7/5

We now express $\Sigma\mathbf{F}$ in terms of its scalar sums and $\delta\mathbf{s}$ in terms of its component virtual displacements in the coordinate directions and write

$$\delta U = \Sigma\mathbf{F}\cdot\delta\mathbf{s} = (\mathbf{i}\Sigma F_x + \mathbf{j}\Sigma F_y + \mathbf{k}\Sigma F_z)\cdot(\mathbf{i}\,\delta x + \mathbf{j}\,\delta y + \mathbf{k}\,\delta z)$$

$$= \Sigma F_x\,\delta x + \Sigma F_y\,\delta y + \Sigma F_z\,\delta z = 0$$

The sum is zero, since $\Sigma\mathbf{F} = \mathbf{0}$ and also $\Sigma F_x = 0$, $\Sigma F_y = 0$, and $\Sigma F_z = 0$. We see therefore that the equation $\delta U = 0$ is an alternative statement of the equilibrium conditions for a particle. This condition of zero virtual work for equilibrium is both necessary and sufficient, since we may apply it to virtual displacements taken one at a time in each of the three mutually perpendicular directions in which case it becomes equivalent to the three known scalar requirements for equilibrium.

The principle of zero virtual work for the equilibrium of a single particle usually does not simplify this already simple problem since $\delta U = 0$ and $\Sigma \mathbf{F} = 0$ provide the same information. The concept of virtual work for a particle is introduced so that it may be applied to systems of particles in the development that follows.

(*b*) *Rigid Body.* We easily extend the principle of virtual work for a particle to a system of rigidly connected particles which form our model of a rigid body. Since the virtual work done on each particle of the body in equilibrium is zero, it follows that the virtual work done on the entire rigid body is zero. Only the virtual work done by *external* forces appears in the evaluation of $\delta U = 0$ for the entire body, since the internal forces occur in pairs of equal, opposite, and collinear forces and the net work done by these forces during any movement is zero. This net work is zero because the displacement components of the two particles along the lines of action of the forces are identical for a rigid body.

Again, as in the case of a particle, we find that the principle of virtual work offers no particular advantage to the solution for a single rigid body in equilibrium. Any assumed virtual displacement defined by a linear or angular movement will appear in each term in $\delta U = 0$ and when canceled will leave us with the same expression as we would have obtained by using one of the force or moment equations of equilibrium directly. This condition is illustrated in Fig. 7/6, where we are asked to determine the reaction R under the roller for the hinged plate of negligible weight under the action of a given force P. A small assumed rotation $\delta\theta$ of the plate about O is consistent with the hinge constraint at O and is taken as the virtual displacement. The work done by P is $-Pa\,\delta\theta$, and the work done by R is $+Rb\,\delta\theta$. Therefore the principle $\delta U = 0$ gives

Figure 7/6

$$-Pa\,\delta\theta + Rb\,\delta\theta = 0$$

Canceling out $\delta\theta$ leaves

$$Pa - Rb = 0$$

which is simply the equation of moment equilibrium about O. Therefore nothing is gained by the use of the virtual-work principle for a single rigid body. Use of the principle will, however, provide us with a decided advantage for interconnected bodies as described in the next section.

(*c*) *Systems of Rigid Bodies.* We now extend the principle of virtual work to describe the equilibrium of an interconnected system of rigid bodies. Our treatment will be limited to so-called *ideal* systems, which are systems composed of two or more rigid members linked together by mechanical connections which are assumed to be frictionless and incapable of absorbing energy through elongation or compression. Figure 7/7*a* shows a simple example of an ideal system where motion between its two parts is possible and where the equilibrium position is determined by the applied external forces **P** and **F**.

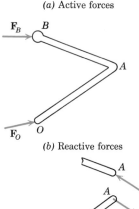

(a) Active forces

(b) Reactive forces

(c) Internal forces

Figure 7/7

For such an interconnected mechanical system we identify three types of forces which act. They are:

(1) *Active forces* are external forces capable of doing virtual work during possible virtual displacements. In Fig. 7/7a forces **P** and **F** are active forces since they would do work as the links move.

(2) *Reactive forces* are forces which act at positions of fixed support where no virtual displacement in the direction of the force takes place. Reactive forces do no work during a virtual displacement. In Fig. 7/7b the horizontal force $\mathbf{F}_B$ exerted on the roller end of the member by the vertical guide can do no work since there can be no horizontal displacement of the roller. The force $\mathbf{F}_O$ exerted on the system by the fixed support at O is also a nonworking reactive force since no displacement of O takes place.

(3) *Internal forces* are forces in the connections between members. During any possible movement of the system or its parts we see that the *net work done by the internal forces at the connections is zero*. This is so because the internal forces always exist in pairs of equal and opposite forces, as indicated for the internal forces $\mathbf{F}_A$ and $-\mathbf{F}_A$ at joint A in Fig. 7/7c, and the work of one force necessarily cancels the work of the other force during their identical displacements.

With the observation that only the external active forces do work during any possible movement of the system, we may now state the principle of virtual work as follows:

The virtual work done by external active forces on an ideal mechanical system in equilibrium is zero for any and all virtual displacements consistent with the constraints.

By constraint we mean restriction of the motion by the supports. In this form the principle finds its greatest use for ideal systems. We state the principle mathematically by the equation

$$\delta U = 0 \qquad\qquad (7/3)$$

where δU stands for the total virtual work done on the system by all active forces during a virtual displacement.

Only now can we see the real advantages of the method of virtual work. There are essentially two. First, it is not necessary for us to dismember ideal systems in order to establish the relations between the active forces, as is generally the case with the equilibrium method based on force and moment summations. Second, we may determine the relations between the active forces directly without reference to the reactive forces. These advantages make the method of virtual work particularly useful in determining the position of equilibrium of a system under known loads. This type of problem is in contrast with the problem of determining the forces acting on a body whose equilibrium position is fixed or specified.

The method of virtual work is especially useful for the purposes mentioned but requires that the internal friction forces do negligible work during any virtual displacement. Consequently, if internal friction in a mechanical system is appreciable, the method of virtual work will produce error when applied to the system as a whole unless the work done by internal friction is accounted for.

In the method of virtual work a diagram which isolates the system under consideration should be drawn. Unlike the free-body diagram, where all forces are shown, the diagram for the method of work need show only the *active forces*, since the reactive forces do not enter into the application of $\delta U = 0$. Such a drawing will be termed an *active-force diagram*. Figure 7/7*a* is an active-force diagram for the system shown.

(*d*) *Degrees of Freedom.* The number of independent coordinates needed to specify completely the configuration of a mechanical system is referred to as the number of *degrees of freedom* for that system. Figure 7/8*a* shows three examples of one-degree-of-freedom systems where only one coordinate is needed to determine the position of each part of the system. The coordinate can be a distance or an angle. Figure 7/8*b* shows three examples of two-degree-of-freedom systems where two independent coordinates are needed to determine the configuration of the system. By adding more links and removing supporting constraints to the linkage in the right-hand figure there is no limit to the number of degrees of freedom which can be achieved.

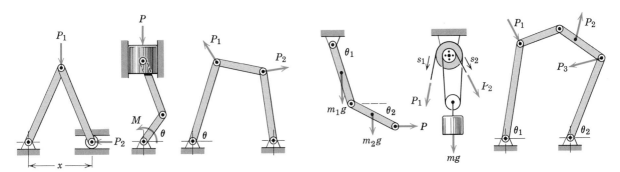

(*a*) Examples of One–Degree–of–Freedom Systems   (*b*) Examples of Two–Degree–of–Freedom Systems

Figure 7/8

The principle of virtual work $\delta U = 0$ may be applied as many times as there are degrees of freedom. With each application we allow only one independent coordinate to change at a time while holding the others constant. In our treatment of virtual work we shall restrict application to one-degree-of-freedom systems.[*]

[*] For examples of solutions to problems of two or more degrees of freedom see Chapter 7 of the author's *Statics, 2nd Edition* 1971 or *SI Version* 1975.

322

Sample Problem   7/1

Each of the two uniform hinged bars has a mass m and a length l, and is supported and loaded as shown. For a given force P determine the angle θ for equilibrium.

Solution.   The active-force diagram for the system composed of the two members is shown separately and includes the weight mg of each bar in addition to the force P. All other forces acting externally on the system are reactive forces which do no work during a virtual movement δx and are not shown.

The principle of virtual work requires that the total work of all external active forces be zero for any virtual displacement consistent with the constraints. Thus for a movement δx the virtual work becomes

① $[\delta U = 0]$ $P\,\delta x + 2mg\,\delta h = 0$

We now express each of these virtual displacements in terms of the variable θ, the required quantity. Hence

$$h = \frac{l}{2}\cos\frac{\theta}{2} \quad \text{and} \quad \delta h = -\frac{l}{4}\sin\frac{\theta}{2}\,\delta\theta$$

② Similarly,

$$x = 2l\sin\frac{\theta}{2} \quad \text{and} \quad \delta x = l\cos\frac{\theta}{2}\,\delta\theta$$

Substitution into the equation of virtual work gives us

$$Pl\cos\frac{\theta}{2}\,\delta\theta - 2mg\frac{l}{4}\sin\frac{\theta}{2}\,\delta\theta = 0$$

from which we get

$$\tan\frac{\theta}{2} = \frac{2P}{mg} \quad \text{or} \quad \theta = 2\tan^{-1}\frac{2P}{mg} \qquad \textit{Ans.}$$

To obtain this result by the principles of force and moment summation, it would be necessary to dismember the frame and take into account all forces acting on each member. Solution by the method of virtual work involves a simpler operation.

① Note carefully that with x positive to the right δx is also positive to the right in the direction of P, so that the virtual work is $P(+\delta x)$. With h positive down δh is also mathematically positive down in the direction of mg, so that the correct mathematical expression for the work is $mg(+\delta h)$. When we express δh in terms of $\delta\theta$ in the next step, δh will have a negative sign, thus bringing our mathematical expression into agreement with the physical observation that the weight mg does negative work as each center of mass moves upward with an increase in x and θ.

② We obtain δh and δx with the same mathematical rules of differentiation with which we may obtain dh and dx.

Sample Problem 7/2

The mass m is brought to an equilibrium position by the application of the couple M to the end of one of the two parallel links that are hinged as shown. The links have negligible mass, and all friction is assumed to be absent. Determine the expression for the equilibrium angle θ assumed by the links with the vertical for a given value of M. Consider the alternative of a solution by force and moment equilibrium.

Solution. The active-force diagram shows the weight mg acting through the center of mass G and the couple M applied to the end of the link. There are no other external active forces or moments which do work on the system during a change in the angle θ.

The vertical position of the center of mass G is designated by the distance h below the fixed horizontal reference line and is $h = b \cos \theta + c$. The work done by mg during a movement δh in the direction of mg is

$$+mg\,\delta h = mg\,\delta(b \cos \theta + c)$$
$$= mg(-b \sin \theta \, \delta\theta + 0)$$
$$= -mgb \sin \theta \, \delta\theta$$

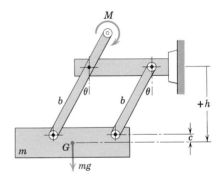

① The minus sign shows that the work is negative for a positive value of $\delta\theta$. The constant c drops out since its variation is zero.

With θ measured positive in the clockwise sense, $\delta\theta$ is also positive clockwise. Thus the work done by the clockwise couple M is $+M\,\delta\theta$. Substitution into the virtual work equation gives us

$[\delta U = 0]$ $\qquad M\,\delta\theta + mg\,\delta h = 0$

which yields

$$M\,\delta\theta = mgb \sin \theta \, \delta\theta$$
$$\theta = \sin^{-1} \frac{M}{mgb} \qquad\qquad Ans.$$

Inasmuch as $\sin \theta$ cannot exceed unity, we see that for equilibrium M is limited to values that do not exceed mgb.

The advantage of the virtual-work solution for this problem is readily seen when we observe what would be involved with a solution by force- and moment-equilibrium. For the latter approach, it would be necessary for us to draw separate free-body diagrams of all of the three moving parts and account for all of the internal reactions at the pin connections. To carry out these steps, it would be necessary for us to include in the analysis the horizontal position of G with respect to the attachment points of the two links, even though reference to this position would finally drop out of the equations when solved. We conclude, then, that the virtual-work method in this problem deals directly with cause and effect and avoids reference to irrelevant quantities.

① Again, as in Sample Problem 7/1, we are consistent mathematically with our definition of work, and we see that the algebraic sign of the resulting expression agrees with the physical change.

Sample Problem 7/3

For link OA in the horizontal position shown, determine the force P on the sliding collar which will prevent OA from rotating under the action of the couple M. Neglect the mass of the moving parts.

Solution. The given sketch serves as the active-force diagram for the system. All other forces are either internal or nonworking reactive forces due to the constraints.

We will give the crank OA a small clockwise angular movement $\delta\theta$ as our virtual displacement and determine the resulting virtual work done by M and P. From the horizontal position of the crank the angular movement gives a downward displacement of A equal to

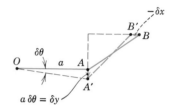

$$\delta y = a\,\delta\theta$$

where $\delta\theta$ is, of course, expressed in radians.

From the right triangle for which link AB is the constant hypotenuse we may write

$$b^2 = x^2 + y^2$$

We now take the differential of the equation and get

$$0 = 2x\,\delta x + 2y\,\delta y \qquad \text{or} \qquad \delta x = -\frac{y}{x}\delta y$$

Thus

$$\delta x = -\frac{y}{x}a\,\delta\theta$$

and the virtual-work equation becomes

$$[\delta U = 0] \qquad M\,\delta\theta + P\,\delta x = 0 \qquad M\,\delta\theta + P\left(-\frac{y}{x}a\,\delta\theta\right) = 0$$

$$P = \frac{Mx}{ya} = \frac{Mx}{ha} \qquad\qquad \textit{Ans.}$$

Again, we observe that the virtual-work method produces a direct relationship between the active force P and the couple M without involving other forces which are irrelevant to this relationship. Solution by the force and moment equations of equilibrium, although fairly simple in this problem, would require accounting for all forces initially and then eliminating the irrelevant ones.

① Note that the displacement $a\,\delta\theta$ of point A would no longer equal δy if the crank OA were not in a horizontal position.

② The length b is constant so that $\delta b = 0$. Notice the negative sign which merely tells us that if one change is positive, the other must be negative.

③ We could just as well use a counterclockwise virtual displacement for the crank which would merely reverse the signs of all terms.

PROBLEMS

(Assume that the negative work of friction is negligible in the following problems unless otherwise indicated.)

7/1 For a given force P determine the angle θ for equilibrium. Neglect the mass of the links.

$$\text{Ans. } \theta = \cos^{-1}\frac{2P}{mg}$$

Problem 7/1

7/2 The toggle mechanism is used to position the mass m in the smooth vertical guides. Determine the expression for the horizontal force P required to support m for any value of θ. Would the action be any more effective if P were applied in a direction other than horizontal?　　Ans. $P = mg \tan \theta$

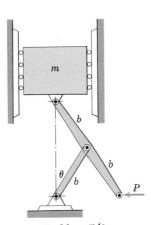

Problem 7/2

7/3 Find the force Q exerted on the paper by the paper punch of Prob. 4/65 repeated here.

Problem 7/3

Problem 7/4

7/4 Determine the torque M on the activating lever of the dump truck necessary to balance the load m with center of mass at G when the dump angle is θ. The polygon $ABDC$ is a parallelogram.

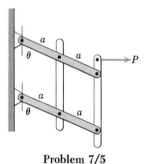

Problem 7/5

7/5 Each of the four uniform links has a mass m. Determine the horizontal force P required to hold them in place in the vertical plane shown.

$$Ans.\ P = \tfrac{5}{2}mg\tan\theta$$

Problem 7/6

7/6 Each of the uniform links of the frame has a mass m and a length b. The equilibrium position of the frame in the vertical plane is determined by the couple M applied to the left-hand link. Express θ in terms of M.

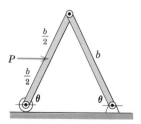

Problem 7/7

7/7 Replace the couple M of Prob. 7/6 by the horizontal force P as shown and determine the equilibrium angle θ.

$$Ans.\ \theta = \tan^{-1}\frac{2mg}{3P}$$

7/8 Determine the force *F* which the patient must apply tangent to the rim of the handwheel of the wheelchair in order to roll up the incline of angle θ. The combined mass of the chair and patient is *m*. (If *s* is the displacement of the center of the wheel measured along the incline and β the corresponding angle in radians through which the wheel turns, it is easily shown that $s = R\beta$ if the wheel rolls without slipping.)

Problem 7/8

7/9 The toggle press of Prob. 4/81 is repeated here. Determine the required force *F* on the handle to produce a compression *R* on the roller for any given value of θ.   Ans. $F = 0.8R\cos\theta$

Problem 7/9

7/10 The hydraulic cylinder is used to spread the linkage and elevate the load *m*. For the position shown determine the compression *C* in the cylinder. Neglect the mass of all parts other than *m*.

Problem 7/10

7/11 The hydraulic cylinder *OA* and link *OB* are arranged to control the tilt of the load which has a mass *m* and a center of mass at *G*. The lower corner *C* is free to roll horizontally as the cylinder linkage elongates. Determine the force *P* in the cylinder necessary to maintain equilibrium at a given angle θ.

$$\text{Ans. } P = \frac{mg}{2}\left(1 - \frac{h}{b}\tan\theta\right)$$

Problem 7/11

Problem 7/12

7/12 Each of the two uniform bars has a mass m with center rollers confined to move in the smooth vertical guide. With the end roller of the lower bar confined to move in the smooth horizontal guide, determine the angle θ for equilibrium under the action of the couple M.

Problem 7/13

7/13 The sketch shows the approximate configuration of one of the four toggle-action hold-down assemblies that clamp the base flange of the Saturn V rocket vehicle to the pedestal of its platform prior to launching. Calculate the preset clamping force F at A if the link CE is under tension produced by a fluid pressure of 2000 lb/in.^{2} acting on the left side of the piston in the hydraulic cylinder. The piston has a net area of 16 in.^{2} The weight of the assembly is considerable, but it is small compared with the clamping force produced and is therefore neglected here. *Ans.* $F = 960,000$ lb

Problem 7/14

7/14 The postal scale consists of a sector of mass m_0 hinged at O and with center of mass at G. The pan and vertical link AB have a mass m_1 and are hinged to the sector at B. End A is hinged to the uniform link AC of mass m_2, which in turn is hinged to the fixed frame. The figure $OBAC$ forms a parallelogram, and the angle GOB is a right angle. Determine the relation between the mass m to be measured and the angle θ assuming that $\theta = \theta_0$ when $m = 0$.

7/15 The claw of the remote-action actuator develops a clamping force C as a result of the tension P in the control rod. Express C in terms of P for the configuration shown where the jaws are parallel.

$$\text{Ans. } C = \frac{P}{2}\frac{e(d+c)}{bc}$$

Problem 7/15

7/16 The cargo box of the food-delivery truck for aircraft servicing has a loaded mass m and is elevated by the application of a torque M on the lower end of the link which is hinged to the truck frame. The horizontal slots allow the linkage to unfold as the cargo box is elevated. Express M as a function of h.

Problem 7/16

7/17 A device for counting the body radiation of a patient is shown. The radiation counter A has a mass m and is positioned by turning the screw of lead L with a torque M which controls the distance BC. Relate the torque M to the load mg for given values of b and θ. Neglect all friction and the mass of the linkage compared with m.

$$\text{Ans. } M = \frac{5mgL}{4\pi}\tan\frac{\theta}{2}$$

Problem 7/17

7/18 The vertical position of the load of mass m is controlled by the adjusting screw which connects joints A and B. The change in the distance between A and B for one revolution of the screw equals the lead L of the screw (advancement per revolution). If a moment M_f is required to overcome friction in the threads and thrust bearing of the screw, determine the expression for the total moment M, applied to the adjusting screw, necessary to raise the load.

Problem 7/18

17.5 mm   40 mm   120 mm   P

B   A

A   P

Problem 7/19

7/19 Determine the torque M exerted by the spaceman's antitorque wrench described in Prob. 4/110 and repeated here. The gripping force on the handles is $P = 150$ N.   *Ans.* $M = 7.88$ N·m

Problem 7/20

7/20 Determine the force F between the jaws of the clamp in terms of a torque M exerted on the handle of the adjusting screw. The screw has a lead (advancement per revolution) L, and friction is to be neglected.

Problem 7/21

7/21 The elevated platform delivers cargo to a cabin opening in an airliner. If the mass of the framework is neglected compared with the mass m of the cargo, determine the compressive force C in the activating hydraulic cylinder AB for any given angle θ.

$$Ans.\ C = \frac{mg\sqrt{1 + 8\sin^2\theta}}{\sin\theta}$$

7/22 Express the compression C in the hydraulic cylinder of the car hoist in terms of the angle θ. The mass of the hoist is neglected compared to the mass m of the vehicle.

$$Ans.\ C = 2mg\ \text{ctn}\ \theta\ \sqrt{1 + \left(\frac{b}{L}\right)^2 - 2\frac{b}{L}\cos\theta}$$

Problem 7/22

7/23 Determine the force P developed at the jaws of the rivet squeezer of Prob. 4/85, repeated here.

$$Ans.\ P = \frac{2Feb}{c(b - a)}$$

Problem 7/23

7/24 The toggle pliers of Prob. 4/71 are repeated here with symbolic dimensions. Determine the clamping force C as a function of α for a given handle gripping force P.

$$Ans.\ C = P\frac{e}{c}\left(\frac{a}{b}\,\text{ctn}\,\alpha - 1\right)$$

Problem 7/24

▶**7/25** The scissors-action military bridge of span L is launched from a tank through the action of the hydraulic linkage AB. If each of the two identical bridge sections has a mass m with center of mass at G, relate the force F in the hydraulic cylinder to the angle θ for equilibrium. An internal mechanism at C maintains the angle 2θ between the two sections. Why must the work of the internal moment at C be evaluated? Discuss the choice of the method of virtual work for this problem.

$$Ans.\ F = \frac{mgL\sqrt{2}}{b}\tan\theta\ \sqrt{1 + \sin\theta}$$

Problem 7/25

7/4 POTENTIAL ENERGY AND STABILITY. In the previous article we have dealt with the equilibrium configuration of mechanical systems composed of individual members that were assumed to be perfectly rigid. We will now extend our method to account for mechanical systems that include elastic elements in the form of springs. For this purpose we find it helpful to introduce the concept of potential energy, which leads us directly to the important problem of determining the stability of equilibrium.

(*a*) *Elastic Potential Energy.* The work done on an elastic member is stored in the member and is called *elastic potential energy V_e*. This energy is potentially available by allowing the member to do work on some other body during the relief of its compression or extension. Consider a spring, Fig. 7/9, which is being compressed by a force F. The spring is assumed to be elastic and linear, that is, the force F is directly proportional to the deflection x. We write this relation as $F = kx$, where k is the *spring constant* or *stiffness* of the spring. The work done on the spring by F during a movement dx is $dU = F\,dx$, so that the elastic potential energy of the spring for a compression x is the total work done on the spring

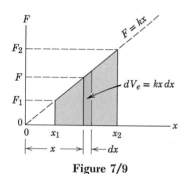

Figure 7/9

$$V_e = \int_0^x F\,dx = \int_0^x kx\,dx$$

or

$$\boxed{V_e = \tfrac{1}{2}kx^2} \tag{7/4}$$

Thus we see that the potential energy of the spring equals the triangular area in the diagram of F versus x from 0 to x.

During a change in the compression of the spring from x_1 to x_2 the work done on the spring equals its *change* in elastic potential energy or

$$\Delta V_e = \int_{x_1}^{x_2} kx\,dx = \tfrac{1}{2}k(x_2^2 - x_1^2)$$

which equals the trapezoidal area from x_1 to x_2.

During a virtual displacement δx of the spring the virtual work done on the spring is the virtual change in elastic potential energy

$$\delta V_e = F\,\delta x = kx\,\delta x$$

If the compressive force is relaxed from $X = x_2$ to $x = x_1$ so that the spring is relieved of its compression, then the *change* (final minus initial) in the potential energy of the spring becomes

$$\Delta V_e = \tfrac{1}{2}k(x_1^2 - x_2^2) = -\tfrac{1}{2}k(x_2^2 - x_1^2)$$

Correspondingly if δx is negative, δV_e will also be negative.

When we have a spring in tension rather than compression, the work and energy relations are the same as those for compression where x now represents the elongation of the spring rather than its compression. While the spring is being stretched, we note that the force again acts in the direction of the displacement doing positive work on the spring and increasing its potential energy.

Since the force acting on the movable end of a spring is the negative of the force exerted by the spring on the body to which its movable end is attached, we conclude that the *work done on the body is the negative of the potential energy change of the spring.*

The units of potential energy are the same as those of work and are expressed in joules (J) in SI units and in foot-pounds (ft-lb) in U.S. customary units.

(*b*) *Gravitational Potential Energy.* In the previous article we treated the work of a gravitational force or weight acting on a body in the same way as the work of any other active force. Thus for an upward displacement δh of the body in Fig. 7/10 the weight $W = mg$ does negative work $\delta U = -mg\,\delta h$. Or, if the body has a downward displacement δh, with h measured positive downward, the weight does positive work $\delta U = +mg\,\delta h$.

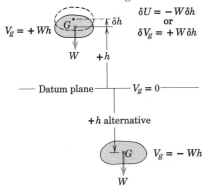

$$\delta U = -W\,\delta h$$
$$\text{or}$$
$$\delta V_g = +W\,\delta h$$

Figure 7/10

We now adopt an alternative to the foregoing treatment by expressing the work done by gravity in terms of a change in potential energy of the body. This alternative treatment is a useful representation when we describe a mechanical system in terms of its total energy. The *gravitational potential energy* V_g of a body is defined simply as the work done on the body by a force equal and opposite to the weight in bringing the body to the position under consideration from some arbitrary datum plane where the potential energy is defined to be zero. The potential energy, then, is the negative of the work done by the weight. When the body is raised, for example, the work done is converted into energy that is potentially available, since the body is capable of doing work on some other body as it returns to its original lower position. If we take V_g to be zero at $h = 0$, Fig. 7/10, then at a height h above the datum plane the gravitational potential energy of the body is

$$\boxed{V_g = mgh} \tag{7/5}$$

If the body is a distance h below the datum plane, its gravitational potential energy is $-mgh$.

It is important to observe that the datum plane for zero potential energy is perfectly arbitrary inasmuch as it is only the *change* in energy with which we are concerned, and this change is the same no matter where we take the datum plane. Also, the gravitational potential energy is independent of the path followed in arriving at the particular level h under consideration. Thus the body of mass m in Fig. 7/11 has the same potential-energy change no matter which path it followed in going from datum plane 1 to datum plane 2 since Δh is the same for both. Our measurement of h, of course, is to the center of mass of the body.

The virtual change in gravitational potential energy is simply

$$\delta V_g = mg\,\delta h$$

where δh is the upward virtual displacement of the mass center of the body. If the mass center should have a downward virtual displacement, then δV_g is negative.

The units of gravitational potential energy are the same as those for work and elastic potential energy, joules (J) in SI units and foot-pounds (ft-lb) in U.S. customary units.

(*c*) *Energy Equation.* In the previous two sections we have noted that the work done *by* a spring *on* the body to which its movable end is attached is the negative of the change in the elastic potential energy of the spring. Also, the work done by the gravitational force or weight mg is the negative of the change in gravitational potential energy. Therefore, when we apply the virtual-work equation to systems with springs and with changes in the vertical position of its members, we may replace the work of the spring and the work of the weight by the negative of the respective potential energy changes. Putting these changes on the right-hand side of Eq. 7/3 enables us to write the virtual-work equation as

$$\boxed{\delta U = \delta V_e + \delta V_g} \quad \text{or} \quad \boxed{\delta U = \delta V} \qquad (7/6)$$

where $V = V_e + V_g$ stands for the total potential energy of the system. The quantity δU now represents the virtual work done on the system during a virtual displacement by all external active forces *other than* spring forces and gravitational forces.

Thus for a mechanical system with elastic members and members that undergo changes in position, we may restate the principle of virtual work as follows:

The virtual work done by external active forces on a mechanical system in equilibrium equals the corresponding change in the total elastic and gravitational potential energy of the system for any and all virtual displacements consistent with the constraints.

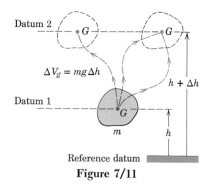

Datum 2

$\Delta V_g = mg\,\Delta h$

Datum 1

$h + \Delta h$

m

h

Reference datum

Figure 7/11

As always, we make the critical decision as to what constitutes our system. With the method of work-energy it is useful to draw the active-force diagram of the system chosen where the boundary of the system clearly distinguishes those members that are a part of the system from other bodies which are not a part of the system. By including an elastic member within the boundary of our system, we see that the forces of interaction between it and the movable member to which it is attached are *internal* to the system and need not be shown. Nor do we show the gravitational forces since their work is accounted for in the V_g-term.

(*d*) *Stability of Equilibrium.* In addition to the replacement of the work of elastic forces and the work of gravitational forces by their corresponding changes in potential energy, we may convert the work of other active forces applied externally to the system to potential energy changes as shown in Fig. 7/12. Here the work done by **F** during a virtual displacement δs of its point of application is equivalent to the change $-mg\,\delta s\cos\alpha$ in potential energy of the mass m for the equivalent system. Hence we may replace the work done by external active forces on a mechanical system by the corresponding potential energy changes with the opposite sign for the equivalent system.

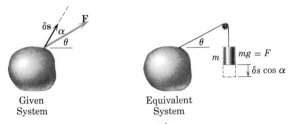

Given System   Equivalent System

Figure 7/12

With the replacement of the work terms by energy terms, the principle of virtual work for a mechanical system without internal kinetic friction, Eq. 7/6, may now be written as

$$\delta(V_e + V_g) = 0 \qquad \text{or} \qquad \delta V = 0 \qquad (7/7)$$

Equation 7/7 expresses the requirement that the equilibrium configuration of a mechanical system is one for which the total potential energy V of the system has a stationary value. For a system of one degree of freedom where the potential energy and its derivatives are continuous functions of the single variable, say x, that describes the configuration, the equilibrium condition $\delta V = 0$ is equivalent mathematically to the requirement

$$\frac{dV}{dx} = 0 \qquad (7/8)$$

Equation 7/8 states that a mechanical system is in equilibrium when the derivative of its total potential energy is zero. For systems with several degrees of freedom the partial derivative of V with respect to each coordinate in turn must be zero for equilibrium.[°]

There are three conditions under which Eq. 7/8 applies, namely, when the total potential energy is a minimum (stable equilibrium), a maximum (unstable equilibrium), or a constant (neutral equilibrium). We see a simple example of these three conditions in Fig. 7/13 where the potential energy of the roller is clearly a minimum in the stable position, a maximum in the unstable position, and a constant in the neutral position.

We may also characterize the stability of a mechanical system by noting that a small displacement away from the stable position results in an increase in potential energy and a tendency to return to the position of lower energy. On the other hand a small displacement away from the unstable position results in a decrease in potential energy and a tendency to move farther away from the equilibrium position to one of lower energy. For the neutral position a small displacement one way or the other results in no change in potential energy and no tendency to move either way.

When a function and its derivatives are continuous, the second derivative is positive at a point of minimum value of the function and negative at a point of maximum value of the function. Thus we may write the mathematical conditions for the equilibrium and stability for a system with a single degree of freedom x as follows:

Stable   Unstable   Neutral

Figure 7/13

$$
\begin{array}{ll}
\text{Equilibrium} & \dfrac{dV}{dx} = 0 \\[2mm]
\text{Stable} & \dfrac{d^2V}{dx^2} > 0 \\[2mm]
\text{Unstable} & \dfrac{d^2V}{dx^2} < 0
\end{array}
\qquad (7/9)
$$

Occasionally we may have a situation where the second derivative of V is also zero at the equilibrium position, in which case we must examine the sign of a higher derivative to ascertain the type of equilibrium. When the order of the lowest remaining nonzero derivative is even, the equilibrium will be stable or unstable according to whether the sign of this lowest even-order derivative is positive or negative. If the order of the lowest remaining nonzero derivative is odd, the equilibrium is classified as unstable, and the plot of V versus x for this case appears as an inflection point in the curve with zero slope at the equilibrium value.

Stability criteria for systems with more than one degree of freedom require more advanced treatment. For two degrees of freedom, for example, we use a Taylor-series expansion for two variables.

[°] For examples of two-degree-of-freedom systems see Art. 43, Chapter 7 of the author's *Statics, 2nd Edition* 1971 or *SI Version* 1975.

Sample Problem 7/4

The 10-kg cylinder is suspended by the spring which has a stiffness of 2 kN/m. Plot the potential energy V of the system and show that it is minimum at the equilibrium position.

$k = 2$ kN/m

$V = 0$

$+x$

$m = 10$ kg

Solution. (Although the equilibrium position in this simple problem is clearly where the force in the spring equals the weight mg, we will proceed as though this fact were unknown in order to illustrate the energy ① relationships in the simplest way.) We choose the datum plane for zero potential energy at the position for which the spring is unextended.

The elastic potential energy for an arbitrary position x is $V_e = \frac{1}{2}kx^2$ and the gravitational potential energy is $-mgx$, so that the total potential energy is

$$[V = V_e + V_g] \qquad V = \tfrac{1}{2}kx^2 - mgx$$

Equilibrium occurs where

$$\left[\frac{dV}{dx} = 0\right] \qquad \frac{dV}{dx} = kx - mg = 0, \qquad x = mg/k$$

Although we know in this simple case that the equilibrium is stable, we prove it by evaluating the sign of the second derivative of V at the equilibrium position. Thus $d^2V/dx^2 = k$, which is positive, proving that the equilibrium is stable.

Substituting numerical values gives

$$V = \tfrac{1}{2}(2000)x^2 - 10(9.81)x$$

expressed in joules, and the equilibrium value of x is

$$x = 10(9.81)/2000 = 0.049 \text{ m or } 49 \text{ mm} \qquad Ans.$$

We calculate V for various values of x and plot V versus x as shown. ② The minimum value of V occurs at $x = 0.049$ m where $dV/dx = 0$ and d^2V/dx^2 is positive.

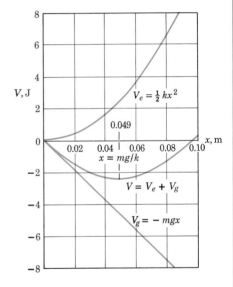

① The choice is arbitrary but simplifies the algebra.

② We could have chosen different datum planes for V_e and V_g without affecting our conclusions. Such a change would merely shift the separate curves for V_e and V_g up or down but would not affect the position of the minimum value of V.

Sample Problem 7/5

The two uniform links, each of mass m, are in the vertical plane and are connected and constrained as shown. As the angle θ between the links increases with the application of the horizontal force P, the light rod, which is connected at A and passes through a pivoted collar at B, compresses the spring of stiffness k. If the spring is uncompressed in the position equivalent to that for which $\theta = 0$, determine the force P which will produce equilibrium at the angle θ.

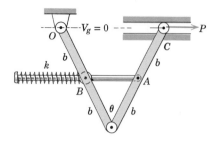

Solution. The given sketch serves as the active-force diagram of the system. The compression x of the spring is the distance that A has moved away from B, which is $x = 2b \sin \dfrac{\theta}{2}$. Thus the elastic potential energy of the spring is

$$[V_e = \tfrac{1}{2}kx^2] \qquad V_e = \tfrac{1}{2}k\left(2b \sin \frac{\theta}{2}\right)^2 = 2kb^2 \sin^2 \frac{\theta}{2}$$

With the datum for zero gravitational potential energy taken through the support at O for convenience, the expression for V_g becomes

$$[V_g = mgh] \qquad V_g = 2mg\left(-b \cos \frac{\theta}{2}\right)$$

The distance between O and C is $4b \sin \dfrac{\theta}{2}$, so that the virtual work done by P is

$$\delta U = P\,\delta\left(4b \sin \frac{\theta}{2}\right) = 2Pb \cos \frac{\theta}{2}\,\delta\theta$$

The virtual-work equation now gives

$$[\delta U = \delta V_e + \delta V_g] \qquad 2Pb \cos \frac{\theta}{2}\,\delta\theta = \delta\left(2kb^2 \sin^2 \frac{\theta}{2}\right) + \delta\left(-2mgb \cos \frac{\theta}{2}\right)$$

$$= 2kb^2 \sin \frac{\theta}{2} \cos \frac{\theta}{2}\,\delta\theta + mgb \sin \frac{\theta}{2}\,\delta\theta$$

Simplifying gives finally

$$P = kb \sin \frac{\theta}{2} + \tfrac{1}{2}mg \tan \frac{\theta}{2} \qquad\qquad \text{Ans.}$$

If we had been asked to express the equilibrium value of θ corresponding to a given force P, we would have difficulty solving explicitly for θ in this particular case. For a numerical problem we could resort to an approximation with a graphical plot of numerical values of the sum of the two functions of θ to determine the value of θ for which the sum equals P.

Sample Problem 7/6

The ends of the uniform bar of mass m slide freely in the horizontal and vertical guides as shown. Examine the stability conditions for the positions of equilibrium. The spring of stiffness k is undeformed when $x = 0$.

Solution. The system consists of the spring and the bar. Since there are no external active forces, the given sketch serves as the active-force diagram. We shall take the x-axis as the datum for zero gravitational potential energy. In the displaced position the elastic and gravitational potential energies are

$$V_e = \tfrac{1}{2}kx^2 = \tfrac{1}{2}kb^2 \sin^2 \theta \quad \text{and} \quad V_g = mg\frac{b}{2}\cos\theta$$

The total potential energy is, then,

$$V = V_e + V_g = \tfrac{1}{2}kb^2 \sin^2 \theta + \tfrac{1}{2}mgb \cos\theta$$

Equilibrium occurs for $dV/d\theta = 0$ so that

$$\frac{dV}{d\theta} = kb^2 \sin\theta \cos\theta - \tfrac{1}{2}mgb \sin\theta = (kb^2 \cos\theta - \tfrac{1}{2}mgb)\sin\theta = 0$$

① The two solutions to this equation are given by

$$\sin\theta = 0 \quad \text{and} \quad \cos\theta = \frac{mg}{2kb}$$

We now determine the stability by examining the sign of the second derivative of V for each of the two equilibrium positions. The second derivative is

$$\frac{d^2V}{d\theta^2} = kb^2(\cos^2 \theta - \sin^2 \theta) - \tfrac{1}{2}mgb \cos\theta$$

$$= kb^2(2\cos^2 \theta - 1) - \tfrac{1}{2}mgb \cos\theta$$

(Solution I) $\sin\theta = 0, \theta = 0$

$$\frac{d^2V}{d\theta^2} = kb^2(2 - 1) - \tfrac{1}{2}mgb = kb^2\left(1 - \frac{mg}{2kb}\right)$$

$$= \text{positive (stable)} \quad \text{if } k > mg/2b$$

$$= \text{negative (unstable)} \quad \text{if } k < mg/2b \qquad \qquad \textit{Ans.}$$

Thus, if the spring is sufficiently stiff, the bar will return to the vertical
② position even though there is no force in the spring at that position.

(Solution II) $\cos\theta = \dfrac{mg}{2kb}, \theta = \cos^{-1}\dfrac{mg}{2kb}$

$$\frac{d^2V}{d\theta^2} = kb^2\left(2\left[\frac{mg}{2kb}\right]^2 - 1\right) - \tfrac{1}{2}mgb\left(\frac{mg}{2kb}\right) = kb^2\left(\left[\frac{mg}{2kb}\right]^2 - 1\right)$$

Ans.

Since the cosine must be less than unity, we see that this solution is limited to the case where $k > mg/2b$, which makes the second derivative
③ of V negative. Hence equilibrium for solution II is never stable. If $k < mg/2b$ we no longer have solution II since the spring will be too weak to maintain equilibrium at a value of θ between $0°$ and $90°$.

① Be careful not to overlook the solution $\theta = 0$ given by $\sin\theta = 0$.

② This result is one that we might not have anticipated without the mathematical analysis of the stability.

③ Again, without the benefit of the mathematical analysis of the stability we might have supposed erroneously that the bar could come to rest in a stable equilibrium position for some value of θ between $0°$ and $90°$.

PROBLEMS

(Assume that the negative work of friction is negligible in the following problems.)

7/26 The bar of mass m with center of mass at G is pivoted about a horizontal axis through O. Prove the stability conditions for the two positions of equilibrium.

Problem 7/26

7/27 Determine the force P required to produce equilibrium of the two hinged links at an angle θ. The uncompressed length of the spring is $2b$. Neglect the mass of the links.

$$Ans.\ P = 4kb(1 - \cos\theta)\tan\theta$$

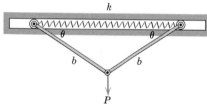

Problem 7/27

7/28 The potential energies of two mechanical systems are given by $V_1 = C_1 x^4$ and $V_2 = C_2 x^3$, where C_1 and C_2 are positive constants and x is the single coordinate expressing the positions of both systems. Specify the stability of each system at the equilibrium position $x = 0$.

7/29 The uniform wheel of mass m is supported in the vertical plane by the light band ABC and the spring of stiffness k. If the wheel is released initially from the position where the force in the spring is zero, determine the clockwise angle θ through which the wheel rotates from the initial position to the final equilibrium position.

$$Ans.\ \theta = \frac{mg}{4kr}$$

Problem 7/29

7/30 The handle is fastened to one of the spring-connected gears, which are mounted in fixed bearings. The spring of stiffness k connects two pins mounted in the faces of the gears. When the handle is in the vertical position, $\theta = 0$ and the spring force is zero. Determine the force P required to maintain equilibrium at an angle θ.

Problem 7/30

7/31 The figure shows the cross section of a uniform
50-kg ventilator door hinged along its upper horizontal edge at O. The door is controlled by the
spring-loaded cable which passes over the small
pulley at A. The spring has a stiffness of 180 N per
meter of stretch and is undeformed when $\theta = 0$.
Determine the angle θ for equilibrium.

Ans. $\theta = 48.6°$

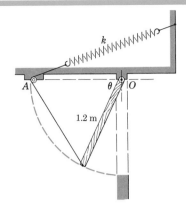

7/32 Determine the equilibrium value of x for the
spring-supported bar. The spring has a stiffness k
and is unstretched when $x = 0$. The force F acts in
the direction of the bar, and the mass of the bar is
negligible.

7/33 When $u = 0$, the spring of stiffness k is uncompressed. As u increases, the rod slides through the
pivoted collar at A and compresses the spring between the collar and the end of the rod. Determine
the force P required to produce a given displacement u. Assume the absence of friction and neglect
the mass of the rod.

$$\text{Ans. } P = \left(1 - \frac{b}{\sqrt{b^2 + u^2}}\right)ku$$

Problem 7/33

Problem 7/34

Problem 7/35

Problem 7/36

Problem 7/37

7/34 Determine the equilibrium value of the coordinate y for the mechanism under the action of the vertical load P. The spring of stiffness k is unstretched when $y = 0$, and the mass of the link is negligible.

7/35 The uniform bar of mass m with center of mass at G is pivoted about O and swings in the vertical plane. Position of the bar is controlled by the light spring-loaded rod that slides through a pivoted collar at A. The spring of stiffness k is uncompressed in the position for which $\theta = 0$. Determine the acute angle θ for equilibrium and prove that the position is stable.

$$Ans. \ \theta = \tan^{-1} \frac{mg}{ka}$$

7/36 The cross section of a trap door hinged at A and having a mass m and a center of mass at G is shown in the figure. The spring is compressed by the rod which is pinned to the lower end of the door and which passes through the swivel block at B. When $\theta = 0$, the spring is undeformed. Show that with the proper stiffness k of the spring, the door will be in equilibrium for any angle θ.

7/37 For a horizontal force F of 50 lb, determine the angle θ for equilibrium of the spring-loaded linkage. The rod DG passes through the swivel at E and compresses the spring, which has a stiffness of 25 lb/in. and an uncompressed condition corresponding to $\theta = 0$. Neglect the weights of the members.   *Ans.* $\theta = 19.5°$

7/38 In the mechanism shown, the rod AB slides through the pivoted collar at C and compresses the spring when a couple M is applied to link DE. The spring has a stiffness k and is uncompressed for the position equivalent to $\theta = 0$. Determine the angle θ for equilibrium. The masses of the parts are negligible.

$$\text{Ans. } \theta = \sin^{-1}\frac{M}{kb^2}$$

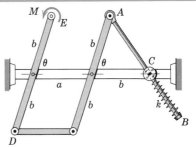

Problem 7/38

7/39 Each of the two gears carries an eccentric mass m and is free to rotate in the vertical plane about its bearing. Determine the values of θ for equilibrium and identify the type of equilibrium for each position.

Problem 7/39

7/40 The center of mass G of the uniform link AB is constrained to move in the smooth vertical guide. Link OA has negligible mass, and the spring of stiffness k is uncompressed for $\theta = 0$. Determine the equilibrium positions and their stability.
   Ans.

$$\theta = \pi, \text{ stable if } k < \frac{mg}{2b} \text{ and unstable if } k > \frac{mg}{2b}$$

$$\theta = 2\sin^{-1}\frac{mg}{2kb} \text{ requires } k > \frac{mg}{2b}, \text{ stable}$$

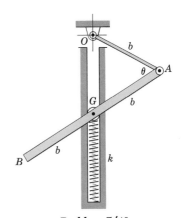

Problem 7/40

7/41 The figure shows the cross section of a container composed of a hemispherical shell of radius r and a cylindrical shell of height h, both made from the same material. Specify the limitation of h for stability in the upright position when the container is placed on the horizontal surface.

Problem 7/41

Problem 7/42

7/42 A uniform rectangular block of height h and mass m is centered in a horizontal position on the fixed circular surface of radius r. Determine the limiting value of h for stability.   *Ans.* $h < 2r$

Problem 7/43

7/43 Predict through calculation whether the homogeneous semicylinder and the half-cylindrical shell will remain in the positions shown or whether they will roll off the lower cylinder.

Problem 7/44

7/44 Determine the maximum height h of the mass m for which the inverted pendulum will be stable in the vertical position shown. Each of the springs has a stiffness k, and they have equal precompressions in this position. Neglect the mass of the remainder of the mechanism.

$$Ans.\ h_{max} = \frac{2kb^2}{mg}$$

Problem 7/45

7/45 One of the critical requirements in the design of an artificial leg for an amputee is to prevent the knee joint from buckling under load when the leg is straight. As a first approximation, simulate the artificial leg by the two light links with a torsion spring at their common joint. The spring develops a torque $M = K\beta$, which is proportional to the angle of bend β at the joint. Determine the minimum value of K that will ensure stability of the knee joint for $\beta = 0$.
Ans. $K_{min} = \frac{1}{2}mgl$

7/46 In the figure is shown a small industrial lift with a foot release. There are four identical springs, two on each side of the central shaft. The stiffness of each pair of springs is $2k$. Specify the value of k that will ensure stable equilibrium when the lift supports a load L in the position shown with no force on the pedal. The springs have an equal initial precompression and may be assumed to act in the horizontal direction at all times.

Problem 7/46

7/47 The figure shows a tilting desk chair together with a detail of the spring-loaded tilting mechanism. The frame of the seat is pivoted about the fixed point O on the base. The increase in distance between A and B as the chair tilts back about O is the increase in compression of the spring. The spring, which has a stiffness of 96 kN/m, is uncompressed when $\theta = 0$. For small angles of tilt it may be assumed with negligible error that the axis of the spring remains parallel to the seat. The center of mass of an 80-kg person who sits in the chair is at G on a line through O perpendicular to the seat. Determine the angle of tilt θ for equilibrium. (*Hint:* The deformation of the spring may be visualized by allowing the base to tilt through the required angle θ about O while the seat is held in a fixed position.)   Ans. $\theta = 11.19°$

Problem 7/47

7/48 The slender bar of length l and mass m is pivoted freely about a horizontal axis through O. The spring has an unstretched length of $l/2$. Determine the equilibrium positions, excluding $\theta = \pi$, and determine the maximum value of the spring stiffness k for stability in the position $\theta = 0$.

$$\text{Ans. } \theta = 0, \ \theta = 2 \cos^{-1} \frac{1/2}{1 - \dfrac{2mg}{kl}}, \ k_{\max} = \frac{4mg}{l}$$

Problem 7/48

Problem 7/49

7/49 The mass m moves in smooth vertical guides and is supported by the four spring-loaded links of negligible mass. The spring of stiffness k is unstretched in the position for which $\theta = 0$. Specify the stability of the system for its equilibrium positions.

Problem 7/50

7/50 The front-end suspension of Prob. 4/82 is repeated here. The frame F must be jacked up so that $h = 350$ mm in order to relieve the compression in the coil springs. Determine the value of h when the jack is removed. Each spring has a stiffness of 120 kN/m. The load L is 12 kN, and the central frame F has a mass of 40 kg. Each wheel and attached link has a mass of 35 kg with a center of mass 680 mm from the vertical centerline.

Ans. h = 265 mm

Problem 7/51

▶**7/51** The platform of mass m is supported by equal legs and braced by the two springs as shown. If the masses of the legs and springs are negligible, determine the minimum stiffness k of each spring that will ensure stability of the platform in the position shown. Each spring has a tensile preset deflection Δ.

$$Ans.\ k_{\min} = \frac{mg}{2b}\left(1 + \frac{b^2}{l^2}\right)$$

Problem 7/52

▶**7/52** The uniform garage door AB shown in section has a mass m and is equipped with two of the spring-loaded mechanisms shown, one on each side of the door. The arm OB has negligible mass, and the upper corner A of the door is free to move horizontally on a roller. The unstretched length of the spring is $r - a$, so that in the top position with $\theta = \pi$ the spring force is zero. In order to ensure smooth action of the door as it reaches the vertical closed position $\theta = 0$, it is desirable that the door be insensitive to movement in this position. Determine the required spring stiffness k.

$$Ans.\ k = \frac{mg(r + a)}{8a^2}$$

7/5 PROBLEM FORMULATION AND REVIEW. When various configurations are possible for a body or a system of interconnected bodies as a result of applied forces, the equilibrium position attained is usually best found by applying the principle of virtual work developed in this chapter. We have seen repeatedly that the only forces which need to be considered when determining the equilibrium position by this method are those which do work (active forces) during the assumed differential movement of the body or system away from its equilibrium position. Those external forces which do no work (reactive forces) need not be involved. For this reason we have constructed the active-force diagram of the body or system (rather than the free-body diagram) to focus attention on only those external forces which do work during the virtual displacements.

Relating the corresponding virtual displacements, linear and angular, of the parts of a mechanical system during a virtual movement consistent with the constraints is often the most difficult part of the analysis. First, the geometric relationships which describe the configuration of the system must be written. Next, the differential changes in the positions of parts of the system are established by the process of differentiation of the geometrical relationship to obtain expressions for the differential virtual movements.

In the method of virtual work we take special note of the fact that a virtual displacement is a first-order differential change in a length or an angle. This change is ficticious in that it is an assumed movement which need not take place in reality. Mathematically, a virtual displacement is treated the same as a differential change in an actual movement. We use the symbol δ for the differential virtual change and the usual symbol d for the differential change in a real movement.

In Chapter 7 we have restricted our attention to mechanical systems for which the positions of the members can be specified by a single variable (single-degree-of-freedom systems). For two or more degrees of freedom, we would apply the virtual-work equation as many times as there are degrees of freedom, allowing one variable to change at a time while holding the remaining ones constant.

We have found that the concept of potential energy, both gravitational (V_g) and elastic (V_e), is useful in solving equilibrium problems where changes in the vertical position of the mass centers of the bodies occur and where corresponding changes in the elongation or compression of elastic members (springs) also result during the virtual displacement. Here we obtain an expression for the total potential energy V of the system in terms of the variable that specifies the possible position of the system. The first and second derivatives of V are used to establish, respectively, the position of equilibrium and the type of stability which exists.

REVIEW PROBLEMS

Plane of each figure is vertical. Size and mass of
each member and applied force are known.

7/53 Identify which of the problems (*a*) through (*f*) are
best solved (*A*) by the force and moment equilibrium equations and (*B*) by virtual work. Outline
briefly the procedure for each solution.

Find *θ* for equilibrium

(*a*)

Find reactions at *A* and *B*

(*b*)

Find *x* for equilibrium

(*c*)

Find forces at *A, B,* and *C*

(*d*)

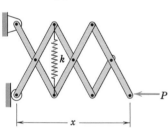

Find *x* for equilibrium

(*e*)

Determine maximum
k for stable equilibrium
at *θ* = 0

(*f*)

Problem 7/53

Problem 7/54

7/54 A "black box" contains a series of interconnected
racks and pinions, gears, and other internal mechanical elements which transfer the linear motion
of the pushrod *A* to produce the linear motion of the
pushrod *B*. For every unit of inward movement of *A*
under the action of force P_1, rod *B* moves outward
from the box one-third of a unit against the action of
force P_2. If $P_1 = 100$ N, calculate P_2 for equilibrium.
Neglect all friction and assume all mechanical components are ideally connected light rigid bodies.

7/55 If the internal mechanism in the box described in Prob. 7/54 is rearranged so that an inward movement of one unit of pushrod A results in a movement of one-third of a unit of rod B also inward, compute the tension T supported by rod B for equilibrium if $P_1 = 100$ N.

7/56 A control mechanism consists of an input shaft at A which is turned by applying a couple M and an output slider B which moves in the x-direction against the action of force P. The mechanism is arranged so that the linear movement of B is proportional to the angular movement of A with x increasing 80 mm for every complete turn of A. If $M = 10$ N·m, determine P for equilibrium. Neglect internal friction and assume all mechanical components are ideally connected rigid bodies.

Ans. P = 785 N

Problem 7/56

7/57 The bar is free to rotate through a complete vertical circle. Verify mathematically the stability conditions that are evident at the two equilibrium positions.

Problem 7/57

7/58 The potential energy of a mechanical system with negligible friction is given by $V = b \sin^2 \theta + c \cos \theta$ where θ is the angular coordinate that defines the position of the system and where b and c are positive constants. Establish the equilibrium position or positions and determine the type of equilibrium for each one.

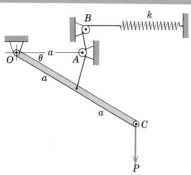

Problem 7/59

7/59 The light bar OC is pivoted at O and swings in the vertical plane. When $\theta = 0$, the spring of stiffness k is unstretched. Determine the equilibrium angle corresponding to a given vertical force P applied to the end of the bar. Neglect the mass of the bar and the diameter of the small pulleys.

$$Ans. \ \theta = \tan^{-1}\frac{2P}{ka}$$

Problem 7/60

7/60 A power-operated loading platform for the back of a truck is shown in the figure. The position of the platform is controlled by the hydraulic cylinder, which applies force at C. The links are pivoted to the truck frame at A, B, and F. Determine the force P supplied by the cylinder in order to support the platform in the position shown. The mass of the platform and links may be neglected compared with that of the 250-kg crate with center of mass at G.

$$Ans. \ P = 3.5 \ kN$$

Problem 7/61

7/61 The figure shows the edge view of a uniform sky-light door of mass m with center of mass midway between A and O. The door is counterbalanced by the action of the spring, which has a stiffness k and which is unstretched when the door is in the vertical plane with $\theta = 0$. Determine the necessary stiffness k to balance the door in an equilibrium position for $\theta > 0$.

$$Ans. \ k = \frac{mg}{2l}, \ \text{independent of } \theta$$

AREA MOMENTS OF INERTIA

A1 INTRODUCTION. When forces are distributed continuously over an area upon which they act, it is often necessary for us to calculate the moment of these forces about some axis either in or perpendicular to the plane of the area. Frequently the intensity of the force (pressure or stress) is proportional to the distance of the force from the moment axis. The elemental force acting on an element of area, then, is proportional to distance times differential area, and the elemental moment is proportional to distance squared times differential area. We see, therefore, that the total moment involves an integral of the form $\int (\text{distance})^2 \, d(\text{area})$. This integral is known as the *moment of inertia* of the area. The integral is a function of the geometry of the area and occurs so frequently in the applications of mechanics that we find it useful to develop its properties in some detail and to have these properties available for ready use when the integral arises.

Figure A1 illustrates the physical origin of these integrals. In the *a*-part of the figure the surface area *ABCD* is subjected to a distributed pressure *p* whose intensity is proportional to the distance *y* from the axis *AB*. This situation was treated in Art. 5/8 of Chapter 5 where we described the action of liquid pressure on a plane surface. The moment about *AB* that is due to the pressure on the element of area dA is $py \, dA = ky^2 \, dA$. Thus the integral in question appears when the total moment $M = k \int y^2 \, dA$ is evaluated.

In Fig. A1*b* we show the distribution of stress acting on a transverse section of a simple elastic beam bent by equal and opposite couples applied to its ends. At any section of the beam a linear distribution of force intensity or stress σ, given by $\sigma = ky$, is present, the stress being positive (tensile) below the axis *O–O* and negative (compressive) above the axis. We see that the elemental moment about the axis *O–O* is $dM = y(\sigma \, dA) = ky^2 \, dA$. Thus the same integral appears when the total moment $M = k \int y^2 \, dA$ is evaluated,

A third example is given in Fig. A1*c*, which shows a circular shaft subjected to a twist or torsional moment. Within the elastic limit of the material this moment is resisted at each cross section of the shaft by a distribution of tangential or shear stress τ which is proportional to the radial distance r from the center. Thus $\tau = kr$,

(a)

(b)

(c)

Figure A/1

and the total moment about the central axis is $M = \int r(\tau\, dA) = k \int r^2\, dA$. Here the integral differs from that in the preceding two examples in that the area is normal instead of parallel to the moment axis and in that r is a radial coordinate instead of a rectangular one.

Although the integral illustrated in the preceding examples is generally called the *moment of inertia* of the area about the axis in question, a more fitting term is the *second moment of area*, since the first moment $y\, dA$ is multiplied by the moment arm y to obtain the second moment for the element dA. The word *inertia* appears in the terminology by reason of the similarity between the mathematical form of the integrals for second moments of areas and those for the resultant moments of the so-called inertia forces in the case of rotating bodies. The moment of inertia of an area is a purely mathematical property of the area and in itself has no physical significance.

A2   DEFINITIONS.   The following definitions of terms form the basis for the analysis of area moments of inertia.

(a) *Rectangular and Polar Moments of Inertia.*   Consider the area A in the x-y plane, Fig. A2. The moments of inertia of the element dA about the x- and y-axes are, by definition, $dI_x = y^2\, dA$ and $dI_y = x^2\, dA$, respectively. Therefore the moments of inertia of A about the same axes are

Figure A/2

$$\boxed{\begin{aligned} I_x &= \int y^2\, dA \\ I_y &= \int x^2\, dA \end{aligned}} \tag{A1}$$

where we carry out the integration over the entire area.

The moment of inertia of dA about the pole O (z-axis) is, by similar definition, $dJ_z = r^2\, dA$, and the moment of inertia of the entire area about O is

$$\boxed{J_z = \int r^2\, dA} \tag{A2}$$

The expressions defined by Eqs. A1 are known as *rectangular* moments of inertia, whereas the expression of Eq. A2 is known as the *polar* moment of inertia. Since $x^2 + y^2 = r^2$, it is clear that

$$\boxed{J_z = I_x + I_y} \tag{A3}$$

A polar moment of inertia for an area whose boundaries are more simply described in rectangular coordinates than in polar coordinates is easily calculated with the aid of Eq. A3.

We note that the moment of inertia of an element involves the square of the distance from the inertia axis to the element. An element whose coordinate is negative contributes as much to the moment of inertia as does an equal element with a positive coordinate of the same magnitude. Consequently we see that the area moment of inertia about any axis is always a positive quantity. In contrast, the first moment of the area, which was involved in the computations of centroids, could be either positive, negative, or zero.

The dimensions of moments of inertia of areas are clearly L^4, where L stands for the dimension of length. Thus the SI units for area moments of inertia are expressed as quartic meters (m^4) or quartic millimeters (mm^4). The U.S. customary units for area moments of inertia are quartic feet (ft^4) or quartic inches $(in.^4)$.

The choice of the coordinates to use for the calculation of moments of inertia is important. Rectangular coordinates should be used for shapes whose boundaries are most easily expressed in these coordinates. Polar coordinates will usually simplify problems involving boundaries that are easily described in r and θ. The choice of an element of area which simplifies the integration as much as possible is also important. These considerations are quite analogous to those we discussed and illustrated in Chapter 5 for the calculation of centroids.

(b) *Radius of Gyration.* Consider an area A, Fig. A3a, which has rectangular moments of inertia I_x and I_y and a polar moment of inertia J_z about O. We now visualize this area to be concentrated into a long narrow strip of area A a distance k_x from the x-axis, Fig. A3b. By definition the moment of inertia of the strip about the x-axis will be the same as that of the original area if $k_x^2 A = I_x$. The distance k_x is known as the *radius of gyration* of the area about the x-axis. A similar relation for the y-axis is written by considering the area to be concentrated into a narrow strip parallel to the y-axis as shown in Fig. A3c. Also, if we visualize the area to be concentrated into a narrow ring of radius k_z as shown in Fig. A3d, we may express the polar moment of inertia as $k_z^2 A = J_z$. In summary we write

$$
\begin{array}{lll}
I_x = k_x^2 A & & k_x = \sqrt{I_x/A} \\
I_y = k_y^2 A & \text{or} & k_y = \sqrt{I_y/A} \\
J_z = k_z^2 A & & k_z = \sqrt{J_z/A}
\end{array}
\qquad (A4)
$$

The radius of gyration, then, is a measure of the distribution of the area from the axis in question. A rectangular or polar moment of inertia may be expressed by specifying the radius of gyration and the area.

When we substitute Eqs. A4 into Eq. A3 we have

$$
k_z^2 = k_x^2 + k_y^2
\qquad (A5)
$$

Thus the square of the radius of gyration about a polar axis equals the sum of the squares of the radii of gyration about the two corresponding rectangular axes.

It is imperative that there be no confusion between the coordinate to the centroid C of an area and the radius of gyration. In Fig. A3a the square of the centroidal distance from the x-axis, for example, is $\bar{y}^2$, which is the square of the mean value of the distances from the

(a)

(b)

(c)

(d)

Figure A/3

elements of the area to the x-axis. The quantity k_x^2, on the other hand, is the mean of the squares of these distances. The moment of inertia is *not* equal to $A\bar{y}^2$, since the square of the mean is less than the mean of the squares.

(*c*) *Transfer of Axes.* The moment of inertia of an area about a noncentroidal axis may be easily expressed in terms of the moment of inertia about a parallel centroidal axis. In Fig. A4 the x_o-y_o axes pass through the centroid C of the area. Let us now determine the moments of inertia of the area about the parallel x-y axes. By definition the moment of inertia of the element dA about the x-axis is

$$dI_x = (y_o + d_x)^2\, dA$$

Expanding and integrating give us

$$I_x = \int y_o^2\, dA + 2d_x \int y_o\, dA + d_x^2 \int dA$$

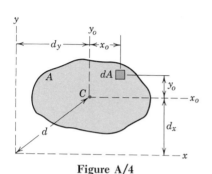

Figure A/4

We see that the first integral is by definition the moment of inertia $\bar{I}_x$ about the centroidal x_o-axis. The second integral is zero, since $\int y_o\, dA = A\bar{y}_o$ and $\bar{y}_o$ is automatically zero with the centroid on the x_o-axis. The third term is simply Ad_x^2. Thus the expression for I_x and the similar expression for I_y become

$$\boxed{\begin{aligned} I_x &= \bar{I}_x + Ad_x^2 \\ I_y &= \bar{I}_y + Ad_y^2 \end{aligned}}$$

(A6)

By Eq. A3 the sum of these two equations gives

$$\boxed{J_z = \bar{J}_z + Ad^2}$$

(A6a)

Equations A6 and A6a are the so-called *parallel-axis theorems.* Two points in particular should be noted. First, the axes between which the transfer is made *must be parallel,* and, second, one of the axes *must pass through the centroid* of the area.

If a transfer is desired between two parallel axes neither one of which passes through the centroid, it is first necessary for us to transfer from one axis to the parallel centroidal axis and then to transfer from the centroidal axis to the second axis.

The parallel-axis theorems also hold for radii of gyration. With substitution of the definition of k into Eqs. A6, the transfer relation becomes

$$\boxed{k^2 = \bar{k}^2 + d^2}$$

(A6b)

where $\bar{k}$ is the radius of gyration about a centroidal axis parallel to the axis about which k applies and d is the distance between the two axes. The axes may be either in the plane or normal to the plane of the area.

A summary of the moment-of-inertia relations for some of the common plane figures is given in Table C3, Appendix C.

Sample Problem A1

Determine the moments of inertia of the rectangular area about the centroidal x_o-y_o axes, the centroidal polar axis z_o through C, the x-axis, and the polar axis z through O.

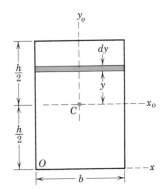

Solution. For the calculation of the moment of inertia $\bar{I}_x$ about the x_o-axis a horizontal strip of area $b\,dy$ is chosen so that all elements of the strip have the same y-coordinate. Thus

$$[I_x = \int y^2\,dA] \qquad \bar{I}_x = \int_{-h/2}^{h/2} y^2 b\,dy = \tfrac{1}{12}bh^3 \qquad \textit{Ans.}$$

By interchanging symbols the moment of inertia about the centroidal y_o-axis is

$$\bar{I}_y = \tfrac{1}{12}hb^3 \qquad \textit{Ans.}$$

The centroidal polar moment of inertia is

$$[J_z = I_x + I_y] \qquad \bar{J}_z = \tfrac{1}{12}(bh^3 + hb^3) = \tfrac{1}{12}A(b^2 + h^2) \qquad \textit{Ans.}$$

By the parallel-axis theorem the moment of inertia about the x-axis is

$$[I_x = \bar{I}_x + Ad_x^{\,2}] \qquad I_x = \tfrac{1}{12}bh^3 + bh\left(\frac{h}{2}\right)^2 = \tfrac{1}{3}bh^3 = \tfrac{1}{3}Ah^2 \qquad \textit{Ans.}$$

We also obtain the polar moment of inertia about O by the parallel-axis theorem which gives us

$$[J_z = \bar{J}_z + Ad^2] \qquad J_z = \tfrac{1}{12}A(b^2 + h^2) + A\left[\left(\frac{b}{2}\right)^2 + \left(\frac{h}{2}\right)^2\right]$$

$$J_z = \tfrac{1}{3}A(b^2 + h^2) \qquad \textit{Ans.}$$

① If we had started with the second-order element $dA = dx\,dy$, integration with respect to x holding y constant amounts simply to multiplication by b and gives us the expression $y^2 b\,dy$ which we chose at the outset.

Sample Problem A2

Determine the moments of inertia of the triangular area about its base and about parallel axes through its centroid and vertex.

Solution. A strip of area parallel to the base is selected as shown in the figure, and it has the area $dA = x\,dy = [(h-y)b/h]\,dy$. By definition

$$[I_x = \int y^2\,dA] \qquad I_x = \int_0^h y^2\frac{h-y}{h}b\,dy = b\left[\frac{y^3}{3} - \frac{y^4}{4h}\right]_0^h = \frac{bh^3}{12} \qquad \textit{Ans.}$$

By the parallel-axis theorem the moment of inertia $\bar{I}$ about an axis through the centroid, a distance $h/3$ above the x-axis, is

$$[\bar{I} = I - Ad^2] \qquad \bar{I} = \frac{bh^3}{12} - \left(\frac{bh}{2}\right)\left(\frac{h}{3}\right)^2 = \frac{bh^3}{36} \qquad \textit{Ans.}$$

A transfer from the centroidal axis to the x'-axis through the vertex gives

$$[I = \bar{I} + Ad^2] \qquad I_{x'} = \frac{bh^3}{36} + \left(\frac{bh}{2}\right)\left(\frac{2h}{3}\right)^2 = \frac{bh^3}{4} \qquad \textit{Ans.}$$

① Here again we choose the simplest possible element. If we had chosen $dA = dx\,dy$, we would have to integrate $y^2\,dx\,dy$ with respect to x first. This gives us $y^2 x\,dy$, which is the expression we chose at the outset.

② Expressing x in terms of y should cause no difficulty if we observe the proportional relationship between the similar triangles.

Sample Problem A3

Calculate the moments of inertia of the area of a circle about a diametral axis and about the polar axis through the center. Specify the radii of gyration.

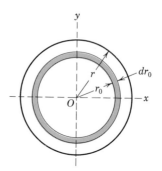

Solution. A differential element of area in the form of a circular ring may be used for the calculation of the moment of inertia about the polar z-axis through O since all elements of the ring are equidistant from O. The elemental area is $dA = 2\pi r_o \, dr_o$, and thus

$$[J_z = \int r^2 \, dA] \qquad J_z = \int_0^r r_o^2 (2\pi r_o \, dr_o) = \frac{\pi r^4}{2} = \tfrac{1}{2}Ar^2 \qquad \text{Ans.}$$

The polar radius of gyration is

$$\left[k = \sqrt{\frac{J}{A}} \right] \qquad k_z = \frac{r}{\sqrt{2}} \qquad \text{Ans.}$$

By symmetry $I_x = I_y$, so that from Eq. A3

$$[J_z = I_x + I_y] \qquad I_x = \tfrac{1}{2}J_z = \frac{\pi r^4}{4} = \tfrac{1}{4}Ar^2 \qquad \text{Ans.}$$

The radius of gyration about the diametral axis is

$$\left[k = \sqrt{\frac{I}{A}} \right] \qquad k_x = \frac{r}{2} \qquad \text{Ans.}$$

The foregoing determination of I_x is the simplest possible. The result may also be obtained by direct integration, using the element of area $dA = r_o \, dr_o \, d\theta$ shown in the lower figure. By definition

$$[I_x = \int y^2 \, dA] \qquad I_x = \int_0^{2\pi} \int_0^r (r_o \sin \theta)^2 r_o \, dr_o \, d\theta$$

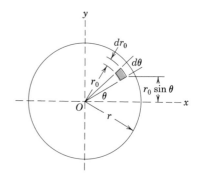

$$= \int_0^{2\pi} \frac{r^4 \sin^2 \theta}{4} \, d\theta = \frac{r^4}{4} \frac{1}{2} \left[\theta - \frac{\sin 2\theta}{2} \right]_0^{2\pi} = \frac{\pi r^4}{4} \qquad \text{Ans.}$$

① Polar coordinates are certainly indicated here. Also, as before, we choose the simplest and lowest-order element possible, which is the differential ring. It should be evident immediately from the definition that the polar moment of inertia of the ring is its area $2\pi r_o \, dr_o$ times r_o^2.

② This integration is straightforward, but the use of Eq. A3 along with the result for J_z is certainly simpler.

Sample Problem A4

Determine the moment of inertia of the area under the parabola about the x-axis. Solve by using (a) a horizontal strip of area and (b) by using a vertical strip of area.

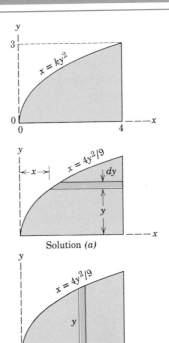

Solution. The constant $k = \frac{4}{9}$ is obtained first by substituting $x = 4$ and $y = 3$ into the equation for the parabola.

(a) *Horizontal strip.* Since all parts of the horizontal strip are the same distance from the x-axis, the moment of inertia of the strip about the x-axis is simply $y^2\,dA$ where $dA = (4 - x)\,dy = 4(1 - y^2/9)\,dy$. Integrating with respect to y gives us

$$\left[I_x = \int y^2\,dA\right] \quad I_x = \int_0^3 4y^2\left(1 - \frac{y^2}{9}\right)dy = \frac{72}{5} = 14.4 \ (\text{units})^4 \qquad Ans.$$

(b) *Vertical strip.* Here all parts of the element are at different distances from the x-axis, so we must use the correct expression for the moment of inertia of the elemental rectangle about its base which, from Sample Problem A1, is $bh^3/3$. For the width dx and the height y the expression becomes

$$dI_x = \tfrac{1}{3}(dx)y^3$$

To integrate with respect to x we must express y in terms of x, which gives $y = 3\sqrt{x}/2$, and the integral becomes

$$I_x = \tfrac{1}{3}\int_0^4 \left(\frac{3\sqrt{x}}{2}\right)^3 dx = \frac{72}{5} = 14.4 \ (\text{units})^4 \qquad Ans.$$

① There is very little preference between Solutions (a) and (b). Solution (b), of course, requires that we know the moment of inertia for a rectangular area about its base, which we should remember because of its frequent use.

Sample Problem A5

Determine the moment of inertia about the x-axis of the semicircular area.

Solution. The moment of inertia of the semicircular area about the x'-axis is one half of that for a complete circle about the same axis. Thus from the results of Sample Problem A3

$$I_{x'} = \frac{1}{2}\frac{\pi r^4}{4} = \frac{20^4\pi}{8} = 2\pi(10^4)\ \text{mm}^4$$

We obtain the moment of inertia $\bar{I}$ about the parallel centroidal axis x_o next. Transfer is made through the distance $\bar{r} = 4r/3\pi = (4)(20)/3\pi = 80/3\pi$ mm by the parallel-axis theorem. Hence

$$[\bar{I} = I - Ad^2] \quad \bar{I} = 2(10^4)\pi - \left(\frac{20^2\pi}{2}\right)\left(\frac{80}{3\pi}\right)^2 = 1.755(10^4)\ \text{mm}^4$$

① Finally, we transfer from the centroidal x_o-axis to the x-axis, which gives

$$[I = \bar{I} + Ad^2] \quad I_x = 1.755(10^4) + \left(\frac{20^2\pi}{2}\right)\left(30 + \frac{80}{3\pi}\right)^2$$

$$= 1.755(10^4) + 93.08(10^4) = 94.8(10^4)\ \text{mm}^4 \qquad Ans.$$

① This problem illustrates the caution we should observe in using a double transfer of axes since neither the x'- nor the x-axis passes through the centroid C of the area. If the circle were complete with the centroid on the x'-axis, only one transfer would be needed.

Sample Problem   A6

Calculate the moment of inertia about the x-axis of the area enclosed between the y-axis and the circular arcs of radius a whose centers are at O and A.

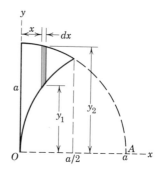

Solution. The choice of a vertical differential strip of area permits one integration to cover the entire area. A horizontal strip would require two integrations with respect to y by virtue of the discontinuity. The moment of inertia of the strip about the x-axis is that of a strip of height y_2 minus that of a strip of height y_1. Thus, from the results of Sample Problem A1 we write

$$dI_x = \tfrac{1}{3}(y_2\,dx)y_2{}^2 - \tfrac{1}{3}(y_1\,dx)y_1{}^2 = \tfrac{1}{3}(y_2{}^3 - y_1{}^3)\,dx$$

The values of y_2 and y_1 are obtained from the equations of the two curves, which are $x^2 + y_2{}^2 = a^2$ and $(x-a)^2 + y_1{}^2 = a^2$ and which give ① $y_2 = \sqrt{a^2 - x^2}$ and $y_1 = \sqrt{a^2 - (x-a)^2}$. Thus

① We choose the positive signs for the radicals here since both y_1 and y_2 lie above the x-axis.

$$I_x = \tfrac{1}{3}\int_0^{a/2} \{(a^2 - x^2)\sqrt{a^2 - x^2} - [a^2 - (x-a)^2]\sqrt{a^2 - (x-a)^2}\}\,dx$$

Simultaneous solution of the two equations which define the two circles gives the x-coordinate of the intersection of the two curves which, by inspection, is $a/2$. Evaluation of the integrals gives

$$\int_0^{a/2} a^2\sqrt{a^2 - x^2}\,dx = \frac{a^4}{4}\left(\frac{\sqrt{3}}{2} + \frac{\pi}{3}\right)$$

$$-\int_0^{a/2} x^2\sqrt{a^2 - x^2}\,dx = \frac{a^4}{16}\left(\frac{\sqrt{3}}{4} - \frac{\pi}{3}\right)$$

$$-\int_0^{a/2} a^2\sqrt{a^2 - (x-a)^2}\,dx = \frac{a^4}{4}\left(\frac{\sqrt{3}}{2} - \frac{2\pi}{3}\right)$$

$$\int_0^{a/2} (x-a)^2\sqrt{a^2 - (x-a)^2}\,dx = \frac{a^4}{8}\left(\frac{\sqrt{3}}{8} + \frac{\pi}{3}\right)$$

Collection of the integrals with the factor of $\tfrac{1}{3}$ gives

$$I_x = \frac{a^4}{96}(9\sqrt{3} - 2\pi) = 0.0969a^4 \qquad\qquad Ans.$$

If we had started from a second-order element $dA = dx\,dy$, we would write $y^2\,dx\,dy$ for the moment of inertia of the element about the x-axis. Integrating from y_1 to y_2 holding x constant produces for the vertical strip

$$dI_x = \left[\int_{y_1}^{y_2} y^2\,dy\right]dx = \tfrac{1}{3}(y_2{}^3 - y_1{}^3)\,dx$$

which is the expression we started with by having the moment-of-inertia result for a rectangle in mind.

PROBLEMS

A1 From the results of Sample Problem A3 write the expression for the polar moment of inertia of the area of the circular sector about its center O.

Problem A/1

A2 The narrow rectangular strip has an area of 6 mm², and its moment of inertia about the y-axis is 170 mm⁴. Obtain a close approximation to the radius of gyration about point O.

$$Ans. \ k_O = 7.30 \ mm$$

Problem A/2

A3 Calculate the moment of inertia of the rectangular area about the x-axis and find the polar moment of inertia about point O.

$$Ans. \ I_x = 21.06(10^6) \ mm^4, \ J_O = 35.64(10^6) \ mm^4$$

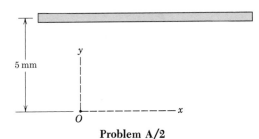

Problem A/3

A4 From the results of Sample Problem A1, state without calculation the moment of inertia of the area of the parallelogram about the x-axis through its base and about a parallel axis through its centroid.

Problem A/4

A5 The moments of inertia of the area A about the y- and y'-axes differ by 0.032 m⁴. Compute the area A, which has its centroid at C.

Problem A/5

Problem A/6

A6 Calculate the moment of inertia of the triangular area about the x-axis. *Ans.* $I_x = 270$ in.^{4}

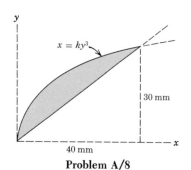

Problem A/7

A7 Obtain the polar moment of inertia of the area of the semicircular ring about point O by direct integration and use the result to find the moment of inertia about the x-axis.

Ans. $I_x = 0.368(10^{-4})$ m^4

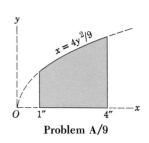

Problem A/8

A8 Calculate the moment of inertia of the shaded area about the x-axis.

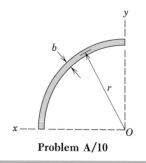

Problem A/9

A9 Calculate the moment of inertia of the shaded area about the x-axis. *Ans.* $I_x = 13.95$ in.^{4}

Problem A/10

A10 The quarter-circular strip of area has a width b which is small compared with its radius r. Determine the polar moment of inertia of the area about its centroid.

A11 Determine the polar moments of inertia of the semicircular area about points A and B.

$$Ans. \; J_A = \tfrac{3}{4}\pi r^4, \; J_B = r^4\left(\frac{3\pi}{4} - \frac{4}{3}\right)$$

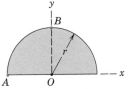

Problem A/11

A12 Calculate by direct integration the moment of inertia of the shaded area about the x-axis. Solve, first, by using a horizontal strip having differential area and, second, by using a vertical strip of differential area.

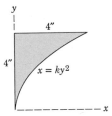

Problem A/12

A13 Determine the polar radius of gyration of the area of the equilateral triangle of side b about its centroid C.

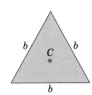

Problem A/13

A14 From considerations of symmetry show that $I_{x'} = I_{y'} = I_x = I_y$ for the semicircular area regardless of the angle α.

Problem A/14

A15 Show that the moment of inertia of the area of the square about any axis x' through its center is the same as that about a central axis x parallel to a side.

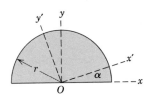

Problem A/15

A16 Determine the moment of inertia about the x-axis of the area under the sine curve.

$$Ans. \; I_x = \frac{4ab^3}{9\pi}$$

Problem A/16

Problem A/17

Problem A/19

Problem A/20

A17 Determine the moment of inertia of the shaded area about the *y*-axis.   *Ans.* $I_y = 27.8(10^4)$ mm^4

A18 The area of a circular ring of inside radius r and outside radius $r + \Delta r$ is approximately equal to the circumference at the mean radius times the thickness Δr. The polar moment of inertia of the ring may be approximated by multiplying this area by the square of the mean radius. What percent error is involved if $\Delta r = r/10$?   *Ans.* Error = 0.226%

A19 Determine the moments of inertia of the area of the circular sector about the *x*- and *y*-axes.

$$Ans.\ I_x = \frac{r^4}{4}\left(\alpha - \frac{\sin 2\alpha}{2}\right)$$

$$I_y = \frac{r^4}{4}\left(\alpha + \frac{\sin 2\alpha}{2}\right)$$

▶**A20** A narrow strip of area of constant width b has the form of a spiral $r = k\theta$. After one complete turn from $\theta = 0$ to $\theta = 2\pi$, the end radius of the spiral is R. Determine the polar moment of inertia and the radius of gyration of the area about O.
    Ans. $J_O = 1.609R^3b$, $k_O = 0.690R$

A3 COMPOSITE AREAS. It is frequently necessary to calculate the moment of inertia of an area that is composed of a number of distinct parts of simple and calculable geometric shape. Since a moment of inertia is the integral or sum of the products of distance squared times element of area, it follows that the moment of inertia of a positive area is always a positive quantity. Therefore the moment of inertia of a composite area about a particular axis is simply the sum of the moments of inertia of its component parts about the same axis. It is often convenient to regard a composite area as being composed of positive and negative parts. We may then treat the moment of inertia of a negative area as a negative quantity.

When a composite area is composed of a large number of parts, it is convenient to tabulate the results for each of the parts in terms of its area A, its centroidal moment of inertia $\bar{I}$, the distance d from its centroidal axis to the axis about which the moment of inertia of the entire section is being computed, and the product Ad^2. For any one of the parts the moment of inertia about the desired axis by the transfer-of-axis theorem is $\bar{I} + Ad^2$. Thus for the entire section the desired moment of inertia becomes $I = \Sigma\bar{I} + \Sigma Ad^2$.

For such an area in the x-y plane, for example, and with the notation of Fig. A4 where I_{x_0} is the same as $\bar{I}_x$ and I_{y_0} is the same as $\bar{I}_y$, the tabulation would include

Part	Area, A	d_x	d_y	$Ad_x^{\,2}$	$Ad_y^{\,2}$	I_{x_0}	I_{y_0}
			Sums	$\Sigma Ad_x^{\,2}$	$\Sigma Ad_y^{\,2}$	ΣI_{x_0}	ΣI_{y_0}

From the sums of the four columns, then, the moments of inertia for the composite area about the x- and y-axes become

$$I_x = \Sigma I_{x_0} + \Sigma Ad_x^{\,2}$$
$$I_y = \Sigma I_{y_0} + \Sigma Ad_y^{\,2}$$

Although we may add the moments of inertia of the individual parts of a composite area about a given axis, we may not add their radii of gyration. The radius of gyration for the composite area about the axis in question is given by $k = \sqrt{I/A}$ where I is the total moment of inertia and A is the total area of the composite figure.

Sample Problem   A7

Calculate the moment of inertia and radius of gyration about the x-axis for the shaded area shown.

Solution. The composite area is composed of the positive area of the rectangle (1) and the negative areas of the quarter circle (2) and triangle (3). For the rectangle the moment of inertia about the x-axis, from Sample Problem A1 (or Table C3), is

$$I_x = \tfrac{1}{3}Ah^2 = \tfrac{1}{3}(80)(60)(60)^2 = 5.76(10^6) \text{ mm}^4$$

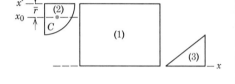

From Sample Problem A3 (or Table C3), the moment of inertia of the negative quarter-circular area about its base axis x′ is

$$I_{x'} = -\frac{1}{4}\left(\frac{\pi r^4}{4}\right) = -\frac{\pi}{16}(30)^4 = -0.1590(10^6) \text{ mm}^4$$

We now transfer this result through the distance $\bar{r} = 4r/3\pi = 4(30)/3\pi = 12.73$ mm by the transfer-of-axis-theorem to get the centroidal moment of inertia of part (2) (or use Table C3 directly)

① $[\bar{I} = I - Ad^2]$ $\quad \bar{I}_x = -0.1590(10^6) - \left[-\frac{\pi(30)^2}{4}(12.73)^2\right]$

$$= -0.0445(10^6) \text{ mm}^4$$

② The moment of inertia of the quarter-circular part about the x-axis is now

$$[I = \bar{I} + Ad^2] \quad I_x = -0.0445(10^6) + \left[-\frac{\pi(30)^2}{4}\right](60 - 12.73)^2$$

$$= -1.624(10^6) \text{ mm}^4$$

Finally, the moment of inertia of the negative triangular area (3) about its base, from Sample Problem A2 (or Table C3), is

$$I_x = -\tfrac{1}{12}bh^3 = -\tfrac{1}{12}(40)(30)^3 = -0.09(10^6) \text{ mm}^4$$

The total moment of inertia about the x-axis of the composite area is, consequently,

③ $\quad I_x = 5.76(10^6) - 1.624(10^6) - 0.09(10^6) = 4.046(10^6) \text{ mm}^4 \quad$ *Ans.*

The net area of the figure is $A = 60(80) - \tfrac{1}{4}\pi(30)^2 - \tfrac{1}{2}(40)(30) = 3493$ mm² so that the radius of gyration about the x-axis is

$$k_x = \sqrt{I_x/A} = \sqrt{4.046(10^6)/3493} = 34.0 \text{ mm} \quad \text{\textit{Ans.}}$$

① Note that we must transfer the moment of inertia for the quarter-circular area to its centroidal axis x_o before we can transfer it to the x-axis, as was done in Sample Problem A5.

② We watch our signs carefully here. Since the area is negative, both $\bar{I}$ and A carry negative signs.

③ If there had been more than the three parts to the composite area, we would have arranged a tabulation of the $\bar{I}$ terms and the Ad^2 terms so as to keep systematic account of the terms and obtain $I = \Sigma\bar{I} + \Sigma Ad^2$.

PROBLEMS

A21 Calculate the polar moment of inertia of the shaded area about point A.   *Ans.* $J_A = 138.9(10^4) \text{ mm}^4$

Problem A/21

A22 Determine the moment of inertia of the shaded area about the x-axis in two different ways.

Problem A/22

A23 Calculate the radius of gyration of the area of the angle section about point A. Note that the width of the legs is small compared to the length of each leg.   *Ans.* $k_A = 10.4 \text{ in.}$

Problem A/23

A24 Find the moment of inertia of the shaded area about the 45° x-axis of symmetry.

Problem A/24

Problem A/25

Problem A/26

Problem A/27

Problem A/28

Problem A/29

A25 Calculate the polar moment of inertia of the shaded area about point O.   Ans. $J_O = 1.265$ m^4

A26 Determine the moments of inertia of the Z-section about its centroidal x_o- and y_o-axes.

A27 Calculate the moment of inertia of the cross section of the beam about its centroidal x_o-axis.
           Ans. $\bar{I}_x = 10.76(10^6)$ mm^4

A28 Approximate the moment of inertia about the x-axis of the semicircular area by dividing it into five horizontal strips of equal width. Treat the moment of inertia of each strip as its area (width times length of its horizontal midline) times the square of the distance from its midline to the x-axis. Compare your result with the exact value.

A29 Compute the polar radius of gyration of the shaded area about point O.   Ans. $k_O = 0.445$ m

A30 Derive the expression for the moment of inertia of the trapezoidal area about its base.

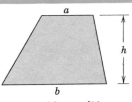

Problem A/30

A31 Develop a formula for the moment of inertia of the regular hexagonal area of side b about its central x-axis.

$$Ans.\ I_x = \frac{5\sqrt{3}}{16}b^4$$

Problem A/31

A32 Determine the expression for the radius of gyration of the hexagonal area of Prob. A31 about a polar axis through its center O.

A33 Calculate the moment of inertia of the shaded area about the x-axis. $Ans.\ I_x = 15.64\ \text{in.}^4$

Problem A/33

A34 Determine the moment of inertia of the rectangular area about its diagonal x-axis.

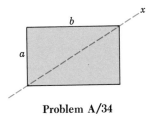

Problem A/34

A35 Calculate the moment of inertia of the area of Prob. A21 about the x-axis. $Ans.\ I_x = 28.3(10^4)\ \text{mm}^4$

Problem A/36

Problem A/37

Problem A/38

A36 Calculate the moment of inertia of the standard 12×4 in. channel section about the centroidal x_o-axis. Neglect the fillets and radii and compare with the handbook value of $I_x = 16.0$ in.^{4}

A37 Determine the moment of inertia of the shaded area about the x-axis.

$$Ans. \ I_x = \left(1 - \frac{5\pi}{16}\right)r^4$$

▶**A38** For the H-beam section, determine the flange width b that will make the moments of inertia about the central x- and y-axes equal. (*Hint:* The solution of a cubic equation is required here. Refer to Appendix B, section B4 for solving a cubic equation or approximate the answer with a numerical-graphical solution.)   *Ans. $b = 161$ mm*

A4 PRODUCTS OF INERTIA AND ROTATION OF AXES.

(*a*) *Definition.* In certain problems involving unsymmetrical cross sections and in the calculation of moments of inertia about rotated axes, an expression

$$dI_{xy} = xy \, dA$$

occurs which has the integrated form

$$\boxed{I_{xy} = \int xy \, dA} \qquad (A7)$$

where x and y are the coordinates of the element of area $dA = dx \, dy$. The quantity I_{xy} is called the *product of inertia* of the area A with respect to the x-y axes. Unlike moments of inertia, which are always positive for positive areas, the product of inertia may be positive, negative, or zero.

The product of inertia is zero whenever either one of the reference axes is an axis of symmetry, such as the x-axis for the area of Fig. A5. Here we see that the sum of the terms $x(-y) \, dA$ and $x(+y) \, dA$ due to symmetrically placed elements vanishes. Since the entire area may be considered composed of pairs of such elements, it follows that the product of inertia I_{xy} for the entire area is zero.

(*b*) *Transfer of Axes.* A transfer-of-axis theorem similar to that for moments of inertia also exists for products of inertia. By definition the product of inertia of the area A in Fig. A4 with respect to the x- and y-axes in terms of the coordinates x_o, y_o to the centroidal axes is

$$I_{xy} = \int (x_o + d_y)(y_o + d_x) \, dA$$

$$= \int x_o y_o \, dA + d_x \int x_o \, dA + d_y \int y_o \, dA + d_x d_y \int dA$$

The first integral is by definition the product of inertia about the centroidal axes, which we write $\bar{I}_{xy}$. The middle two integrals are both zero since the first moment of the area about its own centroid is necessarily zero. The third integral is merely $d_x d_y A$. Thus the transfer-of-axis theorem for products of inertia becomes

$$\boxed{I_{xy} = \bar{I}_{xy} + d_x d_y A} \qquad (A8)$$

(*c*) *Rotation of Axes.* The product of inertia is useful when we need to calculate the moment of inertia of an area about inclined axes. This consideration leads directly to the important problem of determining the axes about which the moment of inertia is a maximum and a minimum.

In Fig. A6 the moments of inertia of the area about the x'- and y'-axes are

$$I_{x'} = \int y'^2 \, dA = \int (y \cos \theta - x \sin \theta)^2 \, dA$$

$$I_{y'} = \int x'^2 \, dA = \int (y \sin \theta + x \cos \theta)^2 \, dA$$

Figure A/5

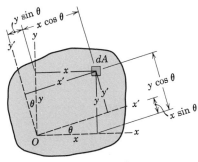

Figure A/6

where x' and y' have been replaced by their equivalent expressions as seen from the geometry of the figure.

Expanding and substituting the trigonometric identities,

$$\sin^2\theta = \frac{1-\cos 2\theta}{2} \qquad \cos^2\theta = \frac{1+\cos 2\theta}{2}$$

and the defining relations for I_x, I_y, I_{xy} give us

$$I_{x'} = \frac{I_x+I_y}{2} + \frac{I_x-I_y}{2}\cos 2\theta - I_{xy}\sin 2\theta$$

$$I_{y'} = \frac{I_x+I_y}{2} - \frac{I_x-I_y}{2}\cos 2\theta + I_{xy}\sin 2\theta$$

$$(A9)$$

In a similar manner we write the product of inertia about the inclined axes as

$$I_{x'y'} = \int x'y'\,dA = \int (y\sin\theta + x\cos\theta)(y\cos\theta - x\sin\theta)\,dA$$

Expanding and substituting the trigonometric identities

$$\sin\theta\cos\theta = \tfrac{1}{2}\sin 2\theta, \qquad \cos^2\theta - \sin^2\theta = \cos 2\theta$$

and the defining relations for I_x, I_y, I_{xy} give us

$$I_{x'y'} = \frac{I_x-I_y}{2}\sin 2\theta + I_{xy}\cos 2\theta \qquad (A9a)$$

Adding Eqs. A9 gives $I_{x'} + I_{y'} = I_x + I_y = J_z$, the polar moment of inertia about O, which checks the results of Eq. A3.

The angle which makes $I_{x'}$ and $I_{y'}$ a maximum or a minimum may be determined by setting the derivative of either $I_{x'}$ or $I_{y'}$ with respect to θ equal to zero. Thus

$$\frac{dI_{x'}}{d\theta} = (I_y - I_x)\sin 2\theta - 2I_{xy}\cos 2\theta = 0$$

Denoting this critical angle by α gives

$$\tan 2\alpha = \frac{2I_{xy}}{I_y - I_x} \qquad (A10)$$

Equation 74 gives two values for 2α which differ by π since $\tan 2\alpha = \tan(2\alpha + \pi)$. Consequently the two solutions for α will differ by $\pi/2$. One value defines the axis of maximum moment of inertia, and the other value defines the axis of maximum moment of inertia. These two rectangular axes are known as the *principal axes of inertia*.

When we substitute Eq. A10 for the critical value of 2θ in Eq. A9a, we see that the product of inertia is zero for the principal axes of inertia. Substitution of $\sin 2\alpha$ and $\cos 2\alpha$, obtained from Eq. A10, for $\sin 2\theta$ and $\cos 2\theta$ in Eqs. A9 gives the magnitudes of the principal moments of inertia as

$$I_{max} = \frac{I_x + I_y}{2} + \frac{1}{2}\sqrt{(I_x - I_y)^2 + 4I_{xy}{}^2}$$

$$I_{min} = \frac{I_x + I_y}{2} - \frac{1}{2}\sqrt{(I_x - I_y)^2 + 4I_{xy}{}^2}$$

(A11)

(d) *Mohr's Circle of Inertia.* We may represent the relations in Eqs. A9, A9a, A10, and A11 graphically by a diagram known as Mohr's circle. For given values of I_x, I_y, and I_{xy} the corresponding values of $I_{x'}$, $I_{y'}$, and $I_{x'y'}$ may be determined from the diagram for any desired angle θ. A horizontal axis for the measurement of moments of inertia and a vertical axis for the measurement of products of inertia are first selected, Fig. A7. Next, point A, which has the

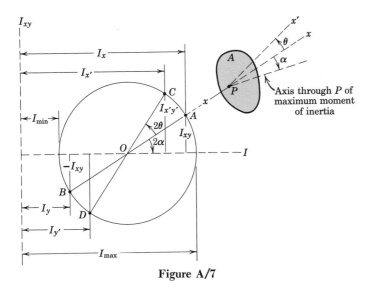

Figure A/7

coordinates (I_x, I_{xy}), and point B, which has the coordinates $(I_y, -I_{xy})$, are located. We now draw a circle with these two points as the extremities of a diameter. The angle from the radius OA to the horizontal axis is 2α or twice the angle from the x-axis of the area in question to the axis of maximum moment of inertia. The angle on the diagram and the angle on the area are both measured in the same sense as shown. The coordinates of any point C are $(I_{x'}, I_{x'y'})$, and those of the corresponding point D are $(I_{y'}, -I_{x'y'})$. Also the angle between OA and OC is 2θ or twice the angle from the x-axis to the x'-axis. Again we measure both angles in the same sense as shown. We may verify from the trigonometry of the circle that Eqs. A9, A9a, and A10 agree with the statements made.

Sample Problem   A8

Write the expression for the product of inertia of the rectangular area with centroid at C with respect to the x-y axes parallel to its sides.

Solution. Since the product of inertia $\bar{I}_{xy}$ about the axes x_o-y_o is zero by symmetry, the transfer-of-axis theorem gives us

$$[I_{xy} = \bar{I}_{xy} + d_x d_y A] \qquad\qquad I_{xy} = d_x d_y bh \qquad\qquad Ans.$$

In this example both d_x and d_y are shown positive. We must be careful to be consistent with the positive directions of d_x and d_y as defined so that their proper signs are observed.

Sample Problem   A9

Determine the product of inertia about the x-y axes for the area under the parabola.

Solution. With the substitution of $x = a$ when $y = b$ the equation of the curve becomes $x = ay^2/b^2$.

Solution I. If we start with the second-order element $dA = dx\,dy$, we have $dI_{xy} = xy\,dx\,dy$. The integral over the entire area then gives us

$$I_{xy} = \int_0^b \int_{ay^2/b^2}^a xy\,dx\,dy = \int_0^b \frac{1}{2}\left(a^2 - \frac{a^2 y^4}{b^4}\right)y\,dy = \tfrac{1}{6}a^2 b^2 \quad Ans.$$

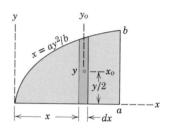

Solution II. Alternatively we can start with a first-order elemental strip and save one integration by using the results of Sample Problem A8. ① Taking a vertical strip $dA = y\,dx$ gives $dI_{xy} = 0 + (\tfrac{1}{2}y)(x)(y\,dx)$ where the distances to the centroidal axes of the elemental rectangle are $d_x = y/2$ and $d_y = x$. Now we have

$$I_{xy} = \int_0^a \frac{y^2}{2}x\,dx = \int_0^a \frac{xb^2}{2a}x\,dx = \frac{b^2}{6a}x^3\Big]_0^a = \tfrac{1}{6}a^2 b^2 \qquad Ans.$$

① If we had chosen a horizontal strip, our expression would have become $dI_{xy} = y\tfrac{1}{2}(a + x)[(a - x)\,dy]$ which when integrated, of course, gives us the same result as before.

Sample Problem   A10

Determine the product of inertia of the semicircular area with respect to the x-y axes.

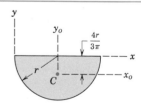

① *Solution.* We may use the transfer-of-axis theorem, Eq. A8, to write

$$[I_{xy} = \bar{I}_{xy} + d_x d_y A] \qquad I_{xy} = 0 + \left(-\frac{4r}{3\pi}\right)(r)\left(\frac{\pi r^2}{2}\right) = -\frac{2r^4}{3} \qquad Ans.$$

where the x- and y-coordinates of the centroid C are $d_y = +r$ and $d_x = -4r/3\pi$. Since one of the centroidal axes is an axis of symmetry, $\bar{I}_{xy} = 0$.

① Proper use of the transfer-of-axis theorem saves a great deal of labor in computing products of inertia.

Sample Problem A11

Determine the orientation of the principal axes of inertia through the centroid of the angle section and determine the corresponding maximum and minimum moments of inertia.

Solution. The location of the centroid C is easily calculated, and its position is shown on the diagram.

Products of inertia. The product of inertia for each rectangle about its own centroidal axes parallel to the x- and y-axes is zero by symmetry. Thus the product of inertia about the x-y axes for part I is

$$[I_{xy} = \bar{I}_{xy} + d_x d_y A] \qquad I_{xy} = 0 + (-12.5)(+7.5)(400) = -3.75(10^4) \text{ mm}^4$$

where

$$d_x = -(7.5 + 5) = -12.5 \text{ mm}$$

and

$$d_y = +(20 - 10 - 2.5) = 7.5 \text{ mm}$$

Likewise for part II,

$$[I_{xy} = \bar{I}_{xy} + d_x d_y A] \qquad I_{xy} = 0 + (12.5)(-7.5)(400) = -3.75(10^4) \text{ mm}^4$$

where

$$d_x = +(20 - 7.5) = 12.5 \text{ mm}$$

and

$$d_y = -(5 + 2.5) = -7.5 \text{ mm}$$

For the complete angle

$$I_{xy} = -3.75(10^4) - 3.75(10^4) = -7.50(10^4) \text{ mm}^4$$

Moments of inertia. The moments of inertia about the x- and y-axes for part I are

$$[I = \bar{I} + Ad^2] \qquad I_x = \tfrac{1}{12}(40)(10)^3 + (400)(12.5)^2 = 6.583(10^4) \text{ mm}^4$$

$$I_y = \tfrac{1}{12}(10)(40)^3 + (400)(7.5)^2 = 7.583(10^4) \text{ mm}^4$$

and the moments of inertia for part II about these same axes are

$$[I = \bar{I} + Ad^2] \qquad I_x = \tfrac{1}{12}(10)(40)^3 + (400)(12.5)^2 = 11.583(10^4) \text{ mm}^4$$

$$I_y = \tfrac{1}{12}(40)(10)^3 + (400)(7.5)^2 = 2.583(10^4) \text{ mm}^4$$

Thus for the entire section we have

$$I_x = 6.583(10^4) + 11.583(10^4) = 18.167(10^4) \text{ mm}^4$$

$$I_y = 7.583(10^4) + 2.583(10^4) = 10.167(10^4) \text{ mm}^4$$

Principal axes. The inclination of the principal axes of inertia is given by Eq. A10, so we have

$$\left[\tan 2\alpha = \frac{2I_{xy}}{I_y - I_x}\right] \qquad \tan 2\alpha = \frac{-2(7.50)}{10.167 - 18.167} = 1.875$$

$$2\alpha = 61.9° \qquad \alpha = 31.0° \qquad \textit{Ans.}$$

We now compute the principal moments of inertia from Eqs. A9 using α for θ and get I_{max} from $I_{x'}$ and I_{min} from $I_{y'}$. Thus

$$I_{max} = \left(\frac{18.167 + 10.167}{2} + \frac{18.167 - 10.167}{2}(0.4705) + (7.50)(0.8824)\right)(10^4)$$

$$= 22.67(10^4) \text{ mm}^4 \qquad \textit{Ans.}$$

$$I_{min} = \left(\frac{18.167 + 10.167}{2} - \frac{18.167 - 10.167}{2}(0.4705) - (7.50)(0.8824)\right)(10^4)$$

$$= 5.67(10^4) \text{ mm}^4 \qquad \textit{Ans.}$$

Mohr's circle. Alternatively we could use Eqs. A11 to obtain the results for I_{max} and I_{min}, or we could construct the Mohr circle from the calculated values of I_x, I_y, and I_{xy}. These values are spotted on the diagram to locate points A and B, which are the extremities of the diameter of the circle. The angle 2α and I_{max} and I_{min} are obtained from the figure as shown.

PROBLEMS

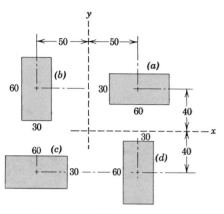

(Dimensions in Millimeters)
Problem A/39

Problem A/40

Problem A/41

Problem A/42

A39 Determine the product of inertia of each of the four areas about the x-y axes.

$$\text{Ans. } (a) \text{ and } (c): I_{xy} = 360(10^4) \text{ mm}^4$$
$$(b) \text{ and } (d): I_{xy} = -360(10^4) \text{ mm}^4$$

A40 Calculate the product of inertia of the triangular area about the x-y axes.

A41 Calculate the product of inertia of the shaded area about the x-y axes.   Ans. $I_{xy} = -128.0(10^6) \text{ mm}^4$

A42 Determine the product of inertia of the area of the right triangle about the x-y axes.

A43 Obtain the product of inertia of the quarter-circular area with respect to the x-y axes and use this result to obtain the product of inertia with respect to the parallel centroidal axes.

Ans. $I_{xy} = r^4/8$, $\bar{I}_{xy} = -0.01647r^4$

Problem A/43

A44 Determine the moments and product of inertia of the area of the square with respect to the x'-y' axes.

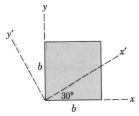

Problem A/44

A45 An area has moments of inertia $I_x = 28$ in.^{4} and $I_y = 12$ in.^{4} about a set of x-y axes. The angle measured clockwise from the x-axis to the axis of maximum moment of inertia through the origin O is $20°$. Determine the minimum moment of inertia of the area about an axis through O.

Ans. $I_{min} = 9.56$ in.^{4}

A46 The products of inertia of the shaded area with respect to the x-y and x'-y' axes are $8(10^6)$ mm^4 and $-42(10^6)$ mm^4, respectively. Compute the area of the figure, whose centroid is C.

Ans. $A = 1.316(10^4)$ mm^2

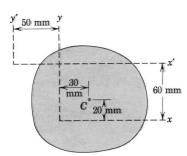

Problem A/46

A47 Where $I_x = I_y$ for an area which is symmetrical about either the x- or the y-axis, prove that the moment of inertia is the same for all axes through the origin.

A48 Determine the proportions of the rectangular area for which the moment of inertia about an x'-axis through the center point C of the base is a constant value regardless of θ. (See Prob. A47.)

Ans. $a = 2b$

Problem A/48

Problem A/49

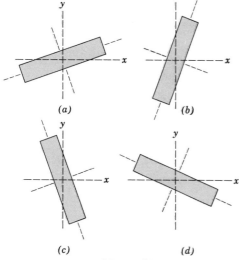

(a)   *(b)*

(c)   *(d)*

Problem A/51

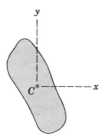

Problem A/52

A49 Determine the relationship between the base b and altitude h of the isosceles triangle which will make the moments of inertia about all axes through the vertex C the same. (See Prob. A47.)

A50 Prove that the magnitude of the product of inertia can be computed from the relation
$$I_{xy} = \sqrt{I_x I_y - I_{\max} I_{\min}}.$$

A51 Sketch the Mohr circle of inertia for each of the four rectangular areas with the proportions and positions shown. Indicate on each diagram point A which has coordinates I_x, I_{xy}, and the angle 2α where α is the angle from the x-axis to the axis of maximum moment of inertia.

A52 The maximum and minimum moments of inertia of the shaded area are $25(10^6)$ mm^4 and $5(10^6)$ mm^4, respectively, about axes passing through the centroid C, and the product of inertia with respect to the x-y axes is $-8(10^6)$ mm^4. From the appropriate equations calculate I_x and the angle α measured counterclockwise from the x-axis to the axis of maximum moment of inertia.

Ans. $I_x = 21(10^6)$ mm^4, $\alpha = 26.6°$

A53 Solve Prob. A52 by constructing the Mohr circle of inertia.

A54 The moments and product of inertia of an area with respect to x-y axes are $I_x = 14$ in.^{4}, $I_y = 24$ in.^{4}, and $I_{xy} = 12$ in.^{4} Construct the Mohr circle of inertia and use it to determine the principal moments of inertia and the angle α from the x-axis to the axis of maximum moment of inertia.

Ans. $I_{\max} = 32$ in.^{4}, $I_{\min} = 6$ in.^{4}, $\alpha = 56.3°$ clockwise

A55 Determine the maximum and minimum moments of inertia with respect to centroidal axes through C for the composite of the four square areas shown. Find the angle α measured from the x-axis to the axis of maximum moment of inertia.

$$Ans. \; I_{max} = a^4(\tfrac{10}{3} + \sqrt{5})$$
$$I_{min} = a^4(\tfrac{10}{3} - \sqrt{5})$$
$$\alpha = 76.7°$$

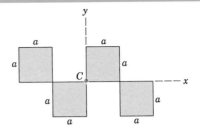

Problem A/55

A56 Calculate the product of inertia of the rectangular area with respect to the x-y axes.

$$Ans. \; I_{xy} = 1225 \text{ mm}^4$$

Problem A/56

A57 Calculate the maximum and minimum moments of inertia of the structural angle about axes through its corner A and find the angle α measured counterclockwise from the x-axis to the axis of maximum inertia. Neglect the small radii and fillet.

$$Ans. \; I_{max} = 1.782(10^6) \text{ mm}^4$$
$$I_{min} = 0.684(10^6) \text{ mm}^4$$
$$\alpha = -13.4°$$

Problem A/57

A58 Calculate the maximum and minimum moments of inertia about centroidal axes for the structural Z-section. Indicate the angle α measured counterclockwise from the x_o-axis to the axis of maximum moment of inertia.

$$Ans. \; I_{max} = 1.820(10^6) \text{ mm}^4$$
$$I_{min} = 0.207(10^6) \text{ mm}^4$$
$$\alpha = 30.1°$$

Problem A/59

SELECTED TOPICS OF MATHEMATICS

B

B1 INTRODUCTION. Appendix B contains an abbreviated summary and reminder of selected topics in basic mathematics which find frequent use in mechanics. The relationships are cited without proof. The student of mechanics will have frequent occasion to use many of these relations, and he will be handicapped if they are not well in hand. Other topics not listed will also be needed from time to time.

As the reader reviews and applies his mathematics, he should bear in mind that mechanics is an applied science descriptive of real bodies and actual motions. Therefore the geometric and physical interpretation of the applicable mathematics should be kept clearly in mind during the development of theory and the formulation and solution of problems.

B2 PLANE GEOMETRY

1. When two intersecting lines are, respectively, perpendicular to two other lines, the angles formed by each pair are equal.

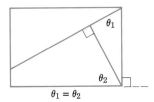

$\theta_1 = \theta_2$

2. Similar triangles

$$\frac{x}{b} = \frac{h - y}{h}$$

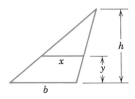

3. Any triangle

Area $= \frac{1}{2}bh$

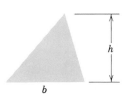

4. Circle

Circumference $= 2\pi r$
Area $= \pi r^2$
Arc length $s = r\theta$
Sector area $= \frac{1}{2}r^2\theta$

5. Every triangle inscribed within a semicircle is a right triangle.

$\theta_1 + \theta_2 = \pi/2$

6. Angles of a triangle

$$\theta_1 + \theta_2 + \theta_3 = 180°$$
$$\theta_4 = \theta_1 + \theta_2$$

B3 SOLID GEOMETRY

1. Sphere

   $\text{Volume} = \frac{4}{3}\pi r^3$

   $\text{Surface Area} = 4\pi r^2$

2. Spherical wedge

   $\text{Volume} = \frac{2}{3}r^3\theta$

3. Right-circular cone

   $\text{Volume} = \frac{1}{3}\pi r^2 h$

   $\text{Lateral area} = \pi r L$

   $L = \sqrt{r^2 + h^2}$

4. Any pyramid or cone

   $\text{Volume} = \frac{1}{3}Bh$

   where $B = $ area of base

B4 ALGEBRA

1. Quadratic equation

   $ax^2 + bx + c = 0$

   $x = \dfrac{-b \pm \sqrt{b^2 - 4ac}}{2a}$, $b^2 > 4ac$ for real roots

2. Logarithms

   $b^x = y$, $x = \log_b y$

   Natural logarithms

   $b = e = 2.718\ 282$

   $e^x = y$, $x = \log_e y = \ln y$

   $\log(ab) = \log a + \log b$

   $\log(a/b) = \log a - \log b$

   $\log(1/n) = -\log n$

   $\log a^n = n \log a$

   $\log 1 = 0$

   $\log_{10} x = 0.4343 \ln x$

3. Determinants
   2nd order

   $\begin{vmatrix} a_1 & b_1 \\ a_2 & b_2 \end{vmatrix} = a_1 b_2 - a_2 b_1$

   3rd order

   $\begin{vmatrix} a_1 & b_1 & c_1 \\ a_2 & b_2 & c_2 \\ a_3 & b_3 & c_3 \end{vmatrix} = \begin{array}{l} +a_1 b_2 c_3 + a_2 b_3 c_1 + a_3 b_1 c_2 \\ -a_3 b_2 c_1 - a_2 b_1 c_3 - a_1 b_3 c_2 \end{array}$

4. Cubic equation

   $x^3 = Ax + B$

   Let $p = A/3$, $q = B/2$

   Case I: $q^2 - p^3$ negative (three roots real and distinct)

   $$\cos u = q/(p\sqrt{p}),\ 0 < u < 180°$$

   $$x_1 = 2\sqrt{p}\cos(u/3)$$

   $$x_2 = 2\sqrt{p}\cos(u/3 + 120°)$$

   $$x_3 = 2\sqrt{p}\cos(u/3 + 240°)$$

   Case II: $q^2 - p^3$ positive (one root real, two roots imaginary).

   $$x_1 = (q + \sqrt{q^2 - p^3})^{1/3} + (q - \sqrt{q^2 - p^3})^{1/3}$$

   Case III: $q^2 - p^3 = 0$ (three roots real, two roots equal)

   $$x_1 = 2q^{1/3},\ x_2 = x_3 = -q^{1/3}$$

   For general cubic equation

   $$x^3 + ax^2 + bx + c = 0$$

   Substitute $x = x_0 - a/3$ and get $x_0{}^3 = Ax_0 + B$. Then proceed as above to find values of x_0 from which $x = x_0 - a/3$.

B5 ANALYTIC GEOMETRY

1. Straight line

$$y = a + mx$$

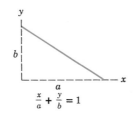

$$\frac{x}{a} + \frac{y}{b} = 1$$

2. Circle

$$x^2 + y^2 = r^2$$

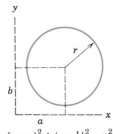

$$(x - a)^2 + (y - b)^2 = r^2$$

3. Parabola

$$y = b\,\frac{x^2}{a^2}$$

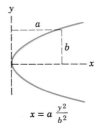

$$x = a\,\frac{y^2}{b^2}$$

4. Ellipse

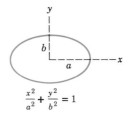

$$\frac{x^2}{a^2} + \frac{y^2}{b^2} = 1$$

5. Hyperbola

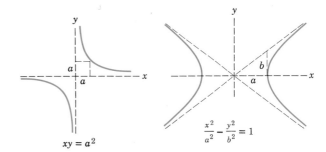

$$xy = a^2$$

$$\frac{x^2}{a^2} - \frac{y^2}{b^2} = 1$$

B6 TRIGONOMETRY

1. Definitions

$$\sin\theta = a/c \qquad \csc\theta = c/a$$
$$\cos\theta = b/c \qquad \sec\theta = c/b$$
$$\tan\theta = a/b \qquad \operatorname{ctn}\theta = b/a$$

2. Signs in the four quadrants

	I	II	III	IV
$\sin\theta$	+	+	−	−
$\cos\theta$	+	−	−	+
$\tan\theta$	+	−	+	−
$\csc\theta$	+	+	−	−
$\sec\theta$	+	−	−	+
$\operatorname{ctn}\theta$	+	−	+	−

3. Miscellaneous relations

$$\sin^2\theta + \cos^2\theta = 1$$
$$1 + \tan^2\theta = \sec^2\theta$$
$$1 + \text{ctn}^2\theta = \csc^2\theta$$

$$\sin\frac{\theta}{2} = \sqrt{\tfrac{1}{2}(1 - \cos\theta)}$$

$$\cos\frac{\theta}{2} = \sqrt{\tfrac{1}{2}(1 + \cos\theta)}$$

$$\sin 2\theta = 2\sin\theta\cos\theta$$
$$\cos 2\theta = \cos^2\theta - \sin^2\theta$$
$$\sin(a \pm b) = \sin a\cos b \pm \cos a\sin b$$
$$\cos(a \pm b) = \cos a\cos b \mp \sin a\sin b$$

4. Law of sines

$$\frac{a}{b} = \frac{\sin A}{\sin B}$$

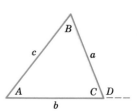

5. Law of cosines

$$c^2 = a^2 + b^2 - 2ab\cos C$$
$$c^2 = a^2 + b^2 + 2ab\cos D$$

B7 VECTOR OPERATIONS

1. *Notation.* Vector quantities are printed in boldface type, and scalar quantities appear in lightface italic type. Thus, the vector quantity **V** has a scalar magnitude V. In longhand work vector quantities should always be consistently indicated by a symbol such as $\underline{V}$ or $\vec{V}$ to distinguish them from scalar quantities.

2. *Addition*

   Triangle addition   **P + Q = R**

   Parallelogram addition   **P + Q = R**

   Commutative law   **P + Q = Q + P**

   Associative law   **P + (Q + R) = (P + Q) + R**

3. *Subtraction*

$$\mathbf{P - Q = P + (-Q)}$$

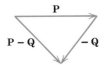

4. *Unit vectors* **i, j, k**

$$\mathbf{V} = \mathbf{i}V_x + \mathbf{j}V_y + \mathbf{k}V_z$$

   where
$$|\mathbf{V}| = V = \sqrt{V_x^{\,2} + V_y^{\,2} + V_z^{\,2}}$$

5. *Direction cosines* l, m, n are the cosines of the angles between **V** and the x-, y-, z-axes. Thus

$$l = V_x/V \qquad m = V_y/V \qquad n = V_z/V$$

   so that
$$\mathbf{V} = V(\mathbf{i}l + \mathbf{j}m + \mathbf{k}n)$$

   and
$$l^2 + m^2 + n^2 = 1$$

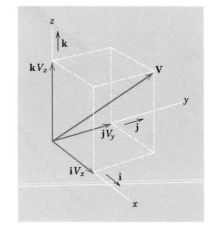

6. *Dot or scalar product*

$$\mathbf{P} \cdot \mathbf{Q} = PQ \cos \theta$$

This product may be viewed as the magnitude of **P** multiplied by the component $Q \cos \theta$ of **Q** in the direction of **P**, or as the magnitude of **Q** multiplied by the component $P \cos \theta$ of **P** in the direction of **Q**.

Commutative law $\mathbf{P} \cdot \mathbf{Q} = \mathbf{Q} \cdot \mathbf{P}$

From the definition of the dot product

$$\mathbf{i} \cdot \mathbf{i} = \mathbf{j} \cdot \mathbf{j} = \mathbf{k} \cdot \mathbf{k} = 1$$

$$\mathbf{i} \cdot \mathbf{j} = \mathbf{j} \cdot \mathbf{i} = \mathbf{i} \cdot \mathbf{k} = \mathbf{k} \cdot \mathbf{i} = \mathbf{j} \cdot \mathbf{k} = \mathbf{k} \cdot \mathbf{j} = 0$$

$$\mathbf{P} \cdot \mathbf{Q} = (\mathbf{i}P_x + \mathbf{j}P_y + \mathbf{k}P_z) \cdot (\mathbf{i}Q_x + \mathbf{j}Q_y + \mathbf{k}Q_z)$$

$$= P_x Q_x + P_y Q_y + P_z Q_z$$

$$\mathbf{P} \cdot \mathbf{P} = P_x^{\,2} + P_y^{\,2} + P_z^{\,2}$$

It follows from the definition of the dot product that two vectors **P** and **Q** are perpendicular when their dot product vanishes, $\mathbf{P} \cdot \mathbf{Q} = 0$.

The angle θ between two vectors $\mathbf{P}_1$ and $\mathbf{P}_2$ may be found from their dot product expression $\mathbf{P}_1 \cdot \mathbf{P}_2 = P_1 P_2 \cos \theta$, which gives

$$\cos \theta = \frac{\mathbf{P}_1 \cdot \mathbf{P}_2}{P_1 P_2} = \frac{P_{1_x} P_{2_x} + P_{1_y} P_{2_y} + P_{1_z} P_{2_z}}{P_1 P_2} = l_1 l_2 + m_1 m_2 + n_1 n_2$$

where l, m, n stand for the respective direction cosines of the vectors. It is also observed that two vectors are perpendicular when their direction cosines obey the relation $l_1 l_2 + m_1 m_2 + n_1 n_2 = 0$.

Distributive law $\mathbf{P} \cdot (\mathbf{Q} + \mathbf{R}) = \mathbf{P} \cdot \mathbf{Q} + \mathbf{P} \cdot \mathbf{R}$

7. *Cross or vector product.* The cross product $\mathbf{P} \times \mathbf{Q}$ of the two vectors **P** and **Q** is defined as a vector with a magnitude

$$|\mathbf{P} \times \mathbf{Q}| = PQ \sin \theta$$

and a direction specified by the right-hand rule as shown. Reversing the order of the vectors and using the right-hand rule gives $\mathbf{Q} \times \mathbf{P} = -\mathbf{P} \times \mathbf{Q}$.

Distributive law $\mathbf{P} \times (\mathbf{Q} + \mathbf{R}) = \mathbf{P} \times \mathbf{Q} + \mathbf{P} \times \mathbf{R}$

From the definition of the cross product, using a *right-handed coordinate system*

$$\mathbf{i} \times \mathbf{j} = \mathbf{k} \qquad \mathbf{j} \times \mathbf{k} = \mathbf{i} \qquad \mathbf{k} \times \mathbf{i} = \mathbf{j}$$

$$\mathbf{j} \times \mathbf{i} = -\mathbf{k} \qquad \mathbf{k} \times \mathbf{j} = -\mathbf{i} \qquad \mathbf{i} \times \mathbf{k} = -\mathbf{j}$$

$$\mathbf{i} \times \mathbf{i} = \mathbf{j} \times \mathbf{j} = \mathbf{k} \times \mathbf{k} = 0$$

With the aid of these identities and the distributive law the vector product may be written

$$\mathbf{P} \times \mathbf{Q} = (\mathbf{i}P_x + \mathbf{j}P_y + \mathbf{k}P_z) \times (\mathbf{i}Q_x + \mathbf{j}Q_y + \mathbf{k}Q_z)$$
$$= \mathbf{i}(P_y Q_z - P_z Q_y) + \mathbf{j}(P_z Q_x - P_x Q_z) + \mathbf{k}(P_x Q_y - P_y Q_x)$$

The cross product may also be expressed by the determinant

$$\mathbf{P} \times \mathbf{Q} = \begin{vmatrix} \mathbf{i} & \mathbf{j} & \mathbf{k} \\ P_x & P_y & P_z \\ Q_x & Q_y & Q_z \end{vmatrix}$$

8. Additional relations

Triple scalar product $(\mathbf{P} \times \mathbf{Q}) \cdot \mathbf{R} = \mathbf{R} \cdot (\mathbf{P} \times \mathbf{Q})$. The dot and cross may be interchanged as long as the order of the vectors is maintained. Parentheses are unnecessary since $\mathbf{P} \times (\mathbf{Q} \cdot \mathbf{R})$ is meaningless since a vector $\mathbf{P}$ cannot be crossed into a scalar $\mathbf{Q} \cdot \mathbf{R}$. Thus the expression may be written

$$\mathbf{P} \times \mathbf{Q} \cdot \mathbf{R} = \mathbf{P} \cdot \mathbf{Q} \times \mathbf{R}$$

The triple scalar product has the determinant expansion

$$\mathbf{P} \times \mathbf{Q} \cdot \mathbf{R} = \begin{vmatrix} P_x & P_y & P_z \\ Q_x & Q_y & Q_z \\ R_x & R_y & R_z \end{vmatrix}$$

Triple vector product $(\mathbf{P} \times \mathbf{Q}) \times \mathbf{R} = -\mathbf{R} \times (\mathbf{P} \times \mathbf{Q})$
$$= \mathbf{R} \times (\mathbf{Q} \times \mathbf{P})$$

Here the parentheses must be used since an expression $\mathbf{P} \times \mathbf{Q} \times \mathbf{R}$ would be ambiguous because it would not identify the vector to be crossed. It may be shown that the triple vector product is equivalent to

$$(\mathbf{P} \times \mathbf{Q}) \times \mathbf{R} = \mathbf{R} \cdot \mathbf{P}\mathbf{Q} - \mathbf{R} \cdot \mathbf{Q}\mathbf{P}$$

or

$$\mathbf{P} \times (\mathbf{Q} \times \mathbf{R}) = \mathbf{P} \cdot \mathbf{R}\mathbf{Q} - \mathbf{P} \cdot \mathbf{Q}\mathbf{R}$$

The first term in the first expression, for example, is the dot product $\mathbf{R} \cdot \mathbf{P}$, a scalar, multiplied by the vector $\mathbf{Q}$.

9. Derivatives of vectors obey the same rules as for scalars.

$$\frac{d\mathbf{P}}{dt} = \dot{\mathbf{P}} = \mathbf{i}\dot{P}_x + \mathbf{j}\dot{P}_y + \mathbf{k}\dot{P}_z$$

$$\frac{d(\mathbf{P}u)}{dt} = \mathbf{P}\dot{u} + \dot{\mathbf{P}}u$$

$$\frac{d(\mathbf{P} \cdot \mathbf{Q})}{dt} = \mathbf{P} \cdot \dot{\mathbf{Q}} + \dot{\mathbf{P}} \cdot \mathbf{Q}$$

$$\frac{d(\mathbf{P} \times \mathbf{Q})}{dt} = \mathbf{P} \times \dot{\mathbf{Q}} + \dot{\mathbf{P}} \times \mathbf{Q}$$

10. *Integration of vectors.* If **V** is a function of x, y, and z and an element of volume is $d\tau = dx\,dy\,dz$, the integral of **V** over the volume may be written as the vector sum of the three integrals of its components. Thus

$$\int \mathbf{V}\,d\tau = \mathbf{i}\int V_x\,d\tau + \mathbf{j}\int V_y\,d\tau + \mathbf{k}\int V_z\,d\tau$$

B8 SERIES

(Expression in brackets following series indicates range of convergence)

$$(1 \pm x)^n = 1 \pm nx + \frac{n(n-1)}{2!}x^2 \pm \frac{n(n-1)(n-2)}{3!}x^3 + \cdots \quad [x^2 < 1]$$

$$\sin x = x - \frac{x^3}{3!} + \frac{x^5}{5!} - \frac{x^7}{7!} + \cdots \qquad\qquad [x^2 < \infty]$$

$$\cos x = 1 - \frac{x^2}{2!} + \frac{x^4}{4!} - \frac{x^6}{6!} + \cdots \qquad\qquad [x^2 < \infty]$$

$$\sinh x = \frac{e^x - e^{-x}}{2} = x + \frac{x^3}{3!} + \frac{x^5}{5!} + \frac{x^7}{7!} + \cdots \qquad [x^2 < \infty]$$

$$\cosh x = \frac{e^x + e^{-x}}{2} = 1 + \frac{x^2}{2!} + \frac{x^4}{4!} + \frac{x^6}{6!} + \cdots \qquad [x^2 < \infty]$$

$$f(x) = \frac{a_0}{2} + \sum_{n=1}^{\infty} a_n \cos\frac{n\pi x}{l} + \sum_{n=1}^{\infty} b_n \sin\frac{n\pi x}{l}$$

where $a_n = \dfrac{1}{l}\displaystyle\int_{-l}^{l} f(x)\cos\frac{n\pi x}{l}dx,\ b_n = \dfrac{1}{l}\displaystyle\int_{-l}^{l} f(x)\sin\frac{n\pi x}{l}dx$

[Fourier expansion for $-l < x < l$]

B9 DERIVATIVES

$$\frac{dx^n}{dx} = nx^{n-1}, \quad \frac{d(uv)}{dx} = u\frac{dv}{dx} + v\frac{du}{dx}, \quad \frac{d\left(\dfrac{u}{v}\right)}{dx} = \frac{v\dfrac{du}{dx} - u\dfrac{dv}{dx}}{v^2}$$

$$\lim_{\Delta x \to 0} \sin \Delta x = \sin dx = \tan dx = dx$$

$$\lim_{\Delta x \to 0} \cos \Delta x = \cos dx = 1$$

$$\frac{d\sin x}{dx} = \cos x, \quad \frac{d\cos x}{dx} = -\sin x, \quad \frac{d\tan x}{dx} = \sec^2 x$$

$$\frac{d\sinh x}{dx} = \cosh x, \quad \frac{d\cosh x}{dx} = \sinh x, \quad \frac{d\tanh x}{dx} = \text{sech}^2 x$$

B10 INTEGRALS

$$\int x^n\, dx = \frac{x^{n+1}}{n+1}$$

$$\int \frac{dx}{x} = \ln x$$

$$\int \sqrt{a+bx}\, dx = \frac{2}{3b}\sqrt{(a+bx)^3}$$

$$\int x\sqrt{a+bx}\, dx = \frac{2}{15b^2}(3bx - 2a)\sqrt{(a+bx)^3}$$

$$\int \frac{dx}{\sqrt{a+bx}} = \frac{2\sqrt{a+bx}}{b}$$

$$\int \frac{x\, dx}{a+bx} = \frac{1}{b^2}[a+bx - a\ln(a+bx)]$$

$$\int \frac{x\, dx}{(a+bx)^n} = \frac{(a+bx)^{1-n}}{b^2}\left(\frac{a+bx}{2-n} - \frac{a}{1-n}\right)$$

$$\int \frac{dx}{a+bx^2} = \frac{1}{\sqrt{ab}}\tan^{-1}\frac{x\sqrt{ab}}{a} \quad \text{or} \quad \frac{1}{\sqrt{-ab}}\tanh^{-1}\frac{x\sqrt{-ab}}{a}$$

$$\int \frac{x\, dx}{a+bx^2} = \frac{1}{2b}\ln\frac{a+bx^2}{b}$$

$$\int \sqrt{x^2 \pm a^2}\, dx = \tfrac{1}{2}[x\sqrt{x^2 \pm a^2} \pm a^2\ln(x+\sqrt{x^2 \pm a^2})]$$

$$\int \sqrt{a^2 - x^2}\, dx = \tfrac{1}{2}\left(x\sqrt{a^2 - x^2} + a^2\sin^{-1}\frac{x}{a}\right)$$

$$\int x\sqrt{a^2 - x^2}\, dx = -\tfrac{1}{3}\sqrt{(a^2 - x^2)^3}$$

$$\int x^2\sqrt{a^2 - x^2}\, dx = -\frac{x}{4}\sqrt{(a^2 - x^2)^3} + \frac{a^2}{8}\left(x\sqrt{a^2 - x^2} + a^2\sin^{-1}\frac{x}{a}\right)$$

$$\int x^3\sqrt{a^2 - x^2}\, dx = -\tfrac{1}{5}(x^2 + \tfrac{2}{3}a^2)\sqrt{(a^2 - x^2)^3}$$

$$\int \frac{dx}{\sqrt{a+bx+cx^2}} = \frac{1}{\sqrt{c}}\ln\left(\sqrt{a+bx+cx^2} + x\sqrt{c} + \frac{b}{2\sqrt{c}}\right) \quad \text{or} \quad \frac{-1}{\sqrt{-c}}\sin^{-1}\left(\frac{b+2cx}{\sqrt{b^2-4ac}}\right)$$

$$\int \frac{dx}{\sqrt{x^2 \pm a^2}} = \ln(x + \sqrt{x^2 \pm a^2})) \qquad \int \frac{x\, dx}{\sqrt{x^2 - a^2}} = \sqrt{x^2 - a^2}$$

$$\int \frac{dx}{\sqrt{a^2 - x^2}} = \sin^{-1}\frac{x}{a} \qquad\qquad \int \frac{x\, dx}{\sqrt{a^2 \pm x^2}} = \pm\sqrt{a^2 \pm x^2}$$

$$\int x\sqrt{x^2 \pm a^2}\, dx = \tfrac{1}{3}\sqrt{(x^2 \pm a^2)^3}$$

$$\int x^2\sqrt{x^2 \pm a^2}\, dx = \frac{x}{4}\sqrt{(x^2 \pm a^2)^3} \mp \frac{a^2}{8}x\sqrt{x^2 \pm a^2} - \frac{a^4}{8}\ln(x + \sqrt{x^2 \pm a^2})$$

$$\int \sin x \, dx = -\cos x$$

$$\int \cos x \, dx = \sin x$$

$$\int \sec x \, dx = \frac{1}{2} \ln \frac{1 + \sin x}{1 - \sin x}$$

$$\int \sin^2 x \, dx = \frac{x}{2} - \frac{\sin 2x}{4}$$

$$\int \cos^2 x \, dx = \frac{x}{2} + \frac{\sin 2x}{4}$$

$$\int \sin x \cos x \, dx = \frac{\sin^2 x}{2}$$

$$\int \sinh x \, dx = \cosh x$$

$$\int \cosh x \, dx = \sinh x$$

$$\int \tanh x \, dx = \ln \cosh x$$

$$\int \ln x \, dx = x \ln x - x$$

$$\int e^{ax} \, dx = \frac{e^{ax}}{a}$$

$$\int x e^{ax} \, dx = \frac{e^{ax}}{a^2}(ax - 1)$$

$$\int e^{ax} \sin px \, dx = \frac{e^{ax}(a \sin px - p \cos px)}{a^2 + p^2}$$

$$\int e^{ax} \cos px \, dx = \frac{e^{ax}(a \cos px + p \sin px)}{a^2 + p^2}$$

$$\int e^{ax} \sin^2 x \, dx = \frac{e^{ax}}{4 + a^2}\left(a \sin^2 x - \sin 2x + \frac{2}{a}\right)$$

$$\int e^{ax} \cos^2 x \, dx = \frac{e^{ax}}{4 + a^2}\left(a \cos^2 x + \sin 2x + \frac{2}{a}\right)$$

$$\int e^{ax} \sin x \cos x \, dx = \frac{e^{ax}}{4 + a^2}\left(\frac{a}{2} \sin 2x - \cos 2x\right)$$

$$\int \sin^3 x \, dx = -\frac{\cos x}{3}(2 + \sin^2 x)$$

$$\int \cos^3 x \, dx = \frac{\sin x}{3}(2 + \cos^2 x)$$

$$\int x \sin x \, dx = \sin x - x \cos x$$

$$\int x \cos x \, dx = \cos x + x \sin x$$

$$\int x^2 \sin x \, dx = 2x \sin x - (x^2 - 2) \cos x$$

$$\int x^2 \cos x \, dx = 2x \cos x + (x^2 - 2) \sin x$$

Radius of curvature
$$\begin{cases} \rho_{xy} = \dfrac{\left[1 + \left(\dfrac{dy}{dx}\right)^2\right]^{3/2}}{\dfrac{d^2y}{dx^2}} \\[30pt] \rho_{r\theta} = \dfrac{\left[r^2 + \left(\dfrac{dr}{d\theta}\right)^2\right]^{3/2}}{r^2 + 2\left(\dfrac{dr}{d\theta}\right)^2 - r\dfrac{d^2r}{d\theta^2}} \end{cases}$$

USEFUL TABLES

C

A. *Density,* ρ

	kg/m^3	lbm/ft^3		kg/m^3	lbm/ft^3
Aluminum	2 690	168	Iron (cast)	7 210	450
Concrete (av.)	2 400	150	Lead	11 370	710
Copper	8 910	556	Mercury	13 570	847
Earth (wet, av.)	1 760	110	Oil (av.)	900	56
(dry, av.)	1 280	80	Steel	7 830	489
Glass	2 590	162	Titanium	3 080	192
Gold	19 300	1205	Water (fresh)	1 000	62.4
Ice	900	56	(salt)	1 030	64
			Wood (soft pine)	480	30
			(hard oak)	800	50

B. *Coefficients of Friction,* μ

(The coefficients in the following table represent typical values under normal working conditions. Actual coefficients for a given situation will depend on the exact nature of the contacting surfaces. A variation of 25 to 100 per cent or more from these values could be expected in an actual application, depending on prevailing conditions of cleanliness, surface finish, pressure, lubrication, and velocity.)

	TYPICAL VALUES OF COEFFICIENT OF FRICTION, μ	
CONTACTING SURFACE	STATIC	KINETIC
Steel on steel (dry)	0.6	0.4
Steel on steel (greasy)	0.1	0.05
Teflon on steel	0.04	0.04
Steel on babbitt (dry)	0.4	0.3
Steel on babbitt (greasy)	0.1	0.07
Brass on steel (dry)	0.5	0.4
Brake lining on cast iron	0.4	0.3
Rubber tires on smooth pavement (dry)	0.9	0.8
Wire rope on iron pulley (dry)	0.2	0.15
Hemp rope on metal	0.3	0.2
Metal on ice	—	0.02

Table C2 Solar System Constants

Universal gravitational constant	$K = 6.673(10^{-11}) \text{ m}^3/(\text{kg} \cdot \text{s}^2)$
	$= 3.439(10^{-8}) \text{ ft}^4/(\text{lbf-s}^4)$
Mass of Earth	$m = 5.976(10^{24}) \text{ kg}$
	$= 4.095(10^{23}) \text{ lbf-s}^2/\text{ft}$
Period of Earth's rotation (1 sidereal day)	$= 23 \text{ h } 56 \text{ min } 4 \text{ s}$
	$= 23.9344 \text{ h}$
Angular velocity of Earth	$\omega = 0.7292(10^{-4}) \text{ rad/s}$
Angular velocity of Earth-Sun line	$\omega' = 0.1991(10^{-6}) \text{ rad/s}$
Mean velocity of Earth's center about Sun	$= 107\,200 \text{ km/h}$
	$= 66{,}610 \text{ mi/h}$

BODY	MEAN DISTANCE TO SUN km (mi)	ECCENTRICITY OF ORBIT e	PERIOD OF ORBIT SOLAR DAYS	MEAN DIAMETER km (mi)	MASS RELATIVE TO EARTH	SURFACE GRAVITATIONAL ACCELERATION m/s² (ft/s²)	ESCAPE VELOCITY km/s (mi/s)
Sun	—	—	—	1 392 000 (865 000)	333 000	274 (898)	616 (383)
Moon	384 398° (238 854)°	0.055	27.32	3 476 (2 160)	0.0123	1.62 (5.32)	2.37 (1.47)
Mercury	57.3 × 10⁶ (35.6 × 10⁶)	0.206	87.97	5 000 (3 100)	0.054	3.47 (11.4)	4.17 (2.59)
Venus	108 × 10⁶ (67.2 × 10⁶)	0.0068	224.70	12 400 (7 700)	0.815	8.44 (27.7)	10.24 (6.36)
Earth	149.6 × 10⁶ (92.96 × 10⁶)	0.0167	365.26	12 742† (7 917)†	1.000	9.821‡ (32. 22)‡	11.18 (6.95)
Mars	227.9 × 10⁶ (141.6 × 10⁶)	0.093	686.98	6 788 (4 218)	0.107	3.73 (12.3)	5.03 (3.13)

° Mean distance to Earth (center-to-center)

† Diameter of sphere of equal volume

    polar diameter $= 12\,713$ km

                   $= 7\,900$ mi

    equatorial diameter $= 12\,755$ km

                        $= 7\,926$ mi

‡ For nonrotating spherical Earth, equivalent to absolute value at sea level and latitude 37.5°

Table C3   Properties of Plane Figures

FIGURE	CENTROID	AREA MOMENTS OF INERTIA
Arc Segment	$\bar{r} = \dfrac{r \sin \alpha}{\alpha}$	—
Quarter and Semicircular Arcs	$\bar{y} = \dfrac{2r}{\pi}$	—
Triangular Area	$\bar{x} = \dfrac{a + b}{3}$ $\bar{y} = \dfrac{h}{3}$	$I_x = \dfrac{bh^3}{12}$ $\bar{I}_x = \dfrac{bh^3}{36}$ $I_{x_1} = \dfrac{bh^3}{4}$
Rectangular Area	—	$I_x = \dfrac{bh^3}{3}$ $\bar{I}_x = \dfrac{bh^3}{12}$ $\bar{J} = \dfrac{bh}{12}(b^2 + h^2)$
Area of Circular Sector	$\bar{x} = \dfrac{2}{3}\dfrac{r \sin \alpha}{\alpha}$	$I_x = \dfrac{r^4}{4}(\alpha - \tfrac{1}{2}\sin 2\alpha)$ $I_y = \dfrac{r^4}{4}(\alpha + \tfrac{1}{2}\sin 2\alpha)$ $J = \tfrac{1}{2}r^4\alpha$
Quarter Circular Area	$\bar{x} = \bar{y} = \dfrac{4r}{3\pi}$	$I_x = I_y = \dfrac{\pi r^4}{16}$ $\bar{I}_x = \bar{I}_y = \left(\dfrac{\pi}{16} - \dfrac{4}{9\pi}\right)r^4$ $J = \dfrac{\pi r^4}{8}$
Area of Elliptical Quadrant Area $A = \dfrac{\pi ab}{4}$	$\bar{x} = \dfrac{4a}{3\pi}$ $\bar{y} = \dfrac{4b}{3\pi}$	$I_x = \dfrac{\pi ab^3}{16},\ \bar{I}_x = \left(\dfrac{\pi}{16} - \dfrac{4}{9\pi}\right)ab^3$ $I_y = \dfrac{\pi a^3b}{16},\ \bar{I}_y = \left(\dfrac{\pi}{16} - \dfrac{4}{9\pi}\right)a^3b$ $J = \dfrac{\pi ab}{16}(a^2 + b^2)$

Table C4. Properties of Homogeneous Solids

(m = mass of body shown)

BODY	MASS CENTER	MOMENTS OF INERTIA
Circular Cylindrical Shell	—	$I_{xx} = \frac{1}{2}mr^2 + \frac{1}{12}ml^2$ $I_{x_1 x_1} = \frac{1}{2}mr^2 + \frac{1}{3}ml^2$ $I_{zz} = mr^2$
Half Cylindrical Shell	$\bar{x} = \dfrac{2r}{\pi}$	$I_{xx} = I_{yy}$ $\quad = \frac{1}{2}mr^2 + \frac{1}{12}ml^2$ $I_{x_1 x_1} = I_{y_1 y_1}$ $\quad = \frac{1}{2}mr^2 + \frac{1}{3}ml^2$ $I_{zz} = mr^2$ $\bar{I}_{zz} = \left(1 - \dfrac{4}{\pi^2}\right)mr^2$
Circular Cylinder	—	$I_{xx} = \frac{1}{4}mr^2 + \frac{1}{12}ml^2$ $I_{x_1 x_1} = \frac{1}{4}mr^2 + \frac{1}{3}ml^2$ $I_{zz} = \frac{1}{2}mr^2$
Semicylinder	$\bar{x} = \dfrac{4r}{3\pi}$	$I_{xx} = I_{yy}$ $\quad = \frac{1}{4}mr^2 + \frac{1}{12}ml^2$ $I_{x_1 x_1} = I_{y_1 y_1}$ $\quad = \frac{1}{4}mr^2 + \frac{1}{3}ml^2$ $I_{zz} = \frac{1}{2}mr^2$ $\bar{I}_{zz} = \left(\dfrac{1}{2} - \dfrac{16}{9\pi^2}\right)mr^2$
Rectangular Parallelepiped	—	$I_{xx} = \frac{1}{12}m(a^2 + l^2)$ $I_{yy} = \frac{1}{12}m(b^2 + l^2)$ $I_{zz} = \frac{1}{12}m(a^2 + b^2)$ $I_{y_1 y_1} = \frac{1}{12}mb^2 + \frac{1}{3}ml^2$

Table C4. *Continued*
(m = mass of body shown)

BODY	MASS CENTER	MOMENTS OF INERTIA
Spherical Shell	—	$I_{zz} = \frac{2}{3}mr^2$
Hemispherical Shell	$\bar{x} = \frac{r}{2}$	$I_{xx} = I_{yy} = I_{zz} = \frac{2}{3}mr^2$ $\bar{I}_{yy} = \bar{I}_{zz} = \frac{5}{12}mr^2$
Sphere	—	$I_{zz} = \frac{2}{5}mr^2$
Hemisphere	$\bar{x} = \frac{3r}{8}$	$I_{xx} = I_{yy} = I_{zz} = \frac{2}{5}mr^2$ $\bar{I}_{yy} = \bar{I}_{zz} = \frac{83}{320}mr^2$
Uniform Slender Rod	—	$I_{yy} = \frac{1}{12}ml^2$ $I_{y_1 y_1} = \frac{1}{3}ml^2$

Table C4.   *Continued*

(m = mass of body shown)

BODY	MASS CENTER	MOMENTS OF INERTIA
Quarter Circular Rod	$\bar{x} = \bar{y}$ $= \dfrac{2r}{\pi}$	$I_{xx} = I_{yy} = \frac{1}{2}mr^2$ $I_{zz} = mr^2$
Elliptical Cylinder	—	$I_{xx} = \frac{1}{4}ma^2 + \frac{1}{12}ml^2$ $I_{yy} = \frac{1}{4}mb^2 + \frac{1}{12}ml^2$ $I_{zz} = \frac{1}{4}m(a^2 + b^2)$ $I_{y_1y_1} = \frac{1}{4}mb^2 + \frac{1}{3}ml^2$
Conical Shell	$\bar{z} = \dfrac{2h}{3}$	$I_{yy} = \frac{1}{4}mr^2 + \frac{1}{2}mh^2$ $I_{y_1y_1} = \frac{1}{4}mr^2 + \frac{1}{6}mh^2$ $I_{zz} = \frac{1}{2}mr^2$ $\bar{I}_{yy} = \frac{1}{4}mr^2 + \frac{1}{18}mh^2$
Half Conical Shell	$\bar{x} = \dfrac{4r}{3\pi}$ $\bar{z} = \dfrac{2h}{3}$	$I_{xx} = I_{yy}$ $\quad = \frac{1}{4}mr^2 + \frac{1}{2}mh^2$ $I_{x_1x_1} = I_{y_1y_1}$ $\quad = \frac{1}{4}mr^2 + \frac{1}{6}mh^2$ $I_{zz} = \frac{1}{2}mr^2$ $\bar{I}_{zz} = \left(\dfrac{1}{2} - \dfrac{16}{9\pi^2}\right)mr^2$
Right Circular Cone	$\bar{z} = \dfrac{3h}{4}$	$I_{yy} = \frac{3}{20}mr^2 + \frac{3}{5}mh^2$ $I_{y_1y_1} = \frac{3}{20}mr^2 + \frac{1}{10}mh^2$ $I_{zz} = \frac{3}{10}mr^2$ $\bar{I}_{yy} = \frac{3}{20}mr^2 + \frac{3}{80}mh^2$

Table C4. *Continued*

(m = mass of body shown)

BODY	MASS CENTER	MOMENTS OF INERTIA
Half Cone	$\bar{x} = \dfrac{r}{\pi}$ $\bar{z} = \dfrac{3h}{4}$	$I_{xx} = I_{yy}$ $= \frac{3}{20}mr^2 + \frac{3}{5}mh^2$ $I_{x_1 x_1} = I_{y_1 y_1}$ $= \frac{3}{20}mr^2 + \frac{1}{10}mh^2$ $I_{zz} = \frac{3}{10}mr^2$ $\bar{I}_{zz} = \left(\dfrac{3}{10} - \dfrac{1}{\pi^2}\right)mr^2$
$\dfrac{x^2}{a^2} + \dfrac{y^2}{b^2} + \dfrac{z^2}{c^2} = 1$ Semiellipsoid	$\bar{z} = \dfrac{3c}{8}$	$I_{xx} = \frac{1}{5}m(b^2 + c^2)$ $I_{yy} = \frac{1}{5}m(a^2 + c^2)$ $I_{zz} = \frac{1}{5}m(a^2 + b^2)$ $\bar{I}_{xx} = \frac{1}{5}m(b^2 + \frac{19}{64}c^2)$ $\bar{I}_{yy} = \frac{1}{5}m(a^2 + \frac{19}{64}c^2)$
$\dfrac{x^2}{a^2} + \dfrac{y^2}{b^2} = \dfrac{z}{c}$ Elliptic Paraboloid	$\bar{z} = \dfrac{2c}{3}$	$I_{xx} = \frac{1}{6}mb^2 + \frac{1}{2}mc^2$ $I_{yy} = \frac{1}{6}ma^2 + \frac{1}{2}mc^2$ $I_{zz} = \frac{1}{6}m(a^2 + b^2)$ $\bar{I}_{xx} = \frac{1}{6}m(b^2 + \frac{1}{3}c^2)$ $\bar{I}_{yy} = \frac{1}{6}m(a^2 + \frac{1}{3}c^2)$
Rectangular Tetrahedron	$\bar{x} = \dfrac{a}{4}$ $\bar{y} = \dfrac{b}{4}$ $\bar{z} = \dfrac{c}{4}$	$I_{xx} = \frac{1}{10}m(b^2 + c^2)$ $I_{yy} = \frac{1}{10}m(a^2 + c^2)$ $I_{zz} = \frac{1}{10}m(a^2 + b^2)$ $\bar{I}_{xx} = \frac{3}{80}m(b^2 + c^2)$ $\bar{I}_{yy} = \frac{3}{80}m(a^2 + c^2)$ $\bar{I}_{zz} = \frac{3}{80}m(a^2 + b^2)$
Half Torus	$\bar{x} = \dfrac{a^2 + 4R^2}{2\pi R}$	$I_{xx} = I_{yy} = \frac{1}{2}mR^2 + \frac{5}{8}ma^2$ $I_{zz} = mR^2 + \frac{3}{4}ma^2$

INDEX

CONVERSION CHARTS BETWEEN SI AND U.S. CUSTOMARY UNITS

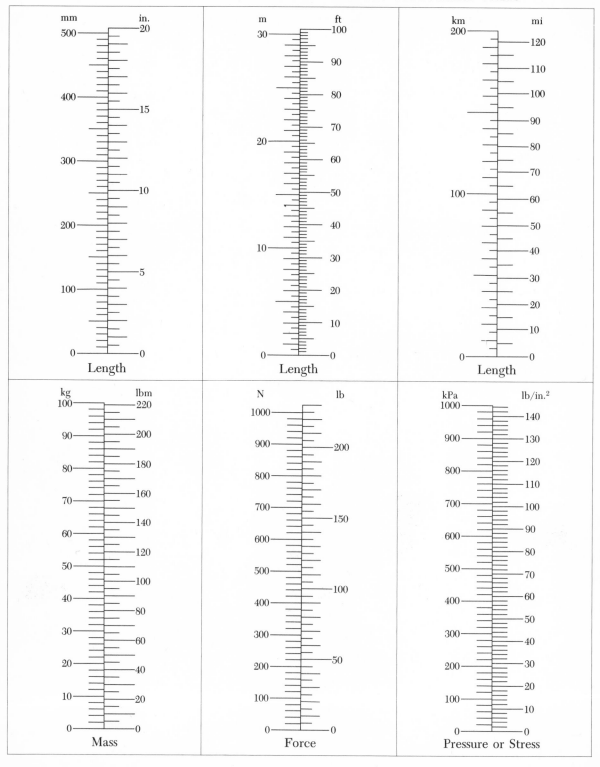